Coronary Calcium

Coronary Calcium

A Comprehensive Understanding of Its Biology, Use in Screening, and Interventional Management

Edited by

Aloke Virmani Finn, MD

Medical Director
Pathology
CVPath Institute
Gaithersburg, Maryland
United States

Associate Clinical Professor
Medicine
University of Maryland
Baltimore, Maryland
United States

ACADEMIC PRESS
An imprint of Elsevier

ELSEVIER

Academic Press is an imprint of Elsevier
125 London Wall, London EC2Y 5AS, United Kingdom
525 B Street, Suite 1650, San Diego, CA 92101, United States
50 Hampshire Street, 5th Floor, Cambridge, MA 02139, United States
The Boulevard, Langford Lane, Kidlington, Oxford OX5 1GB, United Kingdom

Coronary Calcium

Notices

Practitioners and researchers must always rely on their own experience and knowledge in evaluating and using any information, methods, compounds or experiments described herein. Because of rapid advances in the medical sciences, in particular, independent verification of diagnoses and drug dosages should be made. To the fullest extent of the law, no responsibility is assumed by Elsevier, authors, editors or contributors for any injury and/or damage to persons or property as a matter of products liability, negligence or otherwise, or from any use or operation of any methods, products, instructions, or ideas contained in the material herein.

ISBN: 978-0-12-816389-4

Publisher: Stacy Masucci
Acquisition Editor: Katie Chan
Editorial Project Manager: Sam young
Production Project Manager: Kiruthika Govindaraju
Cover Designer: Alan Studholme

Working together
to grow libraries in
developing countries

www.elsevier.com • www.bookaid.org

Contents

*Akiko Maehara, MD, Mitsuaki Matsumura, BS, Ziad A. Ali, MD, DPhil
and Gary S. Mintz, MD*

Paolo Raggi, MD, PhD, Antonio Bellasi, MD, PhD and Nikolaos Alexopoulos, MD

Contributors

Emma Akers, BS
South Australian Health and Medical Research Institute, University of Adelaide, Adelaide, SA, Australia

Nikolaos Alexopoulos, MD
Division of Cardiology, Cardiovascular Imaging, Euroclinic Athens, Athens, Greece

Omar Al Hussein Alawamlh, MD
Department of Radiology and Medicine, Dalio Institute of Cardiovascular Imaging, Weill Cornell Medical College and the New York-Presbyterian Hospital, New York, New York, United States

Ziad A. Ali, MD, DPhil
Director of Intravascular Imaging and Physiology, Center for Interventional Vascular Therapy, New York-Presbyterian Hospital/ Columbia University Medical Center, New York, NY, United States; Director of Angiographic Core Laboratory, Cardiovascular Research Foundation, New York, NY, United States; Department of Cardiology, St. Francis Hospital, Roslyn, NY, United States

Subhi J. Al'Aref, MD
Department of Radiology and Medicine, Dalio Institute of Cardiovascular Imaging, Weill Cornell Medical College and the New York-Presbyterian Hospital, New York, New York, United States

A. Maxim Bax
Department of Radiology and Medicine, Dalio Institute of Cardiovascular Imaging, Weill Cornell Medical College and the New York-Presbyterian Hospital, New York, New York, United States

Antonio Bellasi, MD, PhD
Research, Innovation and Brand Reputation, Ospedale di Bergamo, ASST-Papa Giovanni XXIII, Bergamo, Italy

Manfred Boehm, MD
Translational Vascular Medicine Branch, National Heart, Lung and Blood Institute, National Institutes of Health, Bethesda, MD, United States

Cornelia D. Cudrici, MD
Translational Vascular Medicine Branch, National Heart, Lung and Blood Institute, National Institutes of Health, Bethesda, MD, United States

Belinda Di Bartolo, PhD
Kolling Research Institute, University of Sydney, NSW, Australia

Mhairi Doris
British Heart Foundation Department of Cardiovascular Sciences, Queens Medical Research Institute, University of Edinburgh, Edinburgh, United Kingdom

Marc Dweck, MD, PhD
British Heart Foundation Department of Cardiovascular Sciences, Queens Medical Research Institute, University of Edinburgh, Edinburgh, United Kingdom; BHF Reader in Cardiology, Consultant Cardiologist, Centre for Cardiovascular Science, University of Edinburgh, Edinburgh, United Kingdom

Elisa A. Ferrante, PhD
Translational Vascular Medicine Branch, National Heart, Lung and Blood Institute, National Institutes of Health, Bethesda, MD, United States

Aloke V. Finn, MD
CVPath institute, Gaithersburg, MD, United States; Medical Director, Pathology, CVPath Institute, Gaithersburg, MD, United States; Associate Clinical Professor, School of Medicine, University of Maryland, Baltimore, MD, United States

Rachael Forsythe, MBChB, MRCS
British Heart Foundation Department of Cardiovascular Sciences, Queens Medical Research Institute, University of Edinburgh, Edinburgh, United Kingdom; Speciality Registrar in Vascular Surgery, Edinburgh Vascular Unit, Royal Infirmary of Edinburgh, Edinburgh, United Kingdom

Harvey Hecht, MD
Icahn School of Medicine at Mount Sinai, New York, NY, United States

Hiroyuki Jinnouchi, MD
CVPath institute, Gaithersburg, MD, United States

Daria Larine, BA
Department of Radiology and Medicine, Dalio Institute of Cardiovascular Imaging, Weill Cornell Medical College and the New York-Presbyterian Hospital, New York, New York, United States

Akiko Maehara, MD
Professor, Center for Interventional Vascular Therapy, New York-Presbyterian Hospital/ Columbia University Medical Center, New York, NY, United States; Director of Intravascular Imaging Core Laboratory, Cardiovascular Research Foundation, New York, NY, United States

Petra Zubin Maslov, MD
Icahn School of Medicine at Mount Sinai, New York, NY, United States

Mitsuaki Matsumura, BS
Assistant Director of Intravascular Imaging and Physiology Core Laboratories, Cardiovascular Research Foundation, New York, NY, United States

James K. Min, MD
Department of Radiology and Medicine, Dalio Institute of Cardiovascular
Imaging, Weill Cornell Medical College and the New York-Presbyterian Hospital,
New York, New York, United States

Gary S. Mintz, MD
Senior Medical Advisor, Cardiovascular Research Foundation, New York, NY,
United States

Jagat Narula, MD, PhD, MACC
Icahn School of Medicine at Mount Sinai, New York, NY, United States

David E. Newby, BA, BSc (Hons), PhD, BM, DM, FRCP, FESC, FRSE, FMedSci
British Heart Foundation Department of Cardiovascular Sciences, Queens
Medical Research Institute, University of Edinburgh, Edinburgh, United
Kingdom; Professor, British Heart Foundation Centre for Cardiovascular
Diseases, University of Edinburgh, Edinburgh, United Kingdom

Stephen J. Nicholls, MBBS, PhD
Monash Cardiovascular Research Centre, Monash University, Melbourne,
Australia; Professor of Cardiology, South Australian Health & Medical Research
Institute, Australia; SAHMRI Heart Foundation Heart Disease Theme Leader,
South Australian Health & Medical Research Institute, Australia

Paolo Raggi, MD, PhD
Mazankowski Alberta Heart Institute and University of Alberta, Edmonton, AB,
Canada; Department of Medicine-Division of Cardiology, University of Alberta,
Edmonton, AB, Canada

Atsushi Sakamoto, MD
CVPath Institute, Gaithersburg, MD, United States

Maaz B.J. Syed, MBChB, MSc
British Heart Foundation Department of Cardiovascular Sciences, Queens
Medical Research Institute, University of Edinburgh, Edinburgh, United
Kingdom; Clinical Research Fellow, Department of Cardiovascular Sciences,
University of Edinburgh, Edinburgh, United Kingdom

Sho Torii, MD
CVPath institute, Gaithersburg, MD, United States

Inge J. van den Hoogen, MD
Department of Radiology and Medicine, Dalio Institute of Cardiovascular
Imaging, Weill Cornell Medical College and the New York-Presbyterian Hospital,
New York, New York, United States

Alexander R. van Rosendael, MD
Department of Radiology and Medicine, Dalio Institute of Cardiovascular Imaging, Weill Cornell Medical College and the New York-Presbyterian Hospital, New York, New York, United States

Renu Virmani, MD
CVPath institute, Gaithersburg, MD, United States

Vascular calcification: understanding its clinical significance

Introduction

Coronary artery calcification is concomitant with the development of advanced atherosclerosis and is commonly used by internists and cardiologists in evaluating patients at risk for coronary artery disease. The science of the significance of vascular calcium for risk prediction and even treatment continues to evolve as new knowledge is gained. This book provides a state-of-the-art update for physicians on our current understanding of the significance of vascular calcification. Chapters written by the world's experts on this subject discuss the basic science behind how vascular calcification develops in atherosclerotic lesions, the pathology of calcified coronary lesions in humans and its differential development by race and its relationship to plaque progression, the significance of coronary calcium as detected by modalities such as computed tomography for risk prediction, the potential role of various pharmacologic interventions, the role of intravascular imaging in assisting with coronary interventions in the setting of vascular calcification, clinical perspectives on the meaning at the patient level, and lastly where the field is moving. Each chapter contains summaries with the salient and most important points. This book will serve as a valuable resource for clinicians, researchers, and educators as we continue to learn more about this fascinating area to provide better care for our patients.

Sincerely,
Aloke V. Finn, MD
Gaithersburg, MD

Types and pathology of vascular calcification

Hiroyuki Jinnouchi, MD [1], Atsushi Sakamoto, MD [1], Sho Torii, MD [1], Renu Virmani, MD [1], Aloke V. Finn, MD [1,2,3]

[1]*CVPath institute, Gaithersburg, MD, United States;* [2]*Medical Director, Pathology, CVPath Institute, Gaithersburg, MD, United States;* [3]*Associate Clinical Professor, School of Medicine, University of Maryland, Baltimore, MD, United States*

Introduction

The arterial structure can be categorized into conducting arteries or muscular arteries. The conducting arteries are elastic arteries that include the aorta, iliac, and carotid arteries. These elastic arteries are rich in elastic tissue. The multilayered elastic lamella of elastic arteries is separated by a few smooth muscle cells, proteoglycans, and collagen. The intima usually has a few layers, including smooth muscle cells in combination with proteoglycans and collagen. The adventitia is composed of collagen rich tissue with few elastic fibers and vasa vasorum. Coronary arteries and the peripheral arteries such as femoral, popliteal, and tibial arteries are classified as muscular arteries. They are composed of internal and external elastic membranes, and a media rich in smooth muscle cells, minimal proteoglycans, and collagen. Both coronary and the peripheral arteries can get calcified especially as in the setting of advanced age and atherosclerotic disease progression.

When divided by the location of calcification, vascular calcification is categorized into two forms (intimal or medial). Intimal calcification is the dominant type of calcification in coronary arteries. The peripheral arteries of the lower extremities are mostly influenced by medial calcification but also develop intimal calcification. Medial calcification leads to loss of elasticity and is frequently observed in peripheral vascular disease. We will mainly focus our discussion on calcification in coronary arteries and peripheral arteries of the lower extremities.

Classification of arterial calcification

A variety of mechanisms lead to vascular calcification. These mechanisms suggest the presence of multiple etiologies rather than only one mechanism. Arterial wall calcification is categorized into three different types (i.e., atherosclerotic intimal calcification, medial calcification, and genetic calcification) (Table 1.1) [1]. Atherosclerotic intimal calcification is different from medial calcification and the

Table 1.1 Proposed categorization of common disease-associated pathological calcification.

	Category I	Category II	Category III
Mediators	Inflammatory	Metabolic	Genetic
Commonly associated diseases	Atherosclerosis	Chronic kidney disease; diabetes mellitus	Genetic disorders, e.g., PXE, GACI, ACDC, Marfan syndrome
Arterial site	Intima	Media	Usually media
Regulatory manner	Gain of activators; loss of inhibitors	Loss of inhibitors; gain of activators	Loss of inhibitors; gain of activators
Molecular aspects	Osteogenesis and/or chondrogenesis; matrix Vesicle release	Nucleation in vesicles or elastin; osteogenesis	Defects in various genes and miRs, e.g., ENPP1, ABCC6, NT5E, SLC20A2, MGP, OPG, Ank
Circulating factors	Lipids; cytokines; fetuin	Phosphate; uremic toxins; glucose; soluble RAGE; fetuin	Various, e.g., ATP, pyrophosphate, adenosine, inorganic phosphate
Example local factors	Oxidized lipids; cytokines; matrix metalloproteinases	Elastin; proteases; AGEs; RAGE; transglutaminase; MGP	Elastin; fibrillin; pyrophosphate
Functional effect	Vascular stiffness; plaque vulnerability	Vascular stiffness	Vascular stiffness
Possible analogous ossification	Endochondral ossification	Intramembranous ossification	Variable
X-ray and histological images (H&E)	Intimal calcification	Medial calcification	Genetic calcification

PXE, pseudoxanthoma elasticum; GACI, generalized arterial calcification of infancy; ACDC, arterial calcifications due to deficiency in CD73; miRs, micro-RNAs; ENPP1, ectonucleotide pyrophosphate phosphodiesterase; ABCC6, ATP-binding cassette subfamily C member six; NT5E, 5′-nucleotidase, ecto (CD73); SLC20A2, solute carrier family 20 (phosphate transporter), member two; AGEs, advanced glycation endproducts; RAGE, receptor for AGEs; MGP, matrix gamma-carboxyglutamic acid protein; OPG, osteoprotegerin; Ank, ankylosis protein.

Permission from Demer LL, Tintut Y. Inflammatory, metabolic, and genetic mechanisms of vascular calcification. Arterioscler Thromb Vasc Biol. 2014;34:715−723 mechanisms by which each occurs are thought to be different although some overlap exists. For instance, although medial calcification is not associated with lipid deposition or inflammation, intimal calcification is. In some cases such as in patients with diabetes mellitus (DM) who are likely to have both hypercholesterolemia and chronic kidney disease (CKD), the combination of these risk factors can lead to calcification in intimal and medial wall of the same vessels. There is a strong relationship between medial calcification and osteochondrogenic differentiation factors and ossification. However, medial calcification is also involved with osteoblastic differentiation markers as well as ossification [2]. Additionally, metabolic mediators are also associated with genetic disorders, which can lead to medial and sometimes intimal calcification in arterial wall. Therefore, although types of calcification can be broadly classified into the three different types, it needs to be understood that some overlap in mechanisms likely exists.

Definition of different types of calcification

Arterial calcification is first appreciated in pathological intimal thickening that is the earliest form of progressive lesions [3]. Calcification is classified depending on the size of calcification (Fig. 1.1). Microcalcifications are defined as calcium particles varying in size from ≥ 0.5 μm to less than 15 μm in diameter and are best appreciated in pathologic intimal thickening. Punctate calcification occurs as calcium deposits >15 μm but less than 1 mm and are thought to be the result of macrophage apoptosis and are commonly seen in the deeper areas of the necrotic core. Fragmented calcification occurs when deposits become ≥ 1 mm, and how exactly such phenomenon occur is poorly understood. Sheet calcification is noted when greater than 1 quadrant of the vessel demonstrates calcification and is thought to be consistent with plaque stabilization. Nodular calcification is seen when nodular calcium deposits occur in atherosclerotic lesion remain confined within the atherosclerotic plaque and do not protrude into the lumen. When such a lesion protrudes though the plaque into the lumen, it is called calcified nodule and is accompanied by luminal thrombus. These definitions are applicable to not only coronary arteries but also to peripheral arteries for both intimal and medial calcifications.

Coronary artery calcification
Pathology of coronary calcification

Microcalcifications are generally observed in lipid pools and are thought to result from apoptotic smooth muscle cells and are visualized on histologic assessment. Larger punctate areas of calcifications are thought to originate from apoptotic debris of macrophages and are commonly detected in more-advanced lesions, especially in early fibroatheromas, where macrophages infiltrate the lipid pools. Speckled

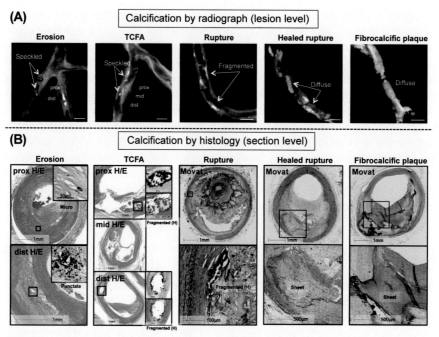

FIGURE 1.1 Type of Calcification by Radiography and Histology.

(A) The typical patterns of calcification by radiography for each plaque type. (B) Corresponding histology with low- and high-power images of sections taken from the areas shown in radiographs as red line. Scale bar (white line) in radiograph indicates 5 mm. Microcalcification by histology is invisible by radiograph in proximal section from a case of coronary erosion. Speckled calcification by radiograph corresponds to punctate or fragmented calcification by histology. TCFA lesions show three sites of sectioning and two spots (proximal and distal) of speckled calcium with the middle section showing TCFA without calcification. Plaque rupture site shows fragmented calcification by radiograph that corresponds to fragmented calcification by histology. Healed rupture and fibrocalcific plaques sites both showing diffuse calcification by radiograph that corresponds to sheet calcification by histology. *Dist*, distal; Fragmented *(H)*, fragmented by histology; *H/E*, hematoxylin-eosin stain; *Mid*, middle; *Prox*, proximal; *TCFA*, thin-cap fibroatheroma.

Permission from Mori H, Torii S, Kutyna M, Sakamoto A, Finn AV, Virmani R. Coronary artery calcification and its progression: what does it really mean? JACC Cardiovasc Imaging. 2018;11:127–142

calcification seen on radiography is <2 mm in diameter, and fragmented calcification is 2−5 mm whereas diffuse is >5 mm of continuous calcification. Histologically, fragmented calcification occurs when microcalcifications coalesce to form larger masses and are observed at the periphery of a necrotic core, or even within lipid pools usually near the media. Sheet or plates of calcium surround intimal smooth muscle cells and collagen regardless of the presence or absence of a necrotic core. Standard noninvasive imaging techniques such as computed tomography

(CT) can visualize these larger fragments of calcium. Nodular calcification is the least frequent and is always seen in coronary arteries that are heavily calcified. Nodular calcification is mostly located in tortuous arteries of older individuals or patients with DM or CKD. The findings of ossification with or without intervening bone marrow are rarely observed in coronary arteries but are more common in the peripheral arteries.

An excellent correlation between the percent of stenosis and mean area of calcification is observed (Fig. 1.2A) [4,5]. The relationship between plaque morphology and percent stenosis cannot fully explain the reason behind the correlation between percent stenosis and calcification. Although fibrocalcific plaques and healed ruptures become more calcified with increasing luminal stenosis, there is no correlation between the necrotic core size and calcification. Indeed, the relationship between coronary calcification and plaque morphology is complex. The presence of coronary calcification is not binary but rather depends on the type of calcification, its location, extent, volume, and density. Whether coronary calcification predicts plaque instability remains a topic of controversy. Pathology studies cannot fully answer this question because of their inherent static nature but can begin to give some clues. The mean area of calcification diverges according to plaque types (Fig. 1.2B). Fibrocalcific plaque, including nodular calcification, tends to have the greatest amount of calcification, followed by healing ruptures, plaque ruptures, and thin cap fibroatheroma, and least calcification is seen in erosions. Although the mean necrotic core area in fibroatheroma, thin-cap fibroatheroma, and rupture increases with percent stenosis, the mean area of calcification did not show a proportionate increase. Spotty calcification correlates with plaque instability, whereas dense calcification correlates with overall plaque burden. On the other hand, calcification of healed ruptures and fibrocalcific plaques outpaces that of necrotic core area as the vessel narrows [6].

There is a correlation between arterial calcification and plaque burden, whereas a correlation between calcification and lumen area has not been shown (Fig. 1.2C) [7]. Although the former is not surprising, the latter finding needs to be interpreted with caution. In general, plaque area is positively correlated with internal elastic lamina (IEL) area, that is, as the plaque increased so did the vessel size. Glagov et al. showed that lumen area does not decrease unless percent plaque reaches up to 40%, that is, vessel positively remodels [8]. Thus, positive remodeling might be the reason why no correlation between calcification and lumen area exists. The association between fibrotic plaque and negative remodeling may be also another possible reason for the lack of correlation between calcification and lumen area [5]. Overall, these data support the idea that calcium burden is greatest in stable plaques compared with unstable plaques with an inverse relationship between necrotic core area and calcification.

The prevalence of coronary calcification

At autopsy, the relationship between coronary calcification and age was shown in individuals dying with severe coronary disease. Radiography showed calcification

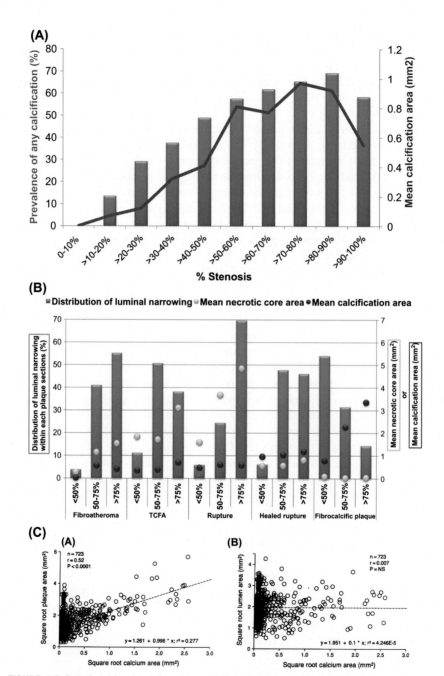

FIGURE 1.2 Relationship of Calcification and Plaque Type and Percent Stenosis.

(A) Graph shows relationship between percent stenosis and the degree of calcification in sudden coronary death victims. Each blue bar represents prevalence of any calcification (%), whereas each red dot represents mean calcification area (mm^2). (B) Blue bar graph shows the percent area luminal narrowing divided into severe (>75% cross-sectional area in 46% of individuals aged <40 years, 79% of those aged 50—60 years, and 100% of those aged >60 years [9].

Gender differences in calcification are well known for a long time. Atherosclerotic coronary artery disease (CAD) in females is delayed by 10—15 years as compared to males. This suggests that estrogen exposure in the premenopausal years likely contributes to providing a protective effect against the development of atherosclerosis [10]. In the CADRE (Coeur Arteres DREpanocytose) study, 108 human hearts from 70 males and 38 females who were victims of sudden cardiac death were enrolled to evaluate calcification by radiography [4]. When divided by decades, the extent of calcification was greater in males than females up to the 60s (Fig. 1.3A). However, in the 70s, the difference between males and females was not obvious, which implies the effects of the postmenopausal state results in the rapid progression of calcification and atherosclerosis in females. When divided by pre- and postmenopausal females, the extent of calcification is three times greater in postmenopausal females versus premenopausal females [11]. The substudy of the Women's Health Initiative trial compared women (50—59 years) over a 7-year period of estrogen therapy with those receiving placebo to evaluate the effect of estrogen therapy on coronary calcification. In the study, calcification was significantly less in those receiving estrogen therapy relative to those receiving placebo [12].

Racial differences in the degree of calcification have also been reported. In the MESA (Multi-Ethnic Study of Atherosclerosis) study, 6814 people aged 45—84 years who were asymptomatic and without clinical cardiovascular disease were

---

narrowing), moderate (50%—75%), and mild (<50%) luminal narrowing stratified by plaque type. Mean necrotic core (yellow circles) and calcification area (blue circles) are also shown for each percent narrowing. The data for fibroatheroma, TCFA, and rupture were obtained ($n = 213$). The data of healed rupture and fibrocalcific plaque were obtained ($n = 30$). TCFA = thin-cap fibroatheroma (C) The graph on the left shows correlation between square root of coronary calcium area values (mm^2) detected by histopathologic and microradiographic analysis and square root of plaque area values (mm^2) for each of the 723 coronary artery segments in humans. The graph on the right shows square root of coronary calcium area (mm^2) detected by histopathologic and microradiographic analysis versus square root of lumen area (mm^2) for each of the 723 coronary artery segments where no relation was identified.

The left and right images are (A) Data are stratified by decades from Burke AP et al. Herz. 2001; 26:239—244; Reproduced with permission from Otsuka F, Sakakura K, Yahagi K, Joner M, Virmani R. Has our understanding of calcification in human coronary atherosclerosis progressed? Arterioscler Thromb Vasc Biol. 2014;34:724—736.

(B) Reproduced with permission from Narula et al. J Am Coll Cardiol 2013; 61:1041—51. Reproduced with permission from Burke AP, Weber DK, Kolodgie FD, Farb A, Taylor AJ, Virmani R. Pathophysiology of calcium deposition in coronary arteries. Herz. 2001;26:239—244. (C) Reproduced with permission from Sangiorgi G, Rumberger JA, Severson A, Edwards WD, Gregoire J, Fitzpatrick LA, Schwartz RS. Arterial calcification and not lumen stenosis is highly correlated with atherosclerotic plaque burden in humans: a histologic study of 723 coronary artery segments using nondecalcifying methodology. J Am Coll Cardiol. 1998;31:126—133

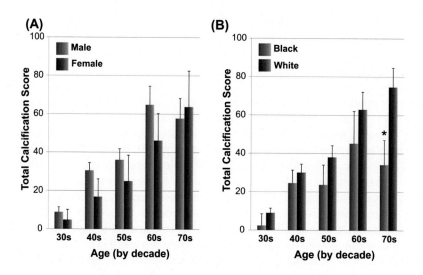

FIGURE 1.3 Relationship between Calcification and Gender and RaceBar.

Bar graphs showing total calcification score in sudden coronary death victims stratified by decade in male and female (A) as well as in black and white (B). Calcification was scored based on one point per mm of epicardial artery involved by calcification by visual inspection of radiographs (density was not assessed).

From Burke AP, et al. Coronary calcification: insights from sudden coronary death victims, Z Kardiol 2000;89:

Suppl 2,II/49-II/53.

enrolled to evaluate the effect of race (White, African American, Hispanic, and Chinese) on calcification [13]. After adjusting for various clinical factors such as age, education, lipids, body mass index, smoking, diabetes, hypertension, gender, treatment for hypercholesterolemia, and scanning center, the relative risk for calcification was significantly less in African Americans (0.78 (95% confidential interval [CI], 0.74−0.82)), Hispanics (0.85 (95% CI, 0.79−0.91)), and Chinese (0.92 (95% CI, 0.85−0.99)) relative to Whites. In our autopsy study enrolling victims of sudden cardiac death, when divided by the decades, Whites showed significantly greater calcification as compared to African Americans regardless of decade, which is consistent with previous clinical studies (Fig. 1.3B) [13−15]. Although the explanation behind the race-based differences in coronary calcium remains unknown, several potential mechanisms have been reported. A previous study reported that bone loss might contribute to abdominal aortic calcification [16]. Whites showed less bone mineral density than African American, suggesting that calcification may be related to lower bone turnover in African Americans [17]. An inherent genetic predisposition might also explain the differences in coronary calcification between Whites and African Americans. The expression of GAB2, a molecular adaptor that associates with the procalcific receptor activator of NF-kB and its ligand RANKL, is more highly expressed in Whites versus African Americans and might provide another explanation for reason why lower calcification is observed in African Americans [18]. The Arg287Gln polymorphisms in the soluble epoxide hydrolase gene were related to lower calcification in African Americans but not Whites, and the minor allele frequency was not significantly different between these races [19]. Although multiple genetic factors are likely to be related to calcification, the precise genetic impact of race on calcification remains to be determined.

Influence of diabetes mellitus on coronary calcification

Patients with DM are more likely to have greater coronary calcification, total plaque burden, and worse clinical outcomes than those without DM [20,21]. Carson et al. evaluated relationship between glycosylated hemoglobin and progression of calcification in total 2076 patients by plain CT at baseline and follow-up of 5 years [22]. Patients with higher glycosylated hemoglobin showed significantly higher incidence of progression of calcification. In our autopsy study, the role of DM on plaque progression was evaluated between patients with DM and without DM (matched for age, gender, and race) who died suddenly [23]. Patients with DM showed greater total plaque burden than those without DM. This result may be related to the larger number of healed plaque ruptures in DM versus non-DM (2.6 vs. 1.9; $P = .04$) [24]. Mean percent calcified area was also higher in subjects with DM versus those without DM (12.1% vs. 9.4%; $P = .05$).

Computed tomography

CT also plays a crucial role in identifying coronary calcification. Calcium scoring has been shown to be helpful in the prediction of the future cardiovascular events on a population wide basis [25]. The most common calcification scoring system is the Agatston score that sums the score of total calcified area and maximum density of calcification (Hounsfield units (HU) > 130). Although other scoring systems have been introduced, Agatston scoring system remains the gold standard because of its simplicity. The relationship between coronary artery calcification and cardiovascular events occurs in a stepwise fashion. Blankstein et al. showed that the presence of any calcification in coronary arteries indicates the risk of the future CAD even if without any atherosclerotic risk factors as compared to those without any calcification at a median follow-up of 5.4 years [26]. Consistent with this study, Joshi et al. showed that a lack of calcification in coronary arteries predicts the future clinical outcomes with a very low rate of coronary events [27]. Gradation in risk becomes obvious when comparing patients divided into percentiles of calcium score. The risk of coronary events is 19 times higher in individuals with a calcium score >75th percentile versus those with <25th percentile [28]. The risk of coronary events is 6.5 times higher in individual patients in the upper risk factor quartile as compared with those in the lowest quartile [28].

The use of multidetector CT may also potentially be valuable in the detection of vulnerable plaques [29]. Spotty calcification on CT is described as one of elements to predict vulnerable plaques along with low plaque density (<30 Hounsfield unit), and the napkin-ring sign. A previous study reported that calcification area of 1.03−1.37 mm^2 is able to be detected by CT in 3−4 contiguous pixels within a plaque area of 5 mm^2 [20,30]. Therefore, by definition microcalcification that is defined as <0.5−15 μm in histology cannot be detected by CT. CT only has capability of detecting dense or sheet calcifications such as are present in fibrocalcific plaques.

Micro-CT has capability of observing more details regarding coronary artery calcification as compared to clinical CT. We have tried to detect different types of plaques in micro-CT. Calcification is seen without any blooming artifact. Micro-CT can even distinguish different types of calcification. Necrotic core as well as cholesterol clefts can be detected by micro-CT. Clinically, if images in clinical CT could be improved to this level, it may be possible to detect features of vulnerable plaques and even distinguish different types of plaques without any invasive procedures. Such advantage might result in raising diagnostic accuracy and make it possible to stratify patients based on risk of future clinical outcomes in clinical practice (Fig. 1.4).

Peripheral artery disease of the lower extremity
Prevalence of PAD

Peripheral artery disease (PAD) contributes to significant morbidity and mortality, affects approximately 202 million adults worldwide, and the global prevalence of PAD has increased by 23.5% during 2000 and 2010 [31]. A previous consensus document showed that the prevalence of asymptomatic PAD is in the range of 3% −10% in individuals 50−69 years, and increasing to 15%−20% in persons over 70 years [32]. PAD is primarily caused by atherosclerotic plaques, and the risk factors for PAD are similar to those for CAD, with diabetes and smoking having a greater impact [33].

PAD in pathology
Above-the-knee artery

A major limitation to our knowledge of PAD of the lower extremities is the paucity of studies evaluating the extent and characteristics of arterial atherosclerosis in this region. In 2006, 1074 individuals aged from 60 to 90 years at autopsy including an equal number of males and females who died from various causes were evaluated for progression of disease in each artery. The progression of disease was the fastest in aorta, followed by coronary, femoral, and the least in the carotid and intracerebral arteries [34]. In 2007, coronary, superficial femoral, and carotid arteries from 100 autopsy cases ranging in age from 20 to 82 years who died from various causes were assessed by Dalager et al. [35]. In this study, the American Heart Association (AHA) classification modified scheme was utilized, which extends the type of

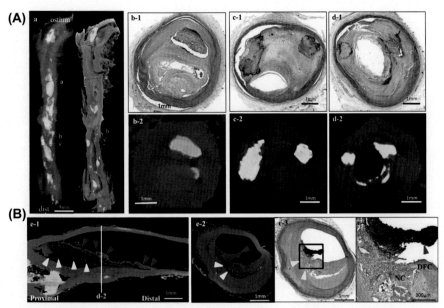

FIGURE 1.4 **Detection of Calcific Morphologies of Human Coronary Artery via High-Resolution Microcomputed Tomography.**

(A) An 81-year old with chronic total occlusion and calcification. a. Left image is a radiography showing diffuse calcification in the mid portion of right coronary artery. Right image is a microcomputed tomography (CT) image showing a longitudinal view of the same vessel. A yellow bracket indicates the extent of chronic total occlusion by micro-CT. b—d. Corresponding images of the histologic sections (1) (Movat staining) and transverse cuts of micro-CT images (2) taken from the areas shown in the radiograph as red lines. (B) A 55-year old male with plaque rupture. e. Longitudinal micro-CT image of the mid left circumflex artery showing irregularly dispersed luminal thrombus (red arrow heads) with an underlying necrotic core (yellow arrow heads). e−2 and e−3. are cross-sections from the micro-CT image and histology. e−4. is high power image corresponding to a black box in e−3. showing rupture site with overlying thrombus and underlying necrotic core (NC). *DFC*, disrupted fibrous cap.

(A) Reproduced with permission from Sakamoto A et al. Current Opinion in Cardiology. 2018 Nov; 33(6): 645–652. (B) Reproduced with permission from Jinnouchi et al. Circ Cardiovasc Imaging. 2018; 11(10): e008331. https://doi.org/10.1161/CIRCIMAGING.118.008331.

plaque to stages VII to IX to separate calcified plaque (VII), fibrous plaque (VIII), and chronic occlusion (IX) from original AHA calcification. Our classification scheme originally described based on coronary experience can also be applied to PAD (Fig. 1.5). Dalager showed that progression of plaque burden was associated with increased age for all vascular beds. He found that atherosclerotic patterns differed depending on the types of artery. Carotid and coronary arteries commonly showed lesions consisting of foam cells and lipid core plaques. These types of

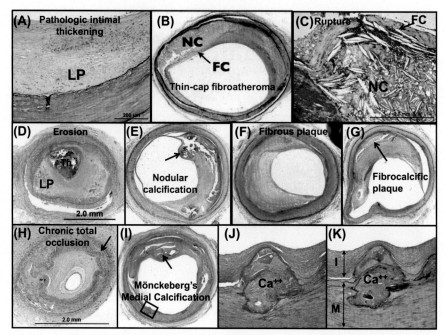

FIGURE 1.5 Histopathology of Atherosclerotic lesions from Human Femoral Arteries.

(A) Accumulation of the extracellular lipid (lipid pool); (B) thin-cap fibroatheroma with necrotic core covered by thin fibrous cap; (C) plaque rupture with a relatively large necrotic core; (D) plaque erosion showing an occlusive organizing thrombus of the femoral artery in an underlying lesion with a necrotic core (NC); (E) nodular calcification (arrow) held within the intima by layers of fibrous tissue. Note the absence of a luminal thrombus; (F) eccentric fibrous plaques rich in smooth muscle cells and collagen (yellow) with the relative absence of inflammation and necrotic core; (G) fibrocalcific plaque rich in smooth muscle cells and collagen, with calcification of the necrotic core (arrow); (H) chronic total occlusion with recanalized thrombi and scalloping of the internal elastic lamina (IEL) (arrow); (I–K) Mönckeberg's medial calcification from asymptomatic patients without peripheral vascular disease. Panel I shows a heavily calcified femoral artery with clinically insignificant stenosis and a large calcified plate seen in the medial layer protruding into the intima (arrow). Panels J and K, higher power images of the area represented by the black box in I, showing a relatively small area of calcification of the IEL (I) and medial (m) layers that protrudes into the intima. Panel J is a histologic section with hematoxylin-eosin staining and all other sections are stained with Movat pentachrome. *Ca++*, calcification; *LP*, lipid pool; *NC*, necrotic core; *FC*, fibrous cap; *Th*, thrombus; *I*, intima; *M*, media.

Reproduced with permission from Otsuka F et al.

plaques were located most frequently in left anterior descending coronary artery and carotid bifurcation. On the other hand, the femoral arteries showed less foam cell lesions. Instead, the femoral arteries were affected by age-dependent atherosclerosis characterized mainly by fibrous plaque. This might imply that atherosclerotic progression in the femoral artery occurs more slowly as compared to coronary and carotid arteries. Severe atherosclerotic lesions in the femoral artery were mainly fibrous plaques with the least likelihood of rupture, although this study did not include symptomatic PAD patients.

Several previous histological studies using endarterectomy specimens were conducted to evaluate the composition of femoral artery plaques. A total of 88 plaques (45 carotid and 43 femoral specimens: mean age, 69.7 ± 1.7 years and 69.2 ± 1.5 years, respectively) retrieved after endarterectomy were investigated by Herisson et al. [36]. The study reported that carotid arteries frequently showed fibroatheromas, whereas fibrocalcific plaques predominated in femoral arteries (93%), and the extent of calcification was greater in femoral arteries than carotid arteries. Derksen et al. reported differences of plaque composition between common femoral artery (CFA) and the superficial femoral artery (SFA). A total 124 CFA and 93 SFA from 217 patients with intermittent claudication (77%) or critical limb ischemia (CLI) (23%) were assessed after endarterectomy. The histological results showed that collagen and smooth muscle cells were more frequently identified in SFA as compared to CFA (collagen, SFA = 69% vs. CFA 31%, $P < .001$; smooth muscle cells, SFA = 64% vs. CFA = 36%, $P < .001$).

Below-the-knee arteries

The most advanced stage of PAD is CLI that is strongly associated with worse cardiovascular outcomes [37]. Females with CLI showed greater risk of amputation and worse outcomes following surgical and endovascular revascularization as compared to males. This might be associated with delayed manifestation of the disease, smaller vessel diameters, and more severe and diffuse atherosclerosis in females versus males [38–40].

Soor et al. assessed histological morphological findings of PAD in 58 patients ranging in age from 43 to 95 years (mean 68.7 ± 12.5 years) following lower extremity amputation (33 [57%] below-the-knee arteries and the rest 25 [43%] above-the-knee arteries). After the peripheral arteries were sectioned at short intervals, only the most advanced atherosclerotic segments were submitted for histological examination. Most of them (218 of 261 segments [84%]) showed atherosclerotic plaques with average luminal stenosis of $68 \pm 29\%$. Severe luminal narrowing (>75%) was observed in 48% of the vessel segments. Medial calcification was seen in 199 segments (76%), which involved $55 \pm 37\%$ of the medial circumference of the vessel, and severe calcification (>75% involvement) was present in only 37% of the vessel segments. Anterior and posterior tibial arteries showed the greatest luminal narrowing due to atherosclerosis (73% and 72%, respectively), followed by peroneal (68%), popliteal (67%), femoral (64%), and the least in the dorsalis pedis (59%) arteries. Medial calcification was also the greatest in anterior and posterior tibial arteries (58% and 59%, respectively), followed by peroneal (55%), dorsalis pedis (54%), popliteal (51%), and the least in the femoral (44%) arteries. Coexistent medial calcification and atherosclerotic disease was observed in 168 (77%) of the 218 arterial segments. Of note, the extent of medial calcification did not correlate with the degree of atherosclerosis [41,42]. Another recent publication by Narula et al. involved 75 patients with CLI underwent 101 amputations, atherosclerosis was more frequent in femoral and popliteal arteries (FEM-POP) compared to infrapopliteal arteries (INFRA-POP) (67.6% vs. 38.5%), and of these only 69% showed >70% narrowing. The presence of chronic luminal thrombi were more frequent in INFRA-POP than FEM-POP arteries with insignificant atherosclerosis, whereas acute thrombi were more frequently seen in FEM-POP than INFRA-POP suggesting atherothromboembolic etiology in INFRA-POP with CLI [43].

In 2018, Torii et al. conducted a systematic pathological characterization of 12 legs from eight victims dying with abundant atherosclerotic risk factors (median age, 82 years; 6 men) [44]. Arteries were sectioned at 3–4 mm intervals from femoral to dorsalis pedis. Atherosclerotic lesions were more frequently observed in above-the-knee arteries (92.9%) versus below-the-knee arteries (56.9%). Calcified nodule was more common in above-the-knee arteries than below-the-knee arteries. Acute thrombi were observed in eight above-the-knee arteries and none below-the-knee arteries. Ten below-the-knee vessels revealed chronic total occlusions, half were embolic in origin and the other half had atherosclerotic lesions. Intimal (75.3%) and medial (86.2%) calcifications were commonly seen (Fig. 1.6). Severity of intimal calcification correlated with percent stenosis in above-the-knee arteries, whereas medial calcification did not show any correlation with present stenosis and severity of medial calcification in above- and below-the-knee arteries was not different (Fig. 1.7).

Mönckeberg's disease

The pathological feature observed in PAD that distinguishes it from that of CAD is medial calcification, which is also called Mönckeberg's disease. Mönckeberg's disease is described as a noninflammatory degenerative disease. The media in small- or medium-sized muscular arteries become calcified with or without associated intimal atherosclerosis. At first, the changes occur in the elastic lamina, and then involve the adjacent medial wall where there are abundant smooth muscle cells. Basically, Mönckeberg's disease does not influence intimal layer. The intimal layer and adventitia are spared in most cases. In general, an inflammatory reaction is not observed in the medial wall. Therefore, medial calcification is not associated with necrotic core, smooth muscle cells, or collagen. An incidental plain radiograph might reveal the appearance of railroad tracks consistent with the outline of the arterial wall that indicates medial calcification [45].

Medial calcification more frequently occurs in elderly patients, those with end-stage renal disease, autonomic neuropathies, osteoporosis, or DM [42,46,47]. Even in the lower extremities of young adults, medial calcification was observed up to 50% [42]. Although the mechanism of calcification remains unknown, medial

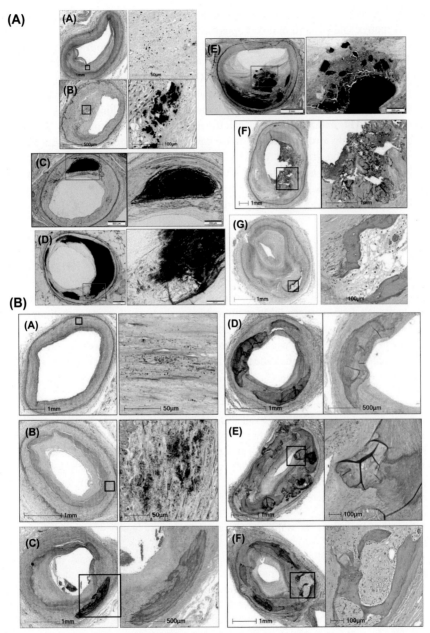

FIGURE 1.6 Representative Histological Images of Intimal.

(A) and Medial (B) Calcium Progression in Human Lower Extremity Peripheral Arteries (A). (A) Intimal microcalcification is observed in areas of lipid pools (boxed area) where there is smooth muscle cell loss from apoptosis, Von Kossa stain from a corresponding calcification is likely associated with lower serum levels of bone-related proteins such as matrix Gla protein and osteopontin, higher levels of alkaline phosphatase, bone sialoprotein, and collagen type II [47]. Mice who lack both osteopontin and matric Gla protein (OPN$-$/$-$, MGP$-$/$-$) show accelerated arterial medial calcification when compared with single knockout (OPN+/+, MGP$-$/$-$) mice [47].

In a recent publication by Kamenskiy et al., 431 superficial femoral arteries (age 13$-$82 years) were examined from legs obtained from organ donors to assess medial calcification. The extent of medial calcification was graded from 0 to 6 with six grades representing circumferential dense calcification. Histologic analysis showed that the calcification grade increased after 40 years and continued to increase with each decade up to the eighth. In addition, the calcification was shown to be related to body mass index, diabetes mellitus, and dyslipidemia, but negatively to smoking. Medial calcification was observed in 46% of patients ($n = 227$) [48].

DM in PAD

Vascular calcification increases with age, risk factors, especially CKD, and type II DM. Ankle brachial index (ABI) has been shown to correlate with PAD. However, early calcification without PAD is not easy to detect by ABI or fluoroscopy [49$-$51]. Kroger et al. did not show any correlation between arterial calcification and Framingham risk factors (age, DM, systolic blood pressure, low-density lipoprotein cholesterol, high-density lipoprotein cholesterol, smoking, and renal function) [52]. In histological study where 36 arteries were enrolled, Vasuri et al. did not find correlations between calcification and any risk factor except hypertension.

---

nondecalcified section showing microcalcification (black dots). The calcifications cannot be seen on Movat pentachrome stained section because of small size. (B) Intimal punctate calcification in the area adjacent to the lipid pool shown in Movat and Von Kossa stained sections (nondecalcified section). (C) Intimal fragment calcification with microcalcification in media (red box) (nondecalcified section). (D) Intimal sheet calcification (arrow) that involves over 50% of the circumference with adjoining two areas of fragments of medial calcifications (arrow heads) (nondecalcified section). (E) Intimal nodular calcification (nondecalcified section). (F) Calcified nodule with luminal fibrin thrombus, note absence of endothelium lining and collagen over the surface of the nodules. (Movat stain, decalcified section) (G) Intimal bone formation. Permission from Torii S et al. JACC imaging in press. (B). Medial microcalcification (black dots) of the media by Von Kossa stain of nondecalcified section in an artery showing adaptive intimal thickening. (B) Medial punctate calcification of the media. Varying degrees (micro to punctate) of calcification is seen circumferentially involving the media. (C) Medial fragment of calcification. (D) Medial sheet calcification involving nearly three quarters of the circumference. (E) Combination of medial nodular and sheet calcification covering the circumference of the medial wall. (F) Bone formation in the presence of near circumferential medial sheet calcification with presence with lacunae containing osteoblasts and bone marrow.

Permission from Torii S et al. JACC imaging in press.

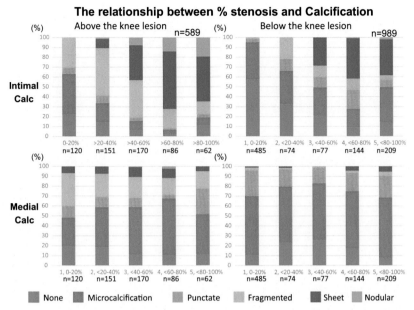

The relationship between % stenosis and Calcification

FIGURE 1.7 Prevalence of Various Types of Intimal and Medial Calcification Morphologies at 20% Incremental Cross-sectional Area Narrowing in Nondecalcified Lesions.

Note: severe intimal calcification (sheet and nodular calcification) was more frequent in AK than BK, beginning as early as 10%—20% cross-sectional area narrowing. On the other hand, severe medial calcification was not as frequent in AK and BK; however, fragment calcification was more frequent in AK lesions compared with BK.

Permission from Torii S et al. JACC imaging in press.

However, the number of cases examined was too few ($n = 143$) and too young (mean of 38 years, range 14—59 years), and only 25% had any calcification. Recently, Kamenskiy et al. reported positive correlations between medial calcification and body mass index, DM, dyslipidemia, CAD, and body mass index as well as hypertension but no correlation with smoking.

In PAD of the lower extremity, intimal and medial calcifications were significantly greater in arteries from DM as compared to those without DM (Fig. 1.8) [44]. In intimal calcification, patients with DM showed 47.9% of sections with calcification characterized as punctate (7.2%), fragmented (15.3%), sheet (23.5%), and nodular (2.0%), whereas in patient without DM 31.0% ($P < .001$) of sectioned showed calcification characterized as punctate (9.9%), fragmented (11.3%), sheet (7.7%), and nodular (2.1%). Similar to intimal calcification, patients with medial calcification and DM also showed higher prevalence of severe calcification (fragmented = 11.5%, sheet = 6.4%, nodular = 0.2%) than those without DM (fragmented = 6.7%, sheet = 0.5%, nodular = 0.1% ($P < .001$).

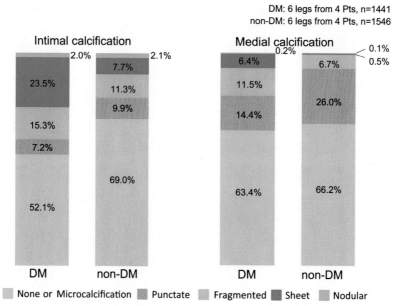

FIGURE 1.8 **Prevalence of Intimal/Medial Calcification in the Lower Limbs Arteries in Patients With Diabetes and Without Diabetes.**

The prevalence of punctate, fragmented, sheet and nodular calcification in both intima and media was significantly higher in patients with diabetes. *DM*, diabetes mellitus.

Permission from Torii S et al. JACC imaging in press.

Correlation of CT with PAD

In our histological study enrolling individuals with PAD, calcification was observed in all legs with inability of imaging to differentiate between intimal and medial calcification. However, ex vivo radiography of the dissected vessels showed higher density of the intimal calcification more as compared to the medial calcification. The extent of calcification was most common in the middle to distal SFA. CT images showed eccentric plaques were predominant rather than concentric plaques. The SFA and popliteal arteries were mostly involved with calcification. The cut-off value of the length and area of calcification for the detection of plaque calcification by CT was determined to be 1.5 mm (sensitivity 97.1%, specificity 96.7%) and 0.5 mm^2 (sensitivity 100%, specificity 96.7%), respectively.

Relationship between CKD and calcification

The pathogenesis of vessel calcification is influenced by many different factors in patients with CKD, and is usually more pronounced than in patients without CKD. The combination of atherosclerotic disease and imbalances in calcium and phosphorus homeostasis results in calcification in the intimal and medial wall in patients with end-stage renal disease. There is strong data supporting a relationship between calcification and CKD. A previous study reported that patients who were younger than 30 years old on hemodialysis showed high prevalence of calcification [53]. Estimated glomerular filtration was correlated inversely with the prevalence of calcification, and increased calcification was observed when transitioning from predialysis to dialysis [54]. These findings are in line with previous other studies [55−57]. Additionally, one of the predictors of an unfavorable outcome in patient undergoing maintenance hemodialysis is coronary artery calcification score [58]. In patients on dialysis, the degree of calcification is proportionate to age, duration of dialysis, presence of DM, abnormalities of mineral metabolism, and use and dose of calcium-based phosphate blinders [53,57,59−62]. In patients with CKD, high serum concentrations of phosphate correlate with degree of calcification, all cause death and cardiovascular death. Even in patients without CKD, all cause death and cardiovascular death are associated with higher level of phosphate [63−65]. Hyperphosphatemia and its treatment can influence clinical outcomes in patients on dialysis. Phosphate-binder treatment has been a focus of interest. The potential mechanism of calcification in CKD is influenced by phosphate stimulation of vascular smooth muscle cell (VSMC) transformation to an osteoblastic phenotype, which releases extracellular matrix vesicles, stimulation of VSMC apoptosis, and inhibition of vessel wall macrophage differentiation to osteoclast-like cells [66−69]. However, a meta-analysis including 77 trials with 12,562 patients with CKD did not show any benefit of phosphate-binder treatment in terms of clinical outcomes, and only small differences were observed in the progression of vascular calcification [70].

Vitamin D has been one of the hot topics for modulation of vascular calcification. Previous studies suggested the usage of high dose of Vitamin D might lead to progression of vascular calcification [71]. On the other hand, proper dose of vitamin D can play a role in protection of calcification [72]. Both low and high concentrations of vitamin D are associated with vascular calcification in patients on dialysis [73]. Calcimimetics also has been a focus of interest, which showed prevention of vascular calcification in in vitro experimental studies [74]. The ADVANCE trial enrolling 165 patients on dialysis treated with cinacalcet plus low-dose vitamin D reported a benefit with less progression of calcification in both arteries and heart valves [75]. However, these treatments did not show any regression of calcification.

Conclusion

Three types of vascular calcification have been described with the most prevalent being atherosclerotic intimal calcification, followed by medial calcification and least frequent is genetic calcification. Although types of calcification can be broadly classified into these three types, it needs to be understood that some overlap in mechanisms likely exists. Atherosclerotic intimal calcification has been liked to bone formation but other factors may also play a role such as loss of calcification-inhibiting pathways and smooth muscle cell and macrophage apoptotic cell death. Factors that govern calcification include alterations in calcium and phosphate balance, especially in patients with CKD, DM, and lipid oxidation products. Despite the undoubted association between coronary atherosclerotic calcification and prediction of future cardiovascular events, there is only weak correlation between calcification-induced coronary plaque progression and luminal narrowing. Calcification is greater in men than women especially in the premenopausal period. African Americans have less calcification than Caucasians. Atherosclerosis of the femoral arteries likely develops more slowly as compared to calcification in the coronary arteries, and is dominated by fibrous plaques with severe calcification. Mönckeberg's medial calcification is seen in peripheral (not coronary) arteries, is a common phenomenon, and is accelerated in patients with diabetes. The media in small or medium-sized muscular arteries become calcified with or without association with atherosclerotic process. The presence of risk factors for atherosclerosis can accelerate calcification of both intima and media. Furthering our understanding of vascular calcification toward better risk stratification to improve outcomes in both CAD and PAD because calcification is associated with poor prognosis.

References

[1] Demer LL, Tintut Y. Inflammatory, metabolic, and genetic mechanisms of vascular calcification. Arterioscler Thromb Vasc Biol 2014;34:715–23.

[2] Qiao JH, Mertens RB, Fishbein MC, Geller SA. Cartilaginous metaplasia in calcified diabetic peripheral vascular disease: morphologic evidence of enchondral ossification. Hum Pathol 2003;34:402–7.

[3] Otsuka F, Sakakura K, Yahagi K, Joner M, Virmani R. Has our understanding of calcification in human coronary atherosclerosis progressed? Arterioscler Thromb Vasc Biol 2014;34:724–36.

[4] Burke AP, Weber DK, Kolodgie FD, Farb A, Taylor AJ, Virmani R. Pathophysiology of calcium deposition in coronary arteries. Herz 2001;26:239–44.

[5] Burke AP, Kolodgie FD, Farb A, Weber D, Virmani R. Morphological predictors of arterial remodeling in coronary atherosclerosis. Circulation 2002;105:297–303.

[6] Mori H, Torii S, Kutyna M, Sakamoto A, Finn AV, Virmani R. Coronary artery calcification and its progression: what does it really mean? JACC Cardiovasc Imaging 2018; 11:127–42.

[7] Sangiorgi G, Rumberger JA, Severson A, Edwards WD, Gregoire J, Fitzpatrick LA, Schwartz RS. Arterial calcification and not lumen stenosis is highly correlated with atherosclerotic plaque burden in humans: a histologic study of 723 coronary artery segments using nondecalcifying methodology. J Am Coll Cardiol 1998;31:126–33.

[8] Glagov S, Weisenberg E, Zarins CK, Stankunavicius R, Kolettis GJ. Compensatory enlargement of human atherosclerotic coronary arteries. N Engl J Med 1987;316: 1371–5.

[9] Burke AP, Virmani R, Galis Z, Haudenschild CC, Muller JE. 34th bethesda conference: task force #2–what is the pathologic basis for new atherosclerosis imaging techniques? J Am Coll Cardiol 2003;41:1874−86.

[10] Williams JK, Adams MR, Klopfenstein HS. Estrogen modulates responses of atherosclerotic coronary arteries. Circulation 1990;81:1680−7.

[11] Burke AP, Farb A, Malcom G, Virmani R. Effect of menopause on plaque morphologic characteristics in coronary atherosclerosis. Am Heart J 2001;141:S58−62.

[12] Manson JE, Allison MA, Rossouw JE, Carr JJ, Langer RD, Hsia J, Kuller LH, Cochrane BB, Hunt JR, Ludlam SE, Pettinger MB, Gass M, Margolis KL, Nathan L, Ockene JK, Prentice RL, Robbins J, Stefanick ML. Estrogen therapy and coronary-artery calcification. N Engl J Med 2007;356:2591−602.

[13] Bild DE, Detrano R, Peterson D, Guerci A, Liu K, Shahar E, Ouyang P, Jackson S, Saad MF. Ethnic differences in coronary calcification: the multi-ethnic study of atherosclerosis (mesa). Circulation 2005;111:1313−20.

[14] Lee TC, O'Malley PG, Feuerstein I, Taylor AJ. The prevalence and severity of coronary artery calcification on coronary artery computed tomography in black and white subjects. J Am Coll Cardiol 2003;41:39−44.

[15] Burke A, Farb A, Kutys R, Zieske A, Weber D, Virmani R. Atherosclerotic coronary plaques in african americans are less likely to calcify than coronary plaques in caucasian americans. Circulation 2002 Nov 5;106(Suppl. 19):II1−766.

[16] Kiel DP, Kauppila LI, Cupples LA, Hannan MT, O'Donnell CJ, Wilson PW. Bone loss and the progression of abdominal aortic calcification over a 25 year period: the framingham heart study. Calcif Tissue Int 2001;68:271−6.

[17] Mauriello A, Servadei F, Zoccai GB, Giacobbi E, Anemona L, Bonanno E, Casella S. Coronary calcification identifies the vulnerable patient rather than the vulnerable plaque. Atherosclerosis 2013;229:124−9.

[18] Huang CC, Lloyd-Jones DM, Guo X, Rajamannan NM, Lin S, Du P, Huang Q, Hou L, Liu K. Gene expression variation between african americans and whites is associated with coronary artery calcification: the multiethnic study of atherosclerosis. Physiol Genom 2011;43:836−43.

[19] Fornage M, Boerwinkle E, Doris PA, Jacobs D, Liu K, Wong ND. Polymorphism of the soluble epoxide hydrolase is associated with coronary artery calcification in african-american subjects: the coronary artery risk development in young adults (cardia) study. Circulation 2004;109:335−9.

[20] Rumberger JA, Simons DB, Fitzpatrick LA, Sheedy PF, Schwartz RS. Coronary artery calcium area by electron-beam computed tomography and coronary atherosclerotic plaque area. A histopathologic correlative study. Circulation 1995;92:2157−62.

[21] Raggi P, Shaw LJ, Berman DS, Callister TQ. Prognostic value of coronary artery calcium screening in subjects with and without diabetes. J Am Coll Cardiol 2004;43: 1663−9.

[22] Carson AP, Steffes MW, Carr JJ, Kim Y, Gross MD, Carnethon MR, Reis JP, Loria CM, Jacobs JrJr, Lewis CE. Hemoglobin A_{1c} and the progression of coronary artery calcification among adults without diabetes. Diabetes Care 2015;38:66−71.

[23] Burke AP, Kolodgie FD, Zieske A, Fowler DR, Weber DK, Varghese PJ, Farb A, Virmani R. Morphologic findings of coronary atherosclerotic plaques in diabetics: a postmortem study. Arterioscler Thromb Vasc Biol 2004;24:1266−71.

[24] Burke AP, Kolodgie FD, Farb A, Weber DK, Malcom GT, Smialek J, Virmani R. Healed plaque ruptures and sudden coronary death: evidence that subclinical rupture has a role in plaque progression. Circulation 2001;103:934−40.

[25] Blaha MJ, Mortensen MB, Kianoush S, Tota-Maharaj R, Cainzos-Achirica M. Coronary artery calcium scoring: is it time for a change in methodology? JACC Cardiovasc Imaging 2017;10:923−37.

[26] Blankstein R, Budoff MJ, Shaw LJ, Goff JrJr, Polak JF, Lima J, Blumenthal RS, Nasir K. Predictors of coronary heart disease events among asymptomatic persons with low low-density lipoprotein cholesterol mesa (multi-ethnic study of atherosclerosis). J Am Coll Cardiol 2011;58:364−74.

[27] Joshi PH, Blaha MJ, Blumenthal RS, Blankstein R, Nasir K. What is the role of calcium scoring in the age of coronary computed tomographic angiography? J Nucl Cardiol 2012;19:1226−35.

[28] Raggi P. Prognostic implications of absolute and relative calcium scores. Herz 2001;26: 252−9.

[29] Hoffmann U, Moselewski F, Nieman K, Jang IK, Ferencik M, Rahman AM, Cury RC, Abbara S, Joneidi-Jafari H, Achenbach S, Brady TJ. Noninvasive assessment of plaque morphology and composition in culprit and stable lesions in acute coronary syndrome and stable lesions in stable angina by multidetector computed tomography. J Am Coll Cardiol 2006;47:1655−62.

[30] Lusis AJ. Atherosclerosis. Nature 2000;407:233−41.

[31] Fowkes FG, Rudan D, Rudan I, Aboyans V, Denenberg JO, McDermott MM, Norman PE, Sampson UK, Williams LJ, Mensah GA, Criqui MH. Comparison of global estimates of prevalence and risk factors for peripheral artery disease in 2000 and 2010: a systematic review and analysis. Lancet 2013;382:1329−40.

[32] Norgren L, Hiatt WR, Dormandy JA, Nehler MR, Harris KA, Fowkes FG, Bell K, Caporusso J, Durand-Zaleski I, Komori K, Lammer J, Liapis C, Novo S, Razavi M, Robbs J, Schaper N, Shigematsu H, Sapoval M, White C, White J, Clement D, Creager M, Jaff M, Mohler 3rd3rd, Rutherford RB, Sheehan P, Sillesen H, Rosenfield K. Inter-society consensus for the management of peripheral arterial disease (Tasc II). Eur J Vasc Endovasc Surg 2007;33(Suppl. 1):S1−75.

[33] Creager MA, Belkin M, Bluth EI, Casey JrJr, Chaturvedi S, Dake MD, Fleg JL, Hirsch AT, Jaff MR, Kern JA, Malenka DJ, Martin ET, Mohler 3rd3rd, Murphy T, Olin JW, Regensteiner JG, Rosenwasser RH, Sheehan P, Stewart KJ, Treat-Jacobson D, Upchurch JrJr, White CJ, Ziffer JA, Hendel RC, Bozkurt B, Fonarow GC, Jacobs JP, Peterson PN, Roger VL, Smith EE, Tcheng JE, Wang T, Weintraub WS. 2012 accf/aha/acr/scai/sir/sts/svm/svn/svs key data elements and definitions for peripheral atherosclerotic vascular disease: a report of the american college of cardiology foundation/american heart association task force on clinical data standards (writing committee to develop clinical data standards for peripheral atherosclerotic vascular disease). Circulation 2012;125:395−467.

[34] Sawabe M, Arai T, Kasahara I, Hamamatsu A, Esaki Y, Nakahara K, Harada K, Chida K, Yamanouchi H, Ozawa T, Takubo K, Murayama S, Tanaka N. Sustained progression and loss of the gender-related difference in atherosclerosis in the very old: a pathological study of 1074 consecutive autopsy cases. Atherosclerosis 2006;186: 374−9.

[35] Dalager S, Paaske WP, Kristensen IB, Laurberg JM, Falk E. Artery-related differences in atherosclerosis expression: implications for atherogenesis and dynamics in intima-media thickness. Stroke 2007;38:2698–705.

[36] Herisson F, Heymann MF, Chetiveaux M, Charrier C, Battaglia S, Pilet P, Rouillon T, Krempf M, Lemarchand P, Heymann D, Goueffic Y. Carotid and femoral atherosclerotic plaques show different morphology. Atherosclerosis 2011;216:348–54.

[37] Diehm N, Silvestro A, Baumgartner I, Do DD, Diehm C, Schmidli J, Dick F. Chronic critical limb ischemia: European experiences. J Cardiovasc Surg 2009;50:647–53.

[38] Nguyen LL, Hevelone N, Rogers SO, Bandyk DF, Clowes AW, Moneta GL, Lipsitz S, Conte MS. Disparity in outcomes of surgical revascularization for limb salvage: race and gender are synergistic determinants of vein graft failure and limb loss. Circulation 2009;119:123–30.

[39] Timaran CH, Stevens SL, Freeman MB, Goldman MH. Predictors for adverse outcome after iliac angioplasty and stenting for limb-threatening ischemia. J Vasc Surg 2002;36: 507–13.

[40] Ortmann J, Nuesch E, Traupe T, Diehm N, Baumgartner I. Gender is an independent risk factor for distribution pattern and lesion morphology in chronic critical limb ischemia. J Vasc Surg 2012;55:98–104.

[41] Soor GS, Vukin I, Leong SW, Oreopoulos G, Butany J. Peripheral vascular disease: who gets it and why? A histomorphological analysis of 261 arterial segments from 58 cases. Pathology 2008;40:385–91.

[42] Kolodgie F, Nakazawa G, Santorgi G, Ladich E, Burke A, Virmani R, Gloviczki P. Differences and commons in pathology and reaction on stents between cardiac and peripheral arteries. In: Chakfé N, durand B, kretz J, editors. European symposium of vascular biomaterials 2007 New Technologies in vascular biomaterials. Strasbourg: EUROPROT; 2007. P49–70.

[43] Narula N, Dannenberg AJ, Olin JW, Bhatt DL, Johnson KW, Nadkarni G, Min J, Torii S, Poojary P, Anand SS, Bax JJ, Yusuf S, Virmani R, Narula J. Pathology of peripheral artery disease in patients with critical limb ischemia. J Am Coll Cardiol 2018;72: 2152–63.

[44] Torii S, Mustapha AJ, Narula J, Mori H, Saab F, Jinnouchi H, Yahagi K, Sakamoto A, Romero EM, Narula N, Kolodgie FD, Virmani R, Finn AV. Histopathologic characterization of peripheral arteries in subjects with abundant risk factors. JACC Cardiovasc Imaging 2018 Dec 6. pii: S1936–878X(18)31008-8. https://doi.org/10.1016/j.jcmg. 2018.08.039. [Epub ahead of print].

[45] Derksen WJ, de Vries JP, Vink A, Velema E, Vos JA, de Kleijn D, Moll FL, Pasterkamp G. Histologic atherosclerotic plaque characteristics are associated with restenosis rates after endarterectomy of the common and superficial femoral arteries. J Vasc Surg 2010;52:592–9.

[46] Lachman AS, Spray TL, Kerwin DM, Shugoll GI, Roberts WC. Medial calcinosis of monckeberg. A review of the problem and a description of a patient with involvement of peripheral, visceral and coronary arteries. Am J Med 1977;63:615–22.

[47] Amos RS, Wright V. Monckeberg's arteriosclerosis and metabolic bone disease. Lancet 1980;2:248–9.

[48] Kamenskiy A, Poulson W, Sim S, Reilly A, Luo J, MacTaggart J. Prevalence of calcification in human femoropopliteal arteries and its association with demographics, risk factors, and arterial stiffness. Arterioscler Thromb Vasc Biol 2018;38:e48–57.

[49] Deas JrJr, Marshall AP, Bian A, Shintani A, Guzman RJ. Association of cardiovascular and biochemical risk factors with tibial artery calcification. Vasc Med 2015;20:326—31.

[50] Shao JS, Cheng SL, Sadhu J, Towler DA. Inflammation and the osteogenic regulation of vascular calcification: a review and perspective. Hypertension 2010;55:579—92.

[51] Moe SM, Chen NX. Pathophysiology of vascular calcification in chronic kidney disease. Circ Res 2004;95:560—7.

[52] Kroger K, Stang A, Kondratieva J, Moebus S, Beck E, Schmermund A, Mohlenkamp S, Dragano N, Siegrist J, Jockel KH, Erbel R. Prevalence of peripheral arterial disease — results of the heinz nixdorf recall study. Eur J Epidemiol 2006;21:279—85.

[53] Goodman WG, Goldin J, Kuizon BD, Yoon C, Gales B, Sider D, Wang Y, Chung J, Emerick A, Greaser L, Elashoff RM, Salusky IB. Coronary-artery calcification in young adults with end-stage renal disease who are undergoing dialysis. N Engl J Med 2000; 342:1478—83.

[54] Sigrist M, Bungay P, Taal MW, McIntyre CW. Vascular calcification and cardiovascular function in chronic kidney disease. Nephrol Dial Transplant 2006;21:707—14.

[55] Baber U, de Lemos JA, Khera A, McGuire DK, Omland T, Toto RD, Hedayati SS. Non-traditional risk factors predict coronary calcification in chronic kidney disease in a population-based cohort. Kidney International 2008;73:615—21.

[56] Block GA, Raggi P, Bellasi A, Kooienga L, Spiegel DM. Mortality effect of coronary calcification and phosphate binder choice in incident hemodialysis patients. Kidney Int 2007;71:438—41.

[57] Raggi P, Boulay A, Chasan-Taber S, Amin N, Dillon M, Burke SK, Chertow GM. Cardiac calcification in adult hemodialysis patients. A link between end-stage renal disease and cardiovascular disease? J Am Coll Cardiol 2002;39:695—701.

[58] Matsuoka M, Iseki K, Tamashiro M, Fujimoto N, Higa N, Touma T, Takishita S. Impact of high coronary artery calcification score (cacs) on survival in patients on chronic hemodialysis. Clin Exp Nephrol 2004;8:54—8.

[59] Oh J, Wunsch R, Turzer M, Bahner M, Raggi P, Querfeld U, Mehls O, Schaefer F. Advanced coronary and carotid arteriopathy in young adults with childhood-onset chronic renal failure. Circulation 2002;106:100—5.

[60] Chertow GM, Raggi P, Chasan-Taber S, Bommer J, Holzer H, Burke SK. Determinants of progressive vascular calcification in haemodialysis patients. Nephrol Dial Transplant 2004;19:1489—96.

[61] Chertow GM, Burke SK, Raggi P. Sevelamer attenuates the progression of coronary and aortic calcification in hemodialysis patients. Kidney Int 2002;62:245—52.

[62] Guerin AP, London GM, Marchais SJ, Metivier F. Arterial stiffening and vascular calcifications in end-stage renal disease. Nephrol Dial Transplant 2000;15:1014—21.

[63] Adeney KL, Siscovick DS, Ix JH, Seliger SL, Shlipak MG, Jenny NS, Kestenbaum BR. Association of serum phosphate with vascular and valvular calcification in moderate ckd. J Am Soc Nephrol 2009;20:381—7.

[64] Kestenbaum B, Sampson JN, Rudser KD, Patterson DJ, Seliger SL, Young B, Sherrard DJ, Andress DL. Serum phosphate levels and mortality risk among people with chronic kidney disease. J Am Soc Nephrol 2005;16:520—8.

[65] Dhingra R, Sullivan LM, Fox CS, Wang TJ, D'Agostino SrSr, Gaziano JM, Vasan RS. Relations of serum phosphorus and calcium levels to the incidence of cardiovascular disease in the community. Arch Intern Med 2007;167:879—85.

[66] Ciceri P, Volpi E, Brenna I, Arnaboldi L, Neri L, Brancaccio D, Cozzolino M. Combined effects of ascorbic acid and phosphate on rat vsmc osteoblastic differentiation. Nephrol Dial Transplant 2012;27:122−7.

[67] Giachelli CM, Speer MY, Li X, Rajachar RM, Yang H. Regulation of vascular calcification: roles of phosphate and osteopontin. Circ Res 2005;96:717−22.

[68] Ciceri P, Elli F, Cappelletti L, Tosi D, Braidotti P, Bulfamante G, Cozzolino M. A new in vitro model to delay high phosphate-induced vascular calcification progression. Mol Cell Biochem 2015;410:197−206.

[69] Mozar A, Haren N, Chasseraud M, Louvet L, Maziere C, Wattel A, Mentaverri R, Morliere P, Kamel S, Brazier M, Maziere JC, Massy ZA. High extracellular inorganic phosphate concentration inhibits rank-rankl signaling in osteoclast-like cells. J Cell Physiol 2008;215:47−54.

[70] Palmer SC, Gardner S, Tonelli M, Mavridis D, Johnson DW, Craig JC, French R, Ruospo M, Strippoli GF. Phosphate-binding agents in adults with ckd: a network meta-analysis of randomized trials. Am J Kidney Dis 2016;68:691−702.

[71] Mizobuchi M, Ogata H, Koiwa F, Kinugasa E, Akizawa T. Vitamin d and vascular calcification in chronic kidney disease. Bone 2009;45(Suppl. 1):S26−9.

[72] Lim K, Lu TS, Molostvov G, Lee C, Lam FT, Zehnder D, Hsiao LL. Vascular klotho deficiency potentiates the development of human artery calcification and mediates resistance to fibroblast growth factor 23. Circulation 2012;125:2243−55.

[73] Cozzolino M, Brandenburg V. Paricalcitol and outcome: a manual on how a vitamin d receptor activator (vdra) can help us get down the "u". Clin Nephrol 2009;71: 593−601.

[74] Ciceri P, Elli F, Brenna I, Volpi E, Brancaccio D, Cozzolino M. The calcimimetic calindol prevents high phosphate-induced vascular calcification by upregulating matrix GLA protein. Nephron Experimental Nephrology 2012;122:75−82.

[75] Raggi P, Chertow GM, Torres PU, Csiky B, Naso A, Nossuli K, Moustafa M, Goodman WG, Lopez N, Downey G, Dehmel B, Floege J. The advance study: a randomized study to evaluate the effects of cinacalcet plus low-dose vitamin d on vascular calcification in patients on hemodialysis. Nephrol Dial Transplant 2011;26:1327−39.

Coronary calcification and atherosclerosis progression

2

Atsushi Sakamoto, MD [1], **Hiroyuki Jinnouchi, MD** [1], **Sho Torii, MD** [1],
Renu Virmani, MD [1], **Aloke V. Finn, MD** [1,2,3]

[1]*CVPath Institute, Gaithersburg, MD, United States;* [2]*CVPath institute, Gaithersburg, MD, United States; Medical Director, Pathology, CVPath Institute, Gaithersburg, MD, United States;* [3]*Associate Clinical Professor, School of Medicine, University of Maryland, Baltimore, MD, United States*

Introduction

The presence of vascular calcium is synonymous with the development of atherosclerosis and has long been used as a surrogate marker for the presence of vascular disease. In general, coronary artery calcium (CAC) correlates with the burden of atherosclerosis and is an established predictor of future cardiac events. Although clinical studies using computed tomography, the most commonly used noninvasive tool to measure coronary calcium, suggest a linear relationship between the amount and extent of calcification and risk of coronary events, this modality is limited by its sensitivity for detecting small amounts of coronary calcium. Indeed, at the pathological level, the relationship between CAC and plaque instability is extremely complex and not fully understood. In this chapter, we will focus on human coronary artery disease from a pathologic standpoint and examine the relationship between different plaque morphologies of coronary calcium and their relationship to plaque progression.

Natural progression of atherosclerosis and calcification
Adaptive intimal thickening and fatty streaks (intimal xanthoma)

According to pathologic observations, nonatherosclerotic intimal lesions, that is, adaptive intimal thickening (AIT) or diffuse intimal thickening (AHA Type I) (Fig. 2.1), are present from birth [1], are often observed in atherosclerosis-prone arteries, and are considered a response to blood flow rather than the actual atherosclerotic process. Nevertheless, AIT is also considered an important transitional nonatherosclerotic plaque that can develop into more advanced atherosclerotic stage, that is, pathologic intimal thickening (PIT) [2], by poorly understood processes.

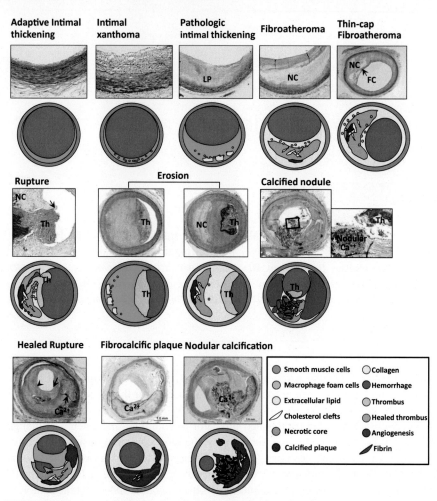

FIGURE 2.1 Spectrum of Atherosclerosis in Human Coronary Arteries.

The two nonprogressive lesions are adaptive intimal thickening (AHA type I) and intimal xanthomas (foam cell collections known as fatty streaks, AHA type II). Pathological intimal thickening (PIT, AHA type III, transitional lesions) marks the first of the progressive plaques as they are the assumed precursors to more advanced fibroatheroma (FA). Thin-cap fibroatheroma (TCFA) are considered precursors to plaque rupture. Essentially missing from the AHA consensus classification are two alternative entities that give rise to coronary thrombosis, namely erosion and the calcified nodule. Erosions can occur on a substrate of PIT or FA that while calcified nodules depict eruptive fragments of calcium that protrude into the lumen causing a thrombotic event. Healed plaque ruptures are lesions with generally smaller necrotic cores and focal areas of calcification where the surface generally shows areas of healing rich in proteoglycans. Multiple healed plaque ruptures are thought responsible for progressive luminal narrowing. Advanced lesions complicated by calcification presenting as "fibrocalcific" and/or "nodular calcification" can arise from FAs and healed rupture plaques. Mechanical factors within heavily calcified tortious vessels can lead to fragmentation giving rise to nodular calcification or calcified nodule with luminal thrombosis. Ca^{2+}, calcification; *FC*, fibrous cap; *LP*, lipid pool; *NC*, necrotic core; *Th*, thrombus.

Modified with permission from Virmani R, Kolodgie FD, Burke AP, et al. Lessons from sudden coronary death: a comprehensive morphological classification scheme for atherosclerotic lesions. Arterioscler Thromb Vasc Biol May 2000;20(5):1262–1275. PubMed PMID: 10807742; Yahagi K et al. Nat Rev Cardiol. 2016 Feb; 13(2): 79–98.

On the other hand, fatty streak or intimal xanthomas (AHA Type II) (Fig. 2.1) are present less frequently than AIT in coronary beds and are primarily composed of infiltrating foamy macrophages and, to a lesser extent, lipid-laden smooth muscle cells (SMC) within the thickened intima. Fatty streaks have been shown to regress over time especially in thoracic aorta [3] and the reason why macrophages emerge and diminish is uncertain. However, these lesions are thought to evolve early and, in children, are seen at similar locations to where more advanced lesions are expected to develop in adults [3]. Histologically, regions of intimal thickening contain SMCs, with proteoglycan-rich extracellular matrix (ECM) without inflammation. Although some studies support the concept that intimal proliferation is a precursor of a more advanced atherosclerotic lesion [3,4], another belief is that intimal proliferation is an adaptive vascular reaction to blood flow in arteries [4].

In these early lesions of AIT and fatty streak, coronary intimal calcification is absent or minimal, and the earliest stage showing calcification is the PIT.

Pathologic intimal thickening—the earliest lesion exhibiting calcification

PIT (AHA Type III) is regarded as the earliest of the progressive atherosclerotic lesions. PIT is characterized by lack or presence of SMC remnants composed predominantly of ECM of several proteoglycans and collagen type III, defined as "lipid pool," and is located near the media (Fig. 2.1) [4]. Lipid pool should not be confused with necrotic core that contains cell debris and is lacking ECM, focally or diffusely. PIT likely represents the earliest phase of atherosclerotic lesion development. Macrophages are absent in the early PITs lesions thus indicating this entity does not originate from fatty streaks. However, Nakagawa and Nakashima have recently shown that mild deposition of extracellular lipids in the middle to the deep layer of the intima is observed in diffuse intimal thickening (DIT) and that at this phase there is no smooth muscle cell loss. Apolipoprotein B (ApoB) and fibrinogen were colocalized with Sudan IV—positive extracellular lipids in lesions of PIT but not in DIT or in fatty streaks. Their findings suggested that extracellular lipids are derived from the plasma, and not from macrophage foam cells [5].

Further, fatty streaks are generally highly cellular lesions while PITs, particularly within areas of the lipid pool, are devoid of nuclei. The "lipid pools" areas are rich in lipids. Nevertheless, several studies have proposed that negatively charged sulfate proteoglycans, such as versican, biglycan, and decorin, contribute to lipoprotein retention that is thought to be the initial proatherogenic step [5–7]. However, no proof exists today of this precise mechanism. Remnants of apoptotic SMCs in PIT are generally visualized by a thickened basement membrane on periodic acid-Schiff (PAS) staining and are thought to contribute to the lipid pool: a process further supporting expansion of the plaque (Fig. 2.2A) [8,9].

PIT is considered as the earliest lesions that exhibit calcification found in lipid pools arising as microcalcification ($\geq 0.5\,\mu m$, but typically $<15\,\mu m$) recognized by special stains for calcium (i.e., von Kossa or Alizarin red) (Figs. 2.2A and 2.3)

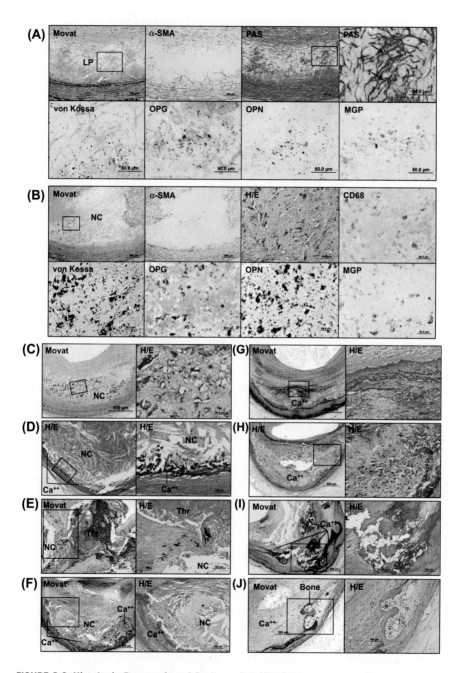

FIGURE 2.2 Histologic Progression of Coronary Calcification.

Nondecalcified arterial segments (A and B) and decalcified segments (C–J) were serially cut for the microscopic assessment. (A) Pathologic intimal thickening (PIT) characterized

[10]. SMC apoptosis or matrix vesicles are considered as origin of early microcalcification, detectable in light microscopy when the size is generally ≥ 1 µm [8,11,12]. Furthermore, electron microscopic examination of vascular wall has shown that the initial calcification occurs in matrix vesicles that vary in size from 100 to 700 nm in diameter [13]. In general, calcification is located in the intima close to the internal elastic lamina. Moreover, calcification detected within the lipid pool revealed coexistence with bone-related proteins such as osteoprotegerin, osteopontin, and matrix Gla protein (Fig. 2.2A).

Fibroatheroma

Fibroatheromas (FA) are characterized by the formation of a acellular necrotic core, which must be distinguished from lipid pool lesions of PIT as they imply further progression of atherosclerosis (AHA Type IV) [10]. In line with the AHA consensus definition of coronary atherosclerotic plaques, FAs are the first clearly

◄―――

by lipid pool (LP) that lacks smooth muscle cells (SMCs; negative for α-smooth muscle actin [α-SMA]) and the presence of apoptotic SMCs that can be identified by prominent basement membrane that stains positive with periodic acid-Schiff (PAS), and the arrows point to in the high-power image (top right corner). Early microcalcification (≥ 0.5 µm, typically <15 µm in diameter) likely results from SMC apoptosis, and calcification is detected by von Kossa staining within the LP (corresponding with a boxed area in the Movat image) where bone-related proteins such as osteoprotegerin (OPG), osteopontin (OPN), and matrix Gla protein (MGP) are detected. Early necrotic core (NC; B) not only lacks SMCs but also is infiltrated by macrophages that eventually undergo apoptosis and calcification, which is observed as punctate (≥ 15 µm) areas of calcification. The microcalcification in early NC shows variable amounts of staining for macrophage CD68 antigen; however, von Kossa staining clearly shows relatively larger punctate areas of calcification resulting from macrophage cell death within the NC as compared with microcalcification of dying SMCs. These calcified macrophages show colocalization of bone-related proteins. Substantial amount of macrophage calcification can be observed in early NC (C), but the degree of calcification in NC typically increases toward the medial wall where fragmented calcifications can be seen (D). Microcalcification resulting from macrophage or SMC deaths can also be detected within a thin fibrous cap and may be associated with plaque rupture (E). Calcification generally progresses into the surrounding area of the NC (F), which leads to the development of sheets of calcification where both collagen matrix (G) and NC itself are calcified (H). Nodular calcification may occur within the plaque in the absence of luminal thrombus and is characterized by breaks in calcified plates with fragments of calcium separated by fibrin (I). Ossification may occur at the edge of an area of calcification especially in nodular calcification (J). *H/E,* hematoxylin and eosin; *Thr,* thrombus.

Reproduced with permission from Otsuka F, Sakakura K, Yahagi K, et al. Has our understanding of calcification in human coronary atherosclerosis progressed? Arterioscler Thromb Vasc Biol April 2014;34(4):724–736. https://doi.org/10.1161/atvbaha.113.302642. PubMed PMID: 24558104; PubMed Central PMCID: PMC4095985.

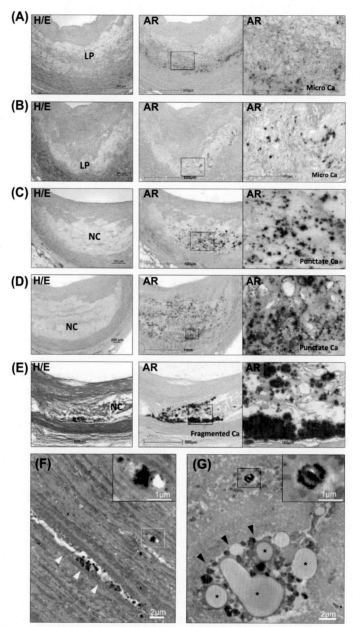

FIGURE 2.3 Coronary Microcalcification in Pathologic Intimal Thickening and Fibroatheroma.

(A—E): Representative histologic nondecalcified sections of pathologic intimal thickening (PIT) with (B) or without (A) macrophage infiltration and early fibroatheroma (C, D) showing progression of micro ($\geq$0.5 μm, typically <15 μm in diameter) and punctate distinguishable plaques consisting of a well-demarcated lipid-rich necrotic core encapsulated by surrounding fibrous tissue (Fig. 2.1) [4,14]. FAs can be further subdivided into early and late stages. The early stage of FA shows infiltration of macrophages into the lipid pool accompanied by focal loss of the matrix of proteoglycans including hyaluronan, versican, and biglycan and other matrix proteins. Focal regions of free cholesterol may also be present in early FA. In contrast, discrete collections of cellular debris and increased amounts of free cholesterol are observed in late FAs. The ECM of necrotic core is almost completely depleted, presumably due to release of matrix metalloproteinases from macrophages, resulting in complete breakdown of proteoglycans and collagen. In addition, large numbers of macrophages are often observed within the necrotic core showing features consistent with apoptosis; thus, free apoptotic bodies are abundant within the necrotic core. The late FA can show substantial luminal stenosis from episodes of intraplaque hemorrhage, probably derived from leaky vasa vasorum characterized by loose endothelial cell—cell junction within the intimal plaque.

The FAs are often accompanied by focal calcification (Figs. 2.2B and 2.3). Micro and punctate calcification in necrotic core may occur not only from SMC remnants but also from macrophage releasing matrix vesicles or from apoptosis [15]. As previously stated, early necrotic core shows infiltration of macrophages that undergo apoptosis and calcification, which is observed as punctate (≥ 15 μm) areas of calcification. Variable amounts of staining for CD68 antigen originating from macrophages are detected in early necrotic core with microcalcifications (Fig. 2.2B). Although SMC apoptosis causes fine microcalcification, macrophage apoptosis exhibits large punctate, blocky appearance. Of note, von Kossa or Alizarin red staining clearly shows relatively larger punctate calcifications resulting from macrophage cell death in the area of necrotic core as compared with microcalcifications from dying SMCs. Additionally, these calcified macrophages also show colocalization of bone-related proteins (Fig. 2.2B). Extensive amounts of macrophage-derived calcification can be observed in early necrotic core and is usually observed within

---

(≥ 15 μm) calcification as assessed by H/E and Alizarin red staining. Histologic section of late fibroatheroma (E) indicating fragmented (≥ 1 mm) calcification located nearby necrotic core. (F—G): Representative transmission electron microscopic images of microcalcification (white rectangle) derived from vascular smooth muscle cell (VSMC) (F) in an area devoid of macrophages and (G) shows macrophage in late fibroatheroma with areas of calcification (black rectangle). Apoptotic VSMC (white arrow head) and macrophage (black arrow head) are observed in each respective image. Macrophage contains lipid droplets (*) showing foam cell phenotype. AR, Alizarin red; Ca, calcification; H/E, hematoxylin and eosin; LP, lipid pool; NC, necrotic core.

Modified and reproduced with permission from Otsuka F, Kramer MC, Woudstra P, et al. Natural progression of atherosclerosis from pathologic intimal thickening to late fibroatheroma in human coronary arteries: a pathology study. Atherosclerosis August 2015;241(2):772—782. https://doi.org/10.1016/j.atherosclerosis.2015.05. 011. PubMed PMID: 26058741; PubMed Central PMCID: PMC4510015.

the lipid pools. These microcalcifications often coalesce into larger masses and involve both the necrotic core and the surrounding collagen-rich ECM to form speckled and fragments of calcification. This specific pattern of calcification also begins from the deeper region of the necrotic core close to the internal elastic lamina (Fig. 2.2C). Further calcification progress and extend from the outer rim of the necrotic core (Fig. 2.2D) into the surrounding collagenous matrix. It is notable that the center of necrotic core either becomes fully calcified or may remain noncalcified at large stages. Nevertheless, further progression of disease results in calcified plaque that forms calcified sheets or plates.

Thin-cap fibroatheroma

During the progression to an advanced fibroatheromatous lesion, the necrotic core has an overlying layer of fibrous tissue, that is, fibrous cap, which is composed mostly of type I and III collagen, proteoglycans, and interspersed SMCs with overlying endothelial cell layer. As the plaque progresses, the fibrous cap begins to thin and results in the formation of a thin fibrous cap fibroatheroma (TCFA) (Fig. 2.1), conceptually often called as a "vulnerable plaque," that may ultimately rupture potentially causing acute coronary syndrome (ACS) and sudden cardiac death. In general, TCFA possesses a relatively large necrotic core (usually >25% of the plaque area) [16] covered with thin fibrous cap that is composed mainly of type I collagen with heavy infiltration by macrophages and, to a lesser extent, T lymphocytes [17−19]. SMCs are quite rare or absent in the fibrous cap due to apoptosis. The thickness of fibrous cap is one of the determining factors of plaque vulnerability and is defined as ≤65 μm, a threshold based on the measurements of thickness of the cap closest to the site of rupture. The mean cap thickness measured 23 ± 19 μm with 95% of caps measuring less than 65-μm within a limit of only two standard deviations [20]. The thinning or weakening of the fibrous cap is an important precursor to plaque rupture [4,21]. Even though there are similar features, TCFAs show several differences from ruptured plaques, such as smaller necrotic cores, fewer macrophages within the fibrous cap, less plaque burden, and less luminal narrowing [19].

Although coronary artery calcification highly correlates with total coronary plaque volume, its effect on plaque instability is still to be elucidated. A previous series of sudden coronary death cases indicated that over 50% of TCFA lesions lacked calcification or had only speckled calcification on postmortem radiographs of coronary arteries. Remaining lesions showed either fragmented or diffuse calcification, meaning a large variation in the degree of calcification within the "vulnerable" plaque [22]. Conversely, 65% of acute rupture lesions contained speckled calcification and remaining 35% showed fragmented or diffuse calcification [22]. When the degree of calcification in TCFA and ruptured plaques was estimated by a semiquantitative method, the mean calcification score was significantly higher in ruptured than in TCFAs, confirming the postmortem radiographic findings. Although the data demonstrated the lack of specificity of calcium patterns in unstable coronary plaques, it suggests that mildly to moderately calcified segments are the most likely to rupture. Contrary to necrotic cores that increase fibrous cap stresses, calcification does not increase fibrous cap stress in typical ruptured or stable human coronary atherosclerotic lesions [23]. In addition, microcalcification within the fibrous cap has been attributed to cap ruptures acting as local tissue stress concentrators leading to a nearly twofold increase in local tissue stress causing interfacial debonding due to mismatch in material properties between microcalcifications and surrounding tissue [24].

Rupture

Plaque rupture is well known as the most frequent cause of myocardial infarction [4]. Pathologically, ruptured plaques (Fig. 2.1) are discriminated from TCFA by the presence of disrupted thin fibrous caps and superimposed thrombus. Underlying necrotic core in ruptured plaques usually is large in size (>30% of the plaque area) [4,11,16]. Similarly, compared with TCFA, greater amount of macrophages and lymphocytes are observed in the fibrous cap of ruptured plaque. These fibrous caps are mostly composed with type I collagen and there are spars smooth muscle cells. As mentioned earlier, the thickness of the fibrous cap at the rupture site is $23 \pm 19\ \mu m$, and 95% of the cap measures $<64\ \mu m$ [20]. It is generally thought that rupture of fibrous cap occurs at structurally the weakest portion, for example, near shoulder zone in most cases; however, an autopsy study using serial sections indicates an equal number of ruptures occur at the middle portion of the fibrous cap, especially in case of rupture after exercise [25]. Thus, we speculate that different processes might lead to the final event of plaque rupture, where select proteases secreted by macrophages [26] might weaken the fibrous cap, and high shear and tensile stress might also be involved [27].

A disruption of the thin fibrous cap of plaque permits direct contact of circulating blood including cellular and noncellular elements with highly thrombogenic components within the necrotic core, for example, tissue factor, and results in thrombus formation. Pathologically, the luminal thrombus at the rupture site contains plenty of platelet and appears as white thrombus on gross inspection, while the thrombus at the proximal and distal propagation sites show a predominance of red thrombus. Thrombus organization is characterized by an infiltration of inflammatory cells, endothelial cells, and smooth muscle cells admixed with extracellular matrix such as proteoglycans and type III collagen.

Although the causes of rupture are poorly understood, matrix metalloproteinases [26], high shear stress [27], and macrophages or smooth muscle cell calcification [24] have been proposed to play an important role in the initial rupture of a TCFA. Pathologically, microcalcification resulting from macrophage or SMC deaths can be detected within a thin fibrous cap (Fig. 2.2E). At the depth of the ruptured plaque, focal calcification (fragmented) and less frequently calcified sheets are observed in ruptured plaques, and are mostly located toward the abluminal surface of the necrotic core [22].

Erosion

Plaque erosion, the second most prevalent cause of ACS, is defined as an acute thrombus in direct contact with the intima in an area of absent endothelium. This pathologic state lacks fibrous cap disruption, distinguished it from plaque rupture (Fig. 2.1) [28]. The underlying plaque lesion of erosion is mostly less advanced than in ruptured plaques, and normally exhibits characteristics of early lesions, that is, PIT or FA usually without an extensive necrotic core, hemorrhage, or calcification. Prior pathologic studies described plenty of smooth muscle cells and proteoglycans, that is, versican, hyaluronan, and type III collagen, exist near the thrombus attached site of erosion, different from ruptured or stable plaques, with the latter being rich in biglycan, decorin, and type I collagen [29]. Different from the vast inflammation of fibrous cap described in ruptures, eroded surfaces contain fewer macrophages (rupture 100% vs. erosion 50%, $P < .0001$) and T-lymphocytes (rupture 75% vs. erosion 32%, $P < .004$) [28,29].

Coronary vasospasm is suspected to play a role in the etiology of plaque erosion because the media is intact at the eroded site while the media at rupture sites is often disrupted together with the internal elastic membrane. An autopsy series of sudden coronary death registry observed that eroded plaques have a greater proportion of women, of younger age, with less severe narrowing, less plaque burden, less thrombus, and less calcification as compared to ruptured plaques [28]. Indeed, plaque erosion accounts for over 80% of thrombi occurring in women <50 years old, which is reduced beyond 50 years.

According to a study of sudden cardiac death autopsy registry, most erosion lesions show no calcification (56%); microcalcification is observed in 40%, whereas fragmented and sheets of calcification are seen in 2%, probably reflecting less advanced lesions than ruptured plaques [30].

Healed plaque rupture and fibrocalcific plaques

Previous pathologic studies provide evidence that plaque progression defined as cross-sectional luminal narrowing occurs following repeated plaque ruptures, which may or may not come to clinical attention. Healed plaque ruptures (HPR) are detected via microscope by the identification of breaks in the fibrous cap with an overlying repair reaction consisting of SMCs surrounded by proteoglycans and/or a type III or type I collagen-rich matrix depending on the phase of healing (Fig. 2.1) [31]. Proteoglycans and type III collagen are the dominant matrix in early-healed lesions with eventually replaced by type I collagen over time. Furthermore, the frequency of HPRs increases along with lumen narrowing. The incidence of HPRs was reported as 8%, 19%, and 73% in plaque with 0%−20%, 21%−50%, and >50% stenosis, respectively [31]. In another report, 61% of hearts from sudden coronary death victims showed HPRs. Indeed, the incidence of HPRs was highest in patients with stable plaques (80%), followed by acute plaque rupture (75%) and least in plaque erosions (9%) [32]. Multiple healed ruptures with layering were more common in segments with acute and healed ruptures and the percent cross-sectional luminal narrowing was dependent on the number of healed repair at the same site. These data indicate that silent plaque rupture is a form of wound healing that results in increased percent stenosis [31]. Although the precise prevalence of silent ruptures in the clinical setting remains unknown, recent development of intracoronary imaging modality, that is, optical coherence tomography, has potential to reveal it and clinical impact of HPRs [33].

Fibrocalcific plaques are defined as lesion with a thick fibrous cap and an extensive deposition of calcification typically in the deeper intimal layers (Fig. 2.1) [4,34]. These lesions are commonly observed in patients with a history of stable angina. Even though coronary calcification is highly correlated with plaque burden, the amount of coronary calcification and plaque instability are not likely to be related in a linear fashion. Because of the minimal or absence of necrotic core, fibrocalcific plaques are not considered as true FAs. Severely narrowed fibrocalcific plaques most likely represent the final end stage of ruptured plaques with healing, and are often considered "burnt-out lesion," marked by dominance of calcification.

Calcification progresses from punctate areas of calcification into the surrounding area of the necrotic core (Fig. 2.2F), to involve the surrounding smooth muscle cells and collagen matrix (Fig. 2.2G and H). Diffuse areas of calcification are described as sheet calcification by histology because of their appearance, are frequently observed in healed ruptures and fibrocalcific plaques, and can be easily identified by radiography, CT, and intravascular imaging. The sheets typically encompass greater than one quadrant of the arterial wall circumference and involve smooth muscle cells as well as collagenous matrix with or without a necrotic core. Calcified sheets may fracture leading to the formation of nodular calcification. Nodular calcification may occur within the plaque in the absence of luminal thrombus and is characterized by breaks in calcified plates with fragments of calcium often separated by fibrin (Fig. 2.2I).

Mechanisms of vascular calcification

Several mechanisms are likely responsible for calcification of an atherosclerotic plaque; some of which include cell death, expression of specific extracellular matrix proteins, and intraplaque hemorrhage [35,36]. Although it is unknown whether processes of calcification share similarities in difference vessel locations, such as carotid lesions, coronary lesions, or other sites, prior pathologic studies indicated that the frequency of calcification is similar in coronary and carotid arteries, with maximum calcification in carotids occurring in lesions with lumens narrowed greater than 70% in cross-sectional area [37].

How calcifications extend and lead to diffuse calcification involving other extracellular matrix protein such as collagen, and proteoglycans, is poorly understood. The eventual transformation into plates of calcification that may appear as diffuse calcification (may be referred as pipestem when the calcification is circumferential), which involves necrotic core, collagen, and inflammatory cells, and in late stages even bone formation may be observed (Fig. 2.2J). Immunohistochemical and gene expression studies have demonstrated that bone morphogenetic protein, osteopontin, bone sialoprotein, and the osteoblast specific transcription factor for bone formation are highly expressed in the calcified arteries as compared to non-calcified arteries. In heavily calcified lesions that are regarded as burnt-out lesions, there is little if any macrophage infiltration and absence of other inflammatory cells. A fair proportion of the calcification is passive, is purely degenerative without biological regulation, and consists of calcium phosphate crystals [38].

In classification by matched radiography and histology, diffuse calcification correlated with sheet calcification by histology and is frequently observed in healed ruptures and fibrocalcific plaques (Figs. 1.1 and 2.4). Conversely, TCFA and rupture have varying degrees of calcification with less diffuse calcification by radiography with greater fragmented calcification and less sheet calcification by histology (Figs. 1.1 and 2.4).

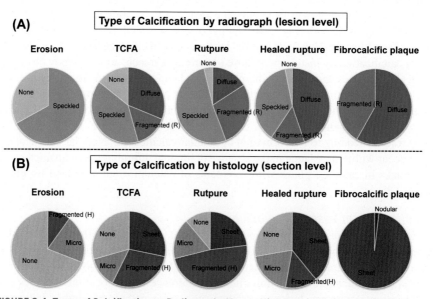

FIGURE 2.4 Types of Calcification on Radiography Versus Histology in Various Plaque Types.

(A) The distribution of calcification types by radiograph. (B) The distribution of calcification by histology. Fragmented (R) = fragmented by radiograph, Fragmented (H) = fragmented by histology, TCFA = thin-cap fibroatheroma.

The data of fibrocalcific plaque in (A) were obtained and reproduced with permission from Yahagi et al. Nat Rev Cardiol 2016;13:79–98; Mori et al. JACC Cardiovasc Imaging 2018;11:127–142. The data shown in (B) was obtained and reproduced with permission from Yahagi et al. Arterioscler Thromb Vasc Biol 2017;37:191–204; Mori et al. JACC Cardiovasc Imaging 2018;11:127–142.

Calcified nodule

Calcified nodule is the least frequent (<5%) morphology seen in cases of ACS [4] and occurs in the state of highly calcified tortuous arteries. Calcified nodule consists of fragmented pieces of calcium that are surrounded by fibrin, and results in disruption of the fibrous cap and endothelium by nodules of dense calcium with overlying thrombus and little or no underlying necrotic core (Fig. 2.1). The eruptive calcified nodule is usually eccentric, protruding into the lumen. There is an absence of endothelium over the nodules of calcium with a platelet-rich "white" thrombus. Fibrin is often present between the calcified spicules, with or without osteoclasts and inflammatory cells. Calcified nodule is generally more prevalent in older men, and in patients with tortuous coronary arteries, diabetes mellitus, or chronic renal failure. The location of this lesion is most frequent in the mid-right coronary artery or left anterior descending artery—sites of maximal tortuosity.

It is important to recognize the difference between "eruptive calcified nodule" and "nodular calcification." Similar to calcified nodules, nodular calcification is also more commonly seen in highly calcified tortuous arteries. The nodular calcification likely also results from breaks in the calcified sheet; however, the pieces of calcium remain within the intima and do not lead to luminal disruption. Nevertheless, medial disruption may be seen with protrusion into the adventitia. Of note, intraplaque fibrin may be observed in nodular calcification, possibly resulting from disruption of surrounding capillaries. Although the etiologies of calcified nodule and nodular calcification are still unclear, calcified sheets crack and break within highly tortuous arteries perhaps during sudden rise of blood pressure [4].

Calcification and plaque stability

Atherosclerotic calcification develops and progress with plaque type as well as with degree of luminal narrowing [10]. Numerous clinical studies support coronary artery calcification as highly related with adverse outcomes in all the populations, and it is a more reliable marker of future events than using risk equations based upon traditional risk factors (Framingham Risk Index) [39]. Nevertheless, it remains unknown whether the coronary calcified plaques cause cardiac events or are just predictors of the presence of coronary artery disease most accurately on a population-wide basis. Moreover, whether the presence of coronary artery calcification predicts plaque instability or stability is a crucial question for clinical practice. Because pathologic studies have fundamental limitations for answering this question, a combination of in vivo coronary imaging and pathology has produced some preliminary suggestions. In brief, one cannot treat coronary artery calcification as an all-or-none variable. The type of calcification, its location, extent, as well as volume and density have differential effects on clinical risk and outcomes [40].

Generally, spotty calcification seems to predict plaque instability while heavy calcification correlates with overall plaque burden. Spotty calcification with positive vessel remodeling and low-attenuation, detected by computed tomography, was more frequent in patients with ACS or likely to develop ACS in short term [41]. Moreover, serial observational studies using intravascular ultrasound have shown that spotty calcification was a predictor for greater progression of plaque volume than noncalcified plaques. On the other hand, heavily calcified plaques were resistant to change in plaques volume in patients with stable coronary artery disease [42,43]. In the pathologic study of human autopsies, the highest histological calcification was observed in fibrocalcific plaque followed by healed plaque rupture, plaque rupture, TCFA, and minimum in plaque erosion (Figs. 1.1 and 2.4) [22]. Moreover, the relationship between percent stenosis in each of these plaque types, mean calcium area, and necrotic core was analyzed. For fibrocalcific plaques and healed plaque ruptures, there was a good relationship between calcification and increasing luminal stenosis; whereas, only mild increase in calcium was observed with luminal narrowing for all other plaque types. In contrast, necrotic core area was greater as stenosis increased in fibroatheroma, TCFA, and plaque rupture [22,44]. Taken together, coronary calcium burden is greater in stable plaques than unstable plaques and exhibit an opposite correlation with necrotic core area.

It is known that microcalcification exists in thin fibrous cap and is not visible by current available in vivo coronary imaging modalities [11,24,45,46]. The prior work by Vengrenyuk et al. proposed microcalcification within the cap as one of the contributors of fibrous cap rupture [24], regarding local tissues stress concentrators within the cap provoking rupture [46]. In the prior pathologic study, calcification was evaluated in 510 coronary artery segments from 17 victims of acute myocardial infarction (AMI), and compared with 450 coronary segments from 15 age-matched control who died from noncardiac causes [47]. Calcification was found in 47% segments of AMI victims and in 25% segments of control. The area of calcification was significantly higher in AMI group ($P = .001$). TCFA and ruptured plaques showed significantly lower degrees of calcification than stable plaques including PIT, FA, and fibrocalcific plaques. The extent of calcification is associated inversely with fibrous cap inflammation and calcification was not identified as an independent determinant of unstable plaque in multivariate analysis. In another study from sudden cardiac death autopsy registry, the amount of overall calcification in whole heart radiographs was divided into four categories, that is, none (0%), mild (<5%), moderate (5%–20%), and severe (>20%), similar to other coronary calcium scoring patterns [48]. Additionally, major plaque types of each patient were decided on the basis of the cause of death and stratified by degree of calcification evaluated by whole heart radiographs (Fig. 2.5). Stable plaques were mainly included in this study. In mild (<5%) and moderate (5%–20%) calcification groups, unstable plaques including rupture and erosion comprised greater proportions. In contrast, the severe calcification (>20%) group showed greatest proportion of stable plaque (PIT, FA, and fibrocalcific plaque). These data indicate that coronary calcium scoring might be helpful to detect general risk of adverse coronary events in a

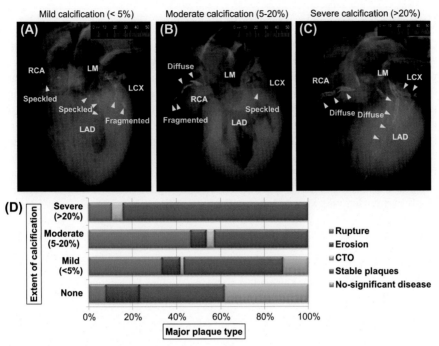

FIGURE 2.5 Extent of Calcification by Plaque Type.

(A–C) are representative images showing different degrees in the extension of CAC as visually assessment in whole heart radiographs. (D) shows the distribution of major plaque types per patient. *CTO,* chronic total occlusion. Stable plaques include PIT, thick-cap fibroatheroma, and fibrocalcific plaque with greater than 75% luminal narrowing. Nonsignificant disease is defined as less than 75% luminal narrowing without thrombus.

The data (n = 120 patients) are obtained and reproduced with permission from Yahagi K, Kolodgie FD, Lutter C, et al. Pathology of human coronary and carotid artery atherosclerosis and vascular calcification in diabetes mellitus. Arterioscler Thromb Vasc Biol February 2017;37(2):191–204. https://doi.org/10.1161/atvbaha.116. 306256. PubMed PMID: 27908890; PubMed Central PMCID: PMC5269516.

population, that is, suggesting the presence of coronary atherosclerotic plaque. However, it is not useful specifically to prospectively identify a culprit lesion of acute coronary syndrome. Furthermore, the relationship between calcification and plaque stability seems to differ according to age [49]. Patients with unstable lesions including ruptured or TCFA in the 4th decade showed greater calcification than those with stable plaques; however, no differences were observed in the 5th and 6th decades. By the 7th decade, greater calcification was detected in stable plaques [50]. Thus, coronary calcification might be more of a determinant of plaque instability in younger individuals.

Conclusion

Coronary artery calcification is a well-recognized feature of coronary artery disease and is commonly used as a risk marker for cardiovascular risk prediction. From a pathologic perspective, there is not a linear relationship between the extent of coronary calcium and plaque progression. Looking at the extent and burden of calcium as a marker for atherosclerosis is not enough to truly understand the relationship of coronary calcium to plaque progression. At the pathologic level, the presence of small, fragmented, or spotty calcium is a better predictor of unstable plaque; however, heavy calcium (diffuse, fibrocalcific plaques, or sheet of calcium) is a better predictor of stable plaque. A greater understanding of this relationship in living patients is needed as well as more sophisticated imaging modalities are needed to successfully translate this knowledge into the care of living patients.

References

[1] Ikari Y, McManus BM, Kenyon J, et al. Neonatal intima formation in the human coronary artery. Arterioscler Thromb Vasc Biol September 1999;19(9):2036—40. PubMed PMID: 10479643.

[2] Stary HC, Chandler AB, Glagov S, et al. A definition of initial, fatty streak, and intermediate lesions of atherosclerosis. A report from the Committee on Vascular Lesions of the Council on Arteriosclerosis, American Heart Association. Circulation May 1994; 89(5):2462—78. PubMed PMID: 8181179.

[3] Velican C. A dissecting view on the role of the fatty streak in the pathogenesis of human atherosclerosis: culprit or bystander? Med Interne 1981 October—December;19(4): 321—37. PubMed PMID: 7038828.

[4] Virmani R, Kolodgie FD, Burke AP, et al. Lessons from sudden coronary death: a comprehensive morphological classification scheme for atherosclerotic lesions. Arterioscler Thromb Vasc Biol May 2000;20(5):1262—75. PubMed PMID: 10807742.

[5] Nakagawa K, Nakashima Y. Pathologic intimal thickening in human atherosclerosis is formed by extracellular accumulation of plasma-derived lipids and dispersion of intimal smooth muscle cells. Atherosclerosis July 2018;274:235—42. https://doi.org/10.1016/j.atherosclerosis.2018.03.039. PubMed PMID: 29622338.

[6] Nakashima Y, Fujii H, Sumiyoshi S, et al. Early human atherosclerosis: accumulation of lipid and proteoglycans in intimal thickenings followed by macrophage infiltration. Arterioscler Thromb Vasc Biol May 2007;27(5):1159—65. https://doi.org/10.1161/atvbaha.106.134080. PubMed PMID: 17303781.

[7] Nakashima Y, Wight TN, Sueishi K. Early atherosclerosis in humans: role of diffuse intimal thickening and extracellular matrix proteoglycans. Cardiovasc Res July, 2008; 79(1):14—23. https://doi.org/10.1093/cvr/cvn099. PubMed PMID: 18430750.

[8] Kockx MM, De Meyer GR, Muhring J, et al. Apoptosis and related proteins in different stages of human atherosclerotic plaques. Circulation June, 1998;97(23):2307—15. PubMed PMID: 9639374.

[9] Otsuka F, Kramer MC, Woudstra P, et al. Natural progression of atherosclerosis from pathologic intimal thickening to late fibroatheroma in human coronary arteries: a

pathology study. Atherosclerosis August 2015;241(2):772—82. https://doi.org/10.1016/j.atherosclerosis.2015.05.011. PubMed PMID: 26058741; PubMed Central PMCID: PMC4510015.

[10] Otsuka F, Sakakura K, Yahagi K, et al. Has our understanding of calcification in human coronary atherosclerosis progressed? Arterioscler Thromb Vasc Biol April 2014;34(4): 724—36. https://doi.org/10.1161/atvbaha.113.302642. PubMed PMID: 24558104; PubMed Central PMCID: PMC4095985.

[11] Kelly-Arnold A, Maldonado N, Laudier D, et al. Revised microcalcification hypothesis for fibrous cap rupture in human coronary arteries. Proc Natl Acad Sci USA June 25, 2013;110(26):10741—6. https://doi.org/10.1073/pnas.1308814110. PubMed PMID: 23733926; PubMed Central PMCID: PMC3696743.

[12] Kapustin AN, Shanahan CM. Calcium regulation of vascular smooth muscle cell-derived matrix vesicles. Trends Cardiovasc Med July 2012;22(5):133—7. https://doi.org/10.1016/j.tcm.2012.07.009. PubMed PMID: 22902179.

[13] Tanimura A, McGregor DH, Anderson HC. Calcification in atherosclerosis. I. Human studies. J Exp Pathol 1986 Summer;2(4):261—73. PubMed PMID: 2946818.

[14] Stary HC, Chandler AB, Dinsmore RE, et al. A definition of advanced types of athero-sclerotic lesions and a histological classification of atherosclerosis. A report from the Committee on Vascular Lesions of the Council on Arteriosclerosis, American Heart Association. Arterioscler Thromb Vasc Biol September 1995;15(9):1512—31. PubMed PMID: 7670967.

[15] New SE, Goettsch C, Aikawa M, et al. Macrophage-derived matrix vesicles: an alterna-tive novel mechanism for microcalcification in atherosclerotic plaques. Circ Res June 2013;113(1):72—7. https://doi.org/10.1161/circresaha.113.301036. PubMed PMID: 23616621; PubMed Central PMCID: PMC3703850.

[16] Narula J, Garg P, Achenbach S, et al. Arithmetic of vulnerable plaques for noninvasive imaging. Nat Clin Pract Cardiovasc Med August 2008;5(Suppl. 2):S2—10. https://doi.org/10.1038/ncpcardio1247. PubMed PMID: 18641603.

[17] Tearney GJ, Regar E, Akasaka T, et al. Consensus standards for acquisition, measure-ment, and reporting of intravascular optical coherence tomography studies: a report from the International Working Group for Intravascular Optical Coherence Tomography Standardization and Validation. J Am Coll Cardiol March 20, 2012;59(12):1058—72. https://doi.org/10.1016/j.jacc.2011.09.079. PubMed PMID: 22421299.

[18] Falk E, Nakano M, Bentzon JF, et al. Update on acute coronary syndromes: the pathol-ogists' view. Eur Heart J March 2013;34(10):719—28. https://doi.org/10.1093/eurheartj/ehs411. PubMed PMID: 23242196.

[19] Kolodgie FD, Burke AP, Farb A, et al. The thin-cap fibroatheroma: a type of vulnerable plaque: the major precursor lesion to acute coronary syndromes. Curr Opin Cardiol September 2001;16(5):285—92. PubMed PMID: 11584167.

[20] Burke AP, Farb A, Malcom GT, et al. Coronary risk factors and plaque morphology in men with coronary disease who died suddenly. N Engl J Med May, 1997;336(18): 1276—82. https://doi.org/10.1056/nejm199705013361802. PubMed PMID: 9113930.

[21] Davies MJ, Thomas AC. Plaque fissuring–the cause of acute myocardial infarction, sud-den ischaemic death, and crescendo angina. Br Heart J April 1985;53(4):363—73. PubMed PMID: 3885978; PubMed Central PMCID: PMC481773.

[22] Burke AP, Weber DK, Kolodgie FD, et al. Pathophysiology of calcium deposition in coronary arteries. Herz June 2001;26(4):239—44. PubMed PMID: 11479935.

[23] Huang H, Virmani R, Younis H, et al. The impact of calcification on the biomechanical stability of atherosclerotic plaques. Circulation February 2001;103(8):1051–6. PubMed PMID: 11222465.

[24] Vengrenyuk Y, Carlier S, Xanthos S, et al. A hypothesis for vulnerable plaque rupture due to stress-induced debonding around cellular microcalcifications in thin fibrous caps. Proc Natl Acad Sci USA October, 2006;103(40):14678–83. https://doi.org/10.1073/pnas.0606310103. PubMed PMID: 17003118; PubMed Central PMCID: PMC1595411.

[25] Burke AP, Farb A, Malcom GT, et al. Plaque rupture and sudden death related to exertion in men with coronary artery disease. Jama March, 1999;281(10):921–6. PubMed PMID: 10078489.

[26] Sukhova GK, Schonbeck U, Rabkin E, et al. Evidence for increased collagenolysis by interstitial collagenases-1 and -3 in vulnerable human atheromatous plaques. Circulation May 1999;99(19):2503–9. PubMed PMID: 10330380.

[27] Gijsen FJ, Wentzel JJ, Thury A, et al. Strain distribution over plaques in human coronary arteries relates to shear stress. Am J Physiol Heart Circ Physiol October 2008; 295(4):H1608–14. https://doi.org/10.1152/ajpheart.01081.2007. PubMed PMID: 18621851.

[28] Farb A, Burke AP, Tang AL, et al. Coronary plaque erosion without rupture into a lipid core. A frequent cause of coronary thrombosis in sudden coronary death. Circulation April, 1996;93(7):1354–63. PubMed PMID: 8641024.

[29] Kolodgie FD, Burke AP, Farb A, et al. Differential accumulation of proteoglycans and hyaluronan in culprit lesions: insights into plaque erosion. Arterioscler Thromb Vasc Biol October, 2002;22(10):1642–8. PubMed PMID: 12377743.

[30] Yahagi K, Zarpak R, Sakakura K, et al. Multiple simultaneous plaque erosion in 3 coronary arteries. JACC Cardiovasc Imaging November 2014;7(11):1172–4. https://doi.org/10.1016/j.jcmg.2014.08.005. PubMed PMID: 25459599.

[31] Mann J, Davies MJ. Mechanisms of progression in native coronary artery disease: role of healed plaque disruption. Heart September 1999;82(3):265–8. PubMed PMID: 10455072; PubMed Central PMCID: PMC1729162.

[32] Burke AP, Kolodgie FD, Farb A, et al. Healed plaque ruptures and sudden coronary death: evidence that subclinical rupture has a role in plaque progression. Circulation February 20, 2001;103(7):934–40. PubMed PMID: 11181466.

[33] Shimokado A, Matsuo Y, Kubo T, et al. In vivo optical coherence tomography imaging and histopathology of healed coronary plaques. Atherosclerosis May, 2018;275:35–42. https://doi.org/10.1016/j.atherosclerosis.2018.05.025. PubMed PMID: 29859471.

[34] Kragel AH, Reddy SG, Wittes JT, et al. Morphometric analysis of the composition of atherosclerotic plaques in the four major epicardial coronary arteries in acute myocardial infarction and in sudden coronary death. Circulation December 1989;80(6): 1747–56. PubMed PMID: 2598434.

[35] Burke AP, Taylor A, Farb A, et al. Coronary calcification: insights from sudden coronary death victims. Zeitschrift fur Kardiologie 2000;89(Suppl. 2):49–53. PubMed PMID: 10769403.

[36] Fischer JW, Steitz SA, Johnson PY, et al. Decorin promotes aortic smooth muscle cell calcification and colocalizes to calcified regions in human atherosclerotic lesions. Arterioscler Thromb Vasc Biol December 2004;24(12):2391–6. https://doi.org/10.1161/01.atv.0000147029.63303.28. PubMed PMID: 15472131.

[37] Kolodgie FD, Nakazawa G, Sangiorgi G, et al. Pathology of atherosclerosis and stenting. Neuroimaging Clinics of North America August 2007;17(3):285–301.

https://doi.org/10.1016/j.nic.2007.03.006. PubMed PMID: 17826632; PubMed Central PMCID: PMC2704337.

[38] Sage AP, Tintut Y, Demer LL. Regulatory mechanisms in vascular calcification. Nat Rev Cardiol September, 2010;7(9):528−36. https://doi.org/10.1038/nrcardio.2010.115. PubMed PMID: 20664518; PubMed Central PMCID: PMC3014092.

[39] Wilson PW, D'Agostino RB, Levy D, et al. Prediction of coronary heart disease using risk factor categories. Circulation May, 1998;97(18):1837−47. PubMed PMID: 9603539.

[40] Shaw LJ, Narula J, Chandrashekhar Y. The never-ending story on coronary calcium: is it predictive, punitive, or protective? J Am Coll Cardiol April 7, 2015;65(13):1283−5. https://doi.org/10.1016/j.jacc.2015.02.024. PubMed PMID: 25835439.

[41] Motoyama S, Kondo T, Sarai M, et al. Multislice computed tomographic characteristics of coronary lesions in acute coronary syndromes. J Am Coll Cardiol July, 2007;50(4): 319−26. https://doi.org/10.1016/j.jacc.2007.03.044. PubMed PMID: 17659199.

[42] Nicholls SJ, Tuzcu EM, Wolski K, et al. Coronary artery calcification and changes in atheroma burden in response to established medical therapies. J Am Coll Cardiol January, 2007;49(2):263−70. https://doi.org/10.1016/j.jacc.2006.10.038. PubMed PMID: 17222740.

[43] Kataoka Y, Wolski K, Uno K, et al. Spotty calcification as a marker of accelerated progression of coronary atherosclerosis: insights from serial intravascular ultrasound. J Am Coll Cardiol May, 2012;59(18):1592−7. https://doi.org/10.1016/j.jacc.2012.03.012. PubMed PMID: 22538329.

[44] Narula J, Nakano M, Virmani R, et al. Histopathologic characteristics of atherosclerotic coronary disease and implications of the findings for the invasive and noninvasive detection of vulnerable plaques. J Am Coll Cardiol March, 2013;61(10):1041−51. https://doi.org/10.1016/j.jacc.2012.10.054. PubMed PMID: 23473409; PubMed Central PMCID: PMC3931303.

[45] Vengrenyuk Y, Cardoso L, Weinbaum S. Micro-CT based analysis of a new paradigm for vulnerable plaque rupture: cellular microcalcifications in fibrous caps. Mol Cell Biomech March 2008;5(1):37−47. PubMed PMID: 18524245.

[46] Maldonado N, Kelly-Arnold A, Vengrenyuk Y, et al. A mechanistic analysis of the role of microcalcifications in atherosclerotic plaque stability: potential implications for plaque rupture. Am J Physiol Heart Circ Physiol September, 2012;303(5):H619−28. https://doi.org/10.1152/ajpheart.00036.2012. PubMed PMID: 22777419; PubMed Central PMCID: PMC3468470.

[47] Mauriello A, Servadei F, Zoccai GB, et al. Coronary calcification identifies the vulnerable patient rather than the vulnerable Plaque. Atherosclerosis July, 2013;229(1):124−9. https://doi.org/10.1016/j.atherosclerosis.2013.03.010. PubMed PMID: 23578355.

[48] Yahagi K, Kolodgie FD, Lutter C, et al. Pathology of human coronary and carotid artery atherosclerosis and vascular calcification in diabetes mellitus. Arterioscler Thromb Vasc Biol February 2017;37(2):191−204. https://doi.org/10.1161/atvbaha.116.306256. PubMed PMID: 27908890; PubMed Central PMCID: PMC5269516.

[49] Mintz GS, Pichard AD, Popma JJ, et al. Determinants and correlates of target lesion calcium in coronary artery disease: a clinical, angiographic and intravascular ultrasound study. J Am Coll Cardiol February 1997;29(2):268−74. PubMed PMID: 9014977.

[50] Otsuka F, Finn AV, Virmani R. Do vulnerable and ruptured plaques hide in heavily calcified arteries? Atherosclerosis July 2013;229(1):34−7. https://doi.org/10.1016/j.atherosclerosis.2012.12.032. PubMed PMID: 23375681.

Basic molecular mechanism of vascular calcification

3

Cornelia D. Cudrici, MD, Elisa A. Ferrante, PhD, Manfred Boehm, MD

Translational Vascular Medicine Branch, National Heart, Lung and Blood Institute, National Institutes of Health, Bethesda, MD, United States

Introduction

Vascular calcification (VC) is a complex, active, and highly regulated biological process associated with the crystallization of calcium crystals within the extracellular matrix and cells that compose the media or intima of the arterial wall. However, VC was once considered a passive and unregulated process linked to aging and a variety of diseases but current data suggest a highly regulated process involved a v;ariety of molecular signaling pathways. Despite considerable progress in our understanding of how multiple signaling pathways regulate VC formation, this complex process is still not completely understood.

In the 19th century, two major forms of ectopic vascular ossifications (an active process) and petrification (a passive process) were reported by Virchow along with a description of two types of VC: one type affecting the media and another the intima of a blood vessel [1]. In the early 1900s, Dr. Mönckeberg, a German pathologist, described for the first time medial calcific sclerosis as a calcification of the muscular middle layer of the arterial wall, which was associated with diabetes mellitus type 2 (DMT2) and end-stage renal disease (ESRD) [2].

For most of the 20th century, VC was considered a degenerative and unregulated process with two major forms of ectopic calcifications described: dystrophic calcifications occurring in soft tissues because of injury/disease state and metastatic calcification associated with widespread ectopic calcification in normal tissues secondary to a systemic mineral imbalance such as hypercalcemia [3,4]. Currently, ectopic calcification is recognized as the pathologic deposition of calcium salts in soft tissues or organs that usually do not calcify, such as the vascular system, valves, heart, lungs, kidney, and brain. These deposits are typically composed of amorphous calcium phosphate or crystals of hydroxyapatite, but can also consist of calcium oxalates, octacalcium phosphate, and other types of calcium salts.

With our evolving understanding of the concept of VC, the idea that this is a highly regulated process and mechanism has reemerged in the late 1990s. At this time, several involved pathways and proteins were reported as well as a mineralization process resembling osteogenesis [5]. VC of the intima is associated with atherosclerotic (ATZ) plaques and is characterized by lipid or cholesterol deposits, inflammation, fibrosis commonly associated with classical cardiovascular disease risk factors (age, sex, obesity, hypertension, smoking, dyslipidemia and diabetes), and it is activated by either oxidative stress or inflammatory pathways. Intimal VC can lead to arterial stiffness and can influence the ATZ plaque composition, potentially predisposing to vulnerability and plaque rupture. The importance of calcification in ATZ plaque instability is influenced by the size of calcifications and the location of calcification [6].

Medial VC (also called Mönckeberg medial sclerosis) represents a group of distinct pathological conditions of differing etiologies, but having as a common final consequence the deposition of calcium hydroxyapatite along the elastic lamina and the extracellular matrix. It is also frequently associated with DMT2 and chronic kidney disease (CKD). Risk factors for medial calcification include reduction of glomerular filtration rate, hyperphosphatemia, hypercalcemia, parathyroid hormone abnormalities, and duration of dialysis (among patients undergoing hemodialysis). However, medial arterial calcification is considered itself a strong, independent marker for future cardiovascular events in patients that present with it. In general, patients with severe medial VC present with arterial stiffness, impaired hemodynamic regulation, and increased cardiac postload [6,7].

VC is a major area of interest in cardiovascular medicine and represents a very dynamic field with many theories emerging over the last few decades. In this chapter, we will review several rare monogenic diseases leading to arterial calcifications for which insight into pathophysiological mechanism may provide a better understanding of molecular and cellular mechanisms of more common forms of VC.

Genetics of vascular calcification

Within recent years, several genes causing diseases associated with severe arterial calcification have been reported and characterized along with their pathogenic molecular mechanisms, shedding some light on the mechanism of arterial calcification. Many of these are primary monogenic deficiencies of *ENPP1* (ectonucleotide pyrophosphatase/phosphodiesterase 1), *CD73* (5′-nucleotidase ecto), *ABCC6* (ATP-binding cassette transporter subfamily C member), *SMGP* (matrix Gla protein), *LMNA* (lamin A/C), and interferon induced with helicase C domain 1 (IFIH1) genes that drive severe arterial calcification in several human diseases such as generalized arterial calcification of infancy (GACI), arterial calcification due to deficiency of CD73 (ACDC), Pseudoxanthoma elasticum (PXE), Keutel syndrome, Hutchinson−Gilford progeria syndrome (HGPS), and Singleton-Merten syndrome (SMS). Characterization of the molecular mechanism and identifying the specific pathways involved in vascular calcification in these rare genetic diseases may lead to strategies in developing new drugs for more common diseases where VC is a major contributor to morbidity and mortality in the general population.

Generalized arterial calcification of infancy

GACI, also known as idiopathic infantile arterial calcification, was initially described in 1899 by Durante and again 2 years later by Bryant [8] as a rare disease in which patients presented with severe arterial calcifications and ectopic mineralization in a variety of tissues, particularly affecting skin, periarticular soft tissues, and multiple visceral organs.

The first mutation responsible for GACI was identified by Rutsch and collaborators in 2003 as a loss of function mutation in the ectonucleotide pyrophosphatase/phosphodiesterase 1 (*ENPP1*) gene. This gene encodes for ENPP1, an extracellular enzyme which hydrolyzes ATP into AMP and inorganic pyrophosphate (PP_i) [9].

GACI is characterized by extreme calcification of medium and large-sized arteries with fragmentation, calcifications of the internal elastic lamina, and myointimal proliferation. Patients can be affected with severe arterial calcification, including severe arterial stenosis in utero or during the first few weeks of life [10]. Mortality is high, up to 70% in these patients and can be caused before birth by polyhydramnios/hydrops fetalis or after birth due to respiratory distress, cardiac heart failure, and cardiac arrhythmias secondary to severe calcifications. Bisphosphonate treatment has been used to reduce the mortality of these patients by as much as 65% [11]. About 70% of cases GACI are caused by biallelic mutations in the ENPP1 gene with more than 40 different mutations described in approximately 200 cases were reported in the scientific literature. However, another identified mutation in ATP-binding cassette, subfamily C member 6 (ABCC6) genes has been identified in about 10% of GACI patients [12]. Patients with GACI carrying mutations in either *ENPP1* or *ABCC6* genes have a similar clinical phenotype with one exception: most *ENPP1*-affected GACI patients eventually develop hypophosphatemic rickets [13].

Mechanistically, ENPP1 is a cell surface glycoprotein enzyme that is present in multiple tissues and is the main source of extracellular PPi by a conversion of extracellular ATP into AMP and PPi (Fig. 3.1). In fact, ENPP1 plays an essential role in sustaining and augmenting extracellular levels of PPi under normal physiologic conditions. PPi is a powerful antimineralization factor, and the tight balance between PPi and inorganic phosphate (Pi) is required to inhibit the precipitation of calcium phosphate complexes in blood vessels and soft connective tissues. PPi is hydrolyzed by tissue nonspecific alkaline phosphatase (TNAP), another cell surface enzyme that generates Pi from PPi hydrolysis (Fig. 3.1). An increase in ENPP1 activity antagonizes the procalcification effect of Pi produced by TNAP. For this reason, loss-of-function mutations in *ENPP1* result in decreased PPi, leading to an imbalanced PPi/Pi ratio. The consequence of this is the formation and growth hydroxyapatite crystals in the vasculature and soft tissues [14,15]. More specifically, patients with GACI have increased TNAP activity, resulting in increased Pi levels and promoting the formation of calcifications.

Bisphosphonates are nonhydrolysable PPi analogs with a high affinity for hydroxyapatite crystals that inhibit crystal formation and, also, mineral and bone resorption by reducing osteoclast activity. Etidronate is a first-generation

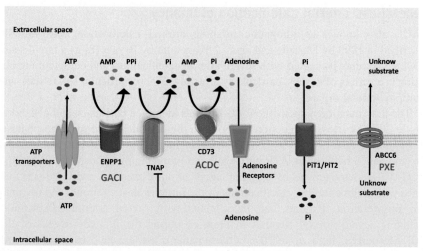

FIGURE 3.1 Overview of Rare Monogenic Diseases with Vascular Calcifications Due to Mutation of the Purine Metabolism Pathway.

ATP gains access via ATP transporters into extracellular space, where it is hydrolyzed by ENPP1 into AMP and PPi. AMP is then hydrolyzed by CD73 into adenosine and Pi while PPi is hydrolyzed by TNAP into inorganic phosphate. PiT1/2 (Na^+-dependent P_i cotransporter) mediates cellular Pi uptake and transport into the cell. PPi is a potent inhibitor of calcification. Adenosine can interact with adenosine receptors and suppress TNAP expression as well as have various effects based on specific receptors. The specific substrate for ABCC6 has yet to be identified. ENPP1, CD73, and ABCC6 deficiencies drive the molecular pathophysiology of several monogenic diseases associated with arterial calcification (GACI, ACDC, and PXE).

nonnitrogen-containing lower antiresorptive potency compared to second- or third-generation bisphosphonates (pamidronate, alendronate, risedronate, ibandronate, and zoledronate) but better inhibitory capacity with respect to crystal formation as compared to newer generation of bisphosphonates. For this reason, etidronate is often used to treat heterotopic calcifications, or prevent vascular calcifications in patients with GACI disease [16,17]. In 2008, Rutsch et al. published a retrospective study of 55 GACI patients and their response to etidronate therapy. Their study resulted in survival of 11 out of 17 patients taking etidronate, while only 8 survived out of 26 patients without treatment. The same study showed that increased survival in this cohort of patients was associated with level of hypophosphatemia and hyperphosphaturia [11].

Both murine models of the disease: Enpp1 knockout mice and homozygous tiptoe-walking mice (ttw/ttw) (homozygous mice carrying a truncated form of Ennp1 protein) present severe arterial and periarticular calcification along with ankyloses of the spinal ligaments, joint fusion and hyperostosis, and, as expected, decreased levels of extracellular PPi [18,19]. Using these murine models, several groups have shown that overexpression of TNAP in vascular endothelial cells leads to arterial calcification [20,21]. Additionally, an increasing amount of data supporting the fine tuning between ENPP-1 and TNAP activity has continued to come from murine models obtained by cross-breeding between ENPP-1 and TNAP knockout mice, which have normal extracellular PPi levels and reduced levels of arterial/ectopic calcifications [14].

In 2015, Albright et al. published the first report in which ENPP1 enzyme replacement therapy was used in ENPP1-asj (ages with stiffened joints) mice that harbor an *ENPP1* mutation with markedly reduced ENPP1 function. Use of this soluble ENPP1 recombinant enzyme (fused to the Fc domain of IgG1) prevented vascular calcification and improved survival in this GACI animal model. However, to prevent formation of auto-antibody-induced immune tolerance, mice were also treated with weekly injections of anti-CD4 antibody. The role of CD4 T cells in arterial calcifications has not yet been established and administration of anti-CD4 with enzymatic replacement therapy may not be entirely feasible in humans [22].

Further work by Rutsch et al. in 2018 showed evidence of a new role for ENPP1 in vascular smooth muscle cells (VSMCs) in Enpp1-deficient ttw/ttw mice, which was also confirmed in GACI patients. This group shows that low levels of ENPP1 in the murine model and patients play a role in increasing VSMC proliferation. Furthermore, myointimal hyperplasia and/or arterial stenosis occurred in approximately 70% of the GACI patient cohort as well as an increase in ATP levels and decrease in PPi levels in both GACI patients and ttw/ttw mice as compared to healthy controls.

A study by Nitschke et al. focused on elucidating the mechanism of intimal proliferation in GACI patients and investigated a potential treatment for GACI with recombinant human (rh) ENPP1-Fc in an ENPP1-deficient mouse model. Results from this study suggest that increased extracellular ATP levels lead to increased VSMC proliferation by suppression of AMPK ($5'$ AMP-activated protein kinase). AMPK is a master regulator of cellular energy homeostasis and plays a major role in controlling cell growth and proliferation signals by inactivating mammalian target of rapamycin (mTOR1) pathway [23]. Recombinant human ENPP1-Fc treatment of human iPSC-derived VSMC in Enpp1-deficient mice inhibited VSMC proliferation.

Pseudoxanthoma elasticum

PXE, or Grönblad—Strandberg syndrome, is an autosomal recessive disease characterized by elastorrhexia or the accumulation of calcium/mineral deposits in elastic fibers, leading to degeneration and ectopic mineralization of skin, retina, and the cardiovascular system. PXE patients experience first symptoms in their early teens, and they are usually diagnosed while in their early 20s. The prevalence of this disease ranges from 1/25,000 to 1/50,000 individuals.

PXE characteristic skin lesions were first described in 1896 as pseudoxantoma—skin lesions that mimic skin lesions of xantoma (yellowish subcutaneous depositions) [24]. These lesions are usually the first manifestation of the disease and appear as a yellow papular lesion that may coalesce into larger plaques in the neck, axilla, inguinal region, knees, and elbows of patients. The retina of PXE patients is also affected, with retinal lesions ranging from less severe angioid streaks and hyperpigmentation to more severe lesions such as subretinal neovascularization and hemorrhage. Vascular lesions are a significant cause of mortality and morbidity in these patients that may result in premature coronary and peripheral artery disease, gastrointestinal hemorrhages, or ischemic stroke. PXE patients present with arterial lesions consisting of elastic laminae fragmentation, elastic fiber mineralization, and diffuse thickening of the wall resulting in blood vessel obstruction. However, arterial compressibility remains relatively well preserved compared to patients with Mönckeberg's calcifications [25].

The most common genetic mutations in PXE patients are reported in the *ABCC6* gene; however, in a small group of patients, PXE can also be caused by mutations in ENPP1 gene [12,26−28]. The *ABCC6 gene* belongs to the ATP-binding cassette [29] superfamily, one of the largest gene families encoding 49 members of ABC transporters and was initially considered to be a group of proteins associated with cell multidrug resistance. All ABC transporters are membrane-bound proteins with two ATP-binding domains and use ATP hydrolysis to translocate various substrates across cell membranes [30].

The ABCC6 transporter substrate is unknown, but ABCC6 protein expression is observed mainly in the liver, followed by the proximal tubules of the kidneys, leading to the hypothesis that PXE may be a metabolic disorder [31,32]. Transport of an unknown "circulating serum factor" by ABCC6 into the systemic circulation has been speculated to play an important role in preventing ectopic calcification in mice, and the effects of the mutation in the ABCC6 transporter may prevent release of this unknown factor and lead to abnormal ectopic calcifications. Multiple substrates for ABCC6 transport including leukotriene C4 (LTC4), BQ-123, an anionic cyclopentapeptide and endothelin receptor antagonist, N-ethylmaleimide S-glutathione, and N-ethylmaleimide S-glutathione have been proposed. However, none of these substrates have been confirmed to inhibit ectopic calcifications [33,34]. Several groups have proposed that ectopic mineralization may not be dependent on the expression of ABCC6 in the liver and that alteration of low levels of ABCC6 in peripheral tissues may be the major driver of ectopic calcifications in PXE (Fig. 3.1) [35−37].

Plasma PPi levels are decreased by as much as 40% both in patients with PXE and ABCC6-deficient mice, and PPi supplementation in the mice delays ectopic calcification [32,33,38,39]. The main pro-osteogenic pathways are upregulated in both PXE and ABCC6 patients and knockout mice model. Activation of the canonical and noncanonical transforming growth factor beta 2 (TGF-β2) pathway is upregulated in human PXE fibroblasts, while increased levels of oxidative stress markers (oxidation protein products, lipid peroxidation derivatives, total circulating antioxidant status, thiol content, and extracellular superoxide dismutase activity) have been reported in PXE patients' blood [40]. On the other hand, fetuin-A and carboxylated matrix Gla protein (MGP), inhibitors of bone morphogenetic protein (BMP) pathway signaling, have been reported to support activation of the BMP-Smad-RUNX2 (Runt-related transcription factor 2) pro-osteogenic pathway, in PXE patients [41,42].

Arterial calcification due to deficiency of CD73

In 2011, our group identified mutations in the 5′-nucleotidase ecto (*NT5E*) gene responsible for a loss-of-function in ecto-5′-nucleotidase, Ecto5′NTase and a deficiency in CD73 enzyme that led to severe arterial and periarticular calcifications in nine patients belonging to three different families [43]. ACDC generally presents in patients when they are young adults with multiple episodes of oligo arthritis, lasting from several days up to several weeks. These episodes usually respond well to treatment with low dose oral corticosteroids or nonsteroidal antiinflammatory drugs. In general after several years since the first presentation of an acute episode of arthritis, patients develop lower extremity claudication secondary to severe calcification of the medial layer in arteries. This results in severe peripheral obstructive arterial large vessel disease and development of extensive networks of collateral circulation. No abnormal coronary calcification has been reported, and none of the patients have any significant atherosclerosis, obesity, impaired kidney function, hypertension, dyslipidemia, or diabetes. In 2015, Zhang et al. reported a novel compound heterozygous *NT5E* mutations encoding for CD73 in a Chinese family with severe upper and lower extremities arterial calcification but also periarticular calcifications [44].

CD73 is a glycosyl phosphatidylinositol (GPI)-anchored glycoprotein with ecto-5′-nucleotidase enzyme activity that converts extracellular AMP to adenosine and inorganic phosphate (Fig. 3.1). The enzyme is localized within a variety of tissues and cells including leukocytes and endothelium. CD73 protein expression is regulated under hypoxic and inflammatory conditions [45].

A study published in 2014 by Fausther et al. reported that mutation in CD73 molecules in a fibroblast-like cell line (COS-7) decreases CD73 expression on the cell surface and reduces endoplasmic reticulum retention, causing a complete loss of catalytic nucleosidase activity in the CD73 molecule [46]. In ACDC-derived fibroblasts, loss of CD73 activity results in decreased adenosine levels and TNAP activity and abundant calcium phosphate crystal formation. Calcification can be rescued in vitro by exogenous adenosine administration in cell culture.

TNAP promotes vascular calcification by regulating the levels of PPi, a potent inhibitor of mineralization (Fig. 3.1). In mice, upregulation of TNAP expression in vascular smooth muscle cells or in endothelial cells is confirmed to lead to severe medial calcification [47,48]. Upregulation of TNAP has also been observed in medial vascular calcification associated with diabetes, chronic kidney disease, and is associated with increased mortality risk factors on patients with coronary artery calcification [49,50].

Studies using induced pluripotent stem cell (iPSC) technology can be a useful tool to study pathological mechanisms of vascular calcification. Under osteogenic stimulation conditions, induced pluripotent stem cell-derived mesenchymal stromal cells (iMSCs) from ACDC patients displayed increased TNAP activity and accelerated calcification as compared to iMSCs from healthy controls. PPi levels produced by iMSCs lacking CD73 expression are also reduced compared to wild-type cells. Inhibition of the Akt and mTOR pathway decreased TNAP activity and inhibited calcification in ACDC-derived fibroblasts. In addition, activation of A2B adenosine receptor reduced calcification via mTOR pathway mechanism.

Rapamycin (an mTOR inhibitor) or etidronate treatment is able to reduce calcification in an ACDC iPSC-derived teratoma animal model in vivo, suggesting that they may be promising potential therapies to prevent and possibly reverse medial vascular calcification [51].

Purinergic signaling in VC

All the diseases described earlier have molecular genetic defects in the ATP metabolism pathway, particularly highlighting the potential role of the purinergic system in the pathophysiology of VC. Identification of these rare disease has been critical in obtaining more insight into the molecular mechanisms of arterial calcification in a variety of conditions. An overview of rare monogenic diseases with mutations in ATP metabolism is depicted in Fig. 3.1.

Within the last few years, increasing evidence suggests that multiple enzymes regulating the purine metabolism are key modulators of cardiovascular homeostasis and VC development [52,53].

Adenosine and purine nucleotides can be released from various cells involved in VC such as endothelial cells (ECs), VSMCs, and platelets. These cellular signals can regulate vascular tone, vascular permeability, proliferation, angiogenesis, and regeneration of damaged vessels [54].

Purinergic signaling is mediated by purine receptors, of which P1 is a G-protein-coupled receptor responding primarily to adenosine and P2 is activated by tri- and dinucleotides. Both receptors play various roles in vascular physiology and pathophysiology. Furthermore, there are four subtypes of P1 receptors: A1, A2a, A2b, and A3 all coupled to G1 proteins [55]. A1 and A3 inhibit the cyclic AMP (cAMP) pathway, while $A2_a$ and $A2_b$ receptors activate cAMP via Gi proteins. As mentioned previously in this text, treatment of $A2_B$ receptors with rapamycin and etidronate successfully reduces calcification [51]. P2 receptors are classified as either P2X or P2Y and are activated by tri- and dinucleotides. P2X receptors (P2X 1−7) are ion-gated channels, while P2Y receptors (P2Y 1−8) are G protein-coupled.

The roles of purine receptors have been extensively explored in relation to heart disease and atherosclerosis. In clinical practice, adenosine (acting via A1 receptors) is used in the treatment of supraventricular tachycardia. ATP signaling is also important in the development of atherosclerosis as it promotes endothelial and smooth muscle cell proliferation and blocking A2b and A3 receptors by specific antagonists reducing ATZ plaque formation [56,57].

The purinergic system regulates numerous cellular responses that either enhance or inhibit a variety of cellular pathways based on the type of purinergic receptors activated. The role of purinergic system in vascular calcification is not completely understood and better characterization of these mechanisms will be essential for the development of new therapies that may target purinergic receptors.

Keutel syndrome

Keutel syndrome is an autosomal recessive disease first described in 1971 as a loss-of-function mutation in the matrix Gla protein *(MGP)* gene that encodes for MGP, which is one of the main inhibitors of the pro-osteogenic BMP-2 pathways in VSMCs and chondrocytes [58,59]. Patients present pulmonary arterial calcifications, diffuse cartilage calcifications including tracheal, bronchial, nasal, auricular hypoplasia of the distal phalanges, facial abnormalities, and hearing loss [60].

The activation of matrix Gla protein is a vitamin K-dependent process, and decreased levels of γ-carboxylated matrix Gla protein are reported in these patients. Contrary to expectations, in one case report of a patient with Keutel syndrome taking daily treatments of vitamin K for 3 months, there was a failure to increase serum γ-carboxylated matrix Gla protein [61].

Mice lacking MGP have extensive arterial calcifications often leading to aortic dissection and premature death. In these mice, it has been shown that osteogenic *trans* differentiation of VSMCs and mesenchymal transition of endothelial cells are important players in calcification of the arterial wall [59]. Furthermore, in MGP-deficient mice, inhibition of the BMP pathway by a small molecule and a recombinant protein that sequesters BMP ligands leads to reduced vascular calcification and improved survival [62].

Hutchinson—Gilford progeria syndrome

HGPS is a rare autosomal dominant disease with only 130 cases described since 1886 after the first case of progeria was described by Dr. Hutchinson. Within the first year of life, children affected by progeria present with severe growth retardation, failure to thrive, alopecia, sclerodermic appearance of the skin, osteoporosis, and dental abnormalities. Life expectancy is around 15 years [63]. Morbidity and mortality of these patients are secondary to cardiovascular disease and stroke.

The underlying vascular disease is characterized by extensive atherosclerotic plaque formation, which results in loss of arterial VSMCs, endothelial dysfunction, vascular calcification, and premature atherosclerosis between ages 5 and 88 in most patients [64].

Progeria is mainly caused by a de novo heterozygotes point mutation in the lamin A/C *(LMNA)* gene, which leads to the synthesis of a truncated variant of prelamin A precursor protein known as progerin or lamin AΔ50. Lamin A is essential for the formation of the nuclear lamina, and *LMNA* mutation affects posttranslational modifications, resulting in a higher proportion of farnesylated and carboxymethylated lamin A (immature lamin A) versus mature lamin A. Lamin A plays an important role in stability of the nucleus, chromatin organization, cell cycle, and signal transduction. Additionally, the *LMNA* mutation results in nuclear envelope blebbing and disruption of nuclear architecture [65−68].

Both HGPS patients and mice with similar mutations in the *LMNA* gene (LmnaG609G/+ knock-in mice) develop severe aortic calcifications. Increased expression of Bmp2, Runx2, osteocalcin, and osteopontin has also been reported in these mice and in patients along with a decrease in ATP and PPi levels and an increase in TNAP activity. In the murine model of HGPS, PPi treatment is able to reduce vascular calcification [69].

Singleton-Merten syndrome

SMS is a rare autosomal dominant disease initially observed in two female patients that had severe calcifications of the aortic valve and thoracic aorta, dental abnormalities, osteoporosis, and skeletal abnormalities [70]. Several additional cases were subsequently reported with more clinical characteristics described by several groups to include psoriasis, glaucoma, hypotonia, scoliosis, cardiac arrhythmia, subluxation of the joints, hyperflexible joint ligaments, acro-osteolysis, widened medullary cavity of the metatarsal bone, and short stature [71]. The aortic calcifications affecting the ascending aorta were generally reported at a very early age along with coronary arteries aortic and mitral valve calcifications that resulted in aortic insufficiency and stenosis, left ventricular hypertrophy, and arrhythmias. Mortality is very high in this group of patients and is mainly driven by their cardiovascular pathology.

In 2015, Rutsch et al. reported the specific gain-of-function mutation (p.Arg822Gln) in four patients with Singleton-Merten syndrome in the interferon induced with helicase C domain 1 (*IFIH1*) gene, encoding for melanoma differentiation-associated protein 5 (MDA5). This mutation is localized in the helicase domain of the MDA5 proteins, enhances MDA5 filament stability, and increases type I IFN immune signaling. Indeed, in SMS patients, the IFN-β1 signature gene pattern is increased in blood [72].

Atypical Singleton-Merten syndrome (or SMS type 2) has also been described by Jang et al. in 11 patients with glaucoma, aortic and valvular calcifications, and skeletal abnormalities but without the dental or facial anomalies reported in typical SMS. SMS type 2 is caused by a mono-allelic mutation (p.Glu373Ala or p.Cys268-Phe) in dead-box polypeptide 58 (*DDX58)* gene [73]. The *DDX58* gene encodes for retinoic-acid-inducible I (RIG- I) protein, a viral cytoplasmic RNA receptor and a member of the RIG-I-like receptor (RLR) family, which includes MDA5. Similar to MDA5, RIG-I recognizes double-stranded RNA viruses and induces antiviral activities by increasing type I interferon production [74,·75].

The RLR receptor family represents homologous double-stranded RNA (dsRNA)-dependent ATPases and include three family members with similar structures: MDA5, RIG-I, and laboratory of genetics and physiology 2 (LGP2). These are cytosolic receptors that recognize viral nucleic acids and activate type I INF production triggering the innate immune system protective defense [76].

MDA 5 and RIG-I have a very similar structure with three distinct domains: an N-terminal tandem caspase for activation and recruitment domains (CARDs)—involved in activating mitochondrial antiviral signaling protein (MAVS); a central DExE/H RNA helicase domain; and a C terminal domain (CTD). The last two domains are responsible for viral or bacterial dsRNA interaction [77]. RIG-I is present in the cytosol in an inactive state but after interaction of the CTD domain with the host or viral nucleic acid, the helicase domain will bind RNAs or DNAs with ATP. This interaction leads to RIG-I activation and oligomerization. The resulting conformational changes allow activation of the CARD domain while interaction with MAVS molecules on the mitochondrial surface results in the increase of INF β production [78].

In contrast with RIG-I, which recognizes the 5′ terminal portion of dsRNA via CTD interaction, MDA5 recognizes the long dsRNA and forms protein-coated filaments that interact with MAVS at the outer mitochondrial membrane via CARD domains. This increases phosphorylation and activation of type I interferon-inducing transcription factors [79]. In classic SMS patients with a p.Arg822Gln mutation in the helicase domain of the *IFIH1* gene, ATP hydrolysis activity of this domain is disrupted because of constitutive activation of MDA5 and increased production of INFβ [79].

On the other hand, in patients with atypical SMS, activation of nuclear factor kappa-light-chain enhancer of activated B cells (NF-kB), activator protein 1 (AP-1), and interferon regulatory factor 3 (IRF3) signaling pathway induces increased expression of interferon-stimulated gene 15 (INF-1β, ISG15) and C—C motif chemokine ligand 5 (CCL5) [73].

In autoimmune disease such as systemic lupus erythematosus dermatomyositis, multiple sclerosis, diabetes mellitus type 1, psoriatic arthritis, and cutaneous psoriasis, type I interferon plays a critical pathogenic role. As such, identification of rare monogenic diseases like SMS is important, because they help to elucidate the specific role of proteins and pathways that are involved.

TGF-β1 and BMP/Smad signaling pathway

The TGF-β family members are involved in multiple cell function processes including embryogenesis, development, and tissue homeostasis [80]. Perturbation of signaling or modification of specific components of the TGF-β signaling pathway can result in a large of spectrum of diseases such as malignancy, autoimmune diseases, tissue fibrosis, and cardiovascular pathology. The TGF-β family is highly conserved throughout species and consists of several members such as BMPs, activins, inhibins, nodals, and antimullerian hormone [81].

TGF-β family members exert their cellular effects via heteromeric serine/threonine kinase receptors composed of type I and type II complexes located at the cell surface. Upon TNF binding, type II receptors phosphorylate and activate the type I receptor. The signal is then transduced intracellularly via a Smad-dependent (canonical pathway) or mitogen-activated protein kinases (MAPK) pathway (non canonical pathway) (Fig. 3.2). TGF-β members are essential for the regulation of angiogenesis and vasculogenesis with mutations often resulting in vascular pathologies such as atherosclerosis and cardiovascular disease.

All members of the TGF-β superfamily bind to a cell surface heterotetrameric receptor complex, composed of two type I receptors and two type II receptors, with common serine/threonine kinases in the cytoplasmic domain. Upon ligand binding, the type II receptors phosphorylate and activate the type I receptor. The activated type I receptor then propagates the signal by phosphorylating a family of transcription factors, the receptor regulated-Smads (R-Smads). The activated phosphorylated R-Smad complex translocates into the nucleus and regulates transcription of target genes, and the pathway is known as Smad canonical pathway [82,83].

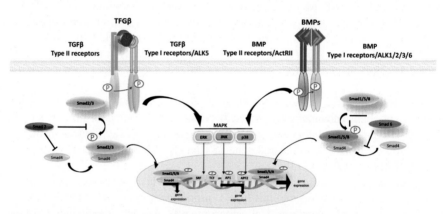

FIGURE 3.2 Signaling Transduction Mediated by TGFβ2 and BMP Pathway.

THFβ and BMPs dimers interact with heteromeric transmembrane serine/threonine kinase type I and II receptors. After being activated, type II receptors will transphosporylate type I receptors and induce activation of regulatory Smad 1/5/7 (R-Smad), which forms heteromeric complexes with Smad4 (common Smad) and translocates to the nucleus where, by interacting with other transcription factors, it will gene transcriptional response (canonical Smad signaling pathway). Inhibitory Smad (I-Smad) 6 and 7 can inhibit activation of R-Smad. Both THFβ and BMPs pathways can activate non-Smad pathways by activation of Extracellular signal-regulated kinase (ERK), c-Jun N-terminal kinase (JNK), and p38 mitogen-activated protein (MAP) kinases, which crosstalk with the Smad pathways. Activated MAPK pathways can translocate to the nucleus and activate several transcription factors such as serum response factor (SRF), ternary complex factor (TCF) family members, activator protein 1 (AP1) complexes, and activating transcription factor 2 (ATF2).

Figure adapted with permission from Cai J, Pardali E, Sánchez-Duffhues G, ten Dijke P. BMP signaling in vascular diseases. FEBS Lett 2012;586:1993–2002.

TGF-β plays an important role in most cell types known to be present in atherosclerotic lesions: smooth muscle cells, ECs, myofibroblasts, monocytes/macrophages, platelets, and T cells. Misregulation of TGF-β signaling, both in human and murine models, has severe consequences of vascular dysfunction and pathologies.

TGF-β and Smad proteins are highly expressed in fibrous atherosclerotic plaques, which increases collagen gene expression and, thus, fibrous content of plaques [84]. In atherosclerosis, TGF-β1 is considered to have antiatherogenic effects by increasing plaque stability through several mechanisms: regulation of vascular (endothelial and smooth muscle) cell proliferation, inhibition of VSMC accumulation in the neointima, increased synthesis of ECM to prevent plaque rupture, decreased expression of adhesion molecules involved in leukocyte recruitment, inhibition of foam cell formation, and suppression of local inflammation [85–88]. In mice, treatment with neutralizing TGF-β1 antibodies decrease TGF-β1 ligand concentration and increase atherosclerosis by reducing VSMC differentiation and EC response to proatherogenic stimuli, resulting in increased vascular inflammation and atherosclerotic plaque instability [89].

TGF-β pathways also play an important role in vascular calcification. TGF-β1 can be released by most cells that make up the atherosclerotic lesion in a process that results in accumulation of fibrotic ECM, promotion of calcium accumulation in a matrix that is prone for mineralization. TGF-β pathways also have an important role in chemotaxis, proliferation, and differentiation of osteoprogenitor cells [90].

In contrast, several groups were reporting the proatherosclerotic effect of TGF-β1. Increased TGF-β1 protein and mRNA levels have been reported in plaques as compared to healthy coronary artery tissue, which is a consequence of various atherogenic stimuli such as oxidized cholesterol and abnormal shear stress on the vessel wall [91,92]. The mechanism of action of TGF-β1 is reviewed in Fig. 3.2.

According to studies on animal models, upregulation of TGF-β1 can be beneficial and protective at early stages of atherosclerosis. However, overproduction of TGF-β1 can ultimately have a detrimental effect of increasing blood vessel fibrosis, suggesting a tight local balance of TGF-β1 signal. It appears that the local concentration of TGF-β1 is more important than systemic concentration, which may imply that examining TGF-β levels in atherosclerotic tissues may be more informative than in blood fractions [93].

BMPs were the first members of the TGF-β family described as critical molecules in the induction of ectopic osteogenesis in vivo [94]. In 1993, the discovery in calcified human atherosclerotic plaques of osteoblast-related BMP2 gave rise to the hypothesis that VC is a regulated process like bone formation [95]. BMPs play an important role in embryogenesis by controlling cellular proliferation, differentiation, and apoptosis but also later in life with control of remodeling processes, tissue homeostasis, and repair. More than 20 members of the BMP subfamily have been identified with several members BMP 2, 4, 5, 6, 7, and 9 reported as important players in osteogenesis.

BMPs are synthetized as dimeric large precursor's proteins within the cytoplasm, and proteolytic cleavage by specific serine endonuclease is required for the generation of the active dimeric mature proteins. Similar to other members of TGF-β family, upon BMP active dimer binding to transmembrane receptors, the type II receptor kinase transphosphorylates the type I receptor serine and threonine residues at the intracellular juxtamembrane site within the glycine-serine-rich domain (Fig. 3.2). The phosphorylated type I receptor causes subsequent phosphorylation of intracellular group of signaling proteins called Smad. There are three distinct subtypes of Smads: receptor-regulated Smads (R-Smads), inhibitory Smads (I-Smads), and common mediator Smads (Co- Smad) [96]. Most BMPs activate Smad 1/5/8 as R-Smads and form heterotrimeric complexes with Smad4, which translocate to the nucleus to regulate the expression of target genes (Fig. 3.2). These complexes can also activate transcription of genes encoding the inhibitory SMADs, creating a negative feedback loop to inhibit Smad signaling [97]. This pathway is called BMP receptor/Smad canonical pathway, but activated BMP receptors can also initiate a noncanonical Smad pathways, which include activation of MAPK kinases (extracellular signal-regulated kinases (ERKs), P38 mitogen-activated protein kinases, and c-Jun N-terminal kinase (JNK)), PI3K/Akt, protein kinase (PKC), and Rho GTPase signaling pathways. Non-SMAD pathways may also modulate canonical SMAD signaling pathways (Fig. 3.2).

Within the BMP subfamily, BMP2 is critical in bone formation and adult skeletal integrity, as it contributes to differentiation of mesenchymal cells into chondrocytes and osteoblasts [98,99]. BMP2 also plays an important role in promoting calcification of vascular media, by upregulating Runt-related transcription factor 2 (Runx2) a key transcription factor involved in osteoblast differentiation. In SMCs, Runx2 promotes osteogenic phenotype transition with a mechanism that remains unclear but has been reported to downregulate two microRNAs: miR-30b (microRNA) and miR-30c [100,101].

Additionally, BMP2 was found to play an important role in atherosclerosis, plaque instability, and vascular calcification by regulating EC inflammation and cell differentiation [99,102,103]. In ECs, BMP-2 can be modulated by proinflammatory and proatherogenic stimuli in a manner that promotes plaque calcification by inducing an osteogenic phenotype in VSMCs [104]. Patients with DMT2 have been shown to have higher circulating levels of BMP-2 than normal controls and their plasma BMP-2 levels correlated positively with plaque burden and calcification [105]. In fact, within calcified coronary arteries, BMP-2 expression has been described to be upregulated along with other inflammatory stimuli such as oxidative stress, hyperglycemia, and hyperlipidemia [95,101,106].

In a murine model for VC, low-density lipoprotein receptor (LDLR$^{-/-}$)-deficient mice fed high-fat diet developed severe vascular calcification. However, a small molecule inhibitor that binds activin receptor-like kinase 3 (ALK3), an inhibitor of BMP pathway, was shown to significantly reduce the development of atheroma, vascular inflammation, osteogenic activity, and calcification [107]. In contrast, the deletion of Smad6, BMP pathway inhibitor, led to severe cardiovascular abnormalities in mice including cardiac valve anomalies, outflow tract defects, and severe aortic calcification [108,109]. In patients, p.C484F mutations in the Smad6 gene result in a lower activity BMP signaling inhibition and cause congenital cardiovascular malformations [110].

Another mutation affecting the BMP pathway has been described in fibrodysplasia ossificans progressive, in which most patients have an identical point mutation, R206H, in the activin A receptor, type 1 (*ACVR1*) gene that encodes for BMP type I receptor ACVR1 (or ALK2). This mutation causes a constitutive activation of the BMP pathway and is characterized by overactive de novo osteogenesis. This results in heterotopic (extraskeletal) ossification, confirming that BMP signaling components are major players in the regulation of extraskeletal bone tissue formation [111].

As discussed here, BMP and TGF pathways are critical processes in VC pathology, and disruption of their pathways is crucial in regulation of vascular function. Future work in this field is essential for elucidating the specific intracellular signaling mechanisms that affect vascular calcification and will facilitate the design of new therapeutic approaches.

Fetuin-A

Fetuin-A is a liver produced glycoprotein with a role as a calcification inhibitor. This protein was initially isolated from bovine serum in 1944 as most abundant globular plasma protein from calf serum and was isolated in humans 20 years later as the Heremans-Schmid glycoprotein [112,113]. In the initial experiments, it was described that fetuin-A can inhibit the precipitation of apatite crystals from solutions supersaturated in calcium and phosphate in cell culture serum. It is hypothesized that fetuin-A inhibits calcification by binding with calciprotein particles (phosphate, calcium phosphate crystals) leading to stabilization of these complexes and decreased inflammatory response after these particles are ingested by local macrophages [114]. Fetuin-A has also been reported to inhibit VC via calcium-dependent uptake in VSMCs [115].

Reduced serum levels of fetuin-A have been reported in dialysis patients and are associated with increased cardiovascular mortality [116−118]. Several other studies report a direct correlation between low serum levels of fetuin-A and presence of vascular disease in lupus patients, or increased arterial stiffness, calcification score, and higher cardiovascular mortality in patients with CKD on hemodialysis or postrenal transplantation [119−122]. On the contrary, high serum levels of fetuin-A have been shown to be correlated with decreased renal function, inflammation, vascular calcification, and increased mortality in patients with CKD [123]. However, a few studies found no correlations between fetuin-A serum levels and vascular calcification or mortality secondary to cardiovascular causes [124−126]. Despite the discrepancy between various studies, data suggest that fetuin-A protects against VSMC calcifications and low levels may be correlated with increased VC.

Matrix Gla-protein

MGP is a vitamin-K-dependent activated protein and an important factor in VC pathogenesis. MGP is expressed by multiple cells involved in vascular calcification including VSMCs, chondrocytes, ECs, and fibroblasts and is considered an inhibitor of VC via the BMP-2 pathway. For activation from the inactive Glu-MGP to the active Gla-MGP form, MGP requires posttranslation modification by an enzyme called γ-carboxyglutamyl carboxylase (GGCX). MGP γ-carboxylation takes place in the Endoplasmic Reticulum (ER) and it is essential in the inhibition of vascular calcification, most likely due to the ability of Gla protein to binding calcium ions and prevents precipitation/mineralization of calcium salts [127,128]. VSMCs are capable of synthesizing carboxylated MGP, which inhibits the BMP-2 pathways and prevents VMSCs differentiation toward an osteochondrogenic phenotype. As previously described in this chapter, MGP deficiency in humans leads to Keutel syndrome.

MGP KO mice usually die in the first 2 months of life due to spontaneous rupture of aortic or other large vessel secondary to calcification of the elastic layer of the tunica media [59]. Khavandgar at all have shown that medial calcification in MGP-deficient mice is linked with alteration of ECM, and that elastin haploinsufficiency reduces arterial calcification in this strain. This suggests that MGP may protect from elastin calcification and that MGP deficiency may lead to alterations of vascular ECM [129].

Vitamin K deficiency or treatments with vitamin K antagonists such as warfarin may increase in VC both in humans as well as animal models via an MGP-dependent mechanism.

Generalized VC is common in patients with CKD, and multiple studies report an inverse relationship between levels of γ-carboxylated MGP and decreased glomerular filtration rate (GFR) in a potentially renoprotective manner [130−132].

In patients with DMT2, higher levels of desphospho-uncarboxylated MGP were associated with higher risk of developing cardiovascular disease and higher arterial stiffness [133,134].

Osteoprotegerin—RANK/RANKL

The late 1990s discovery and characterization of receptor activator of nuclear factor kappa-B ligand (RANKL)—receptor activator of nuclear factor κ B (RANK)—osteoprotegerin (OPG) system as a key player in bone remodeling, has led to significant advances in our understanding of the relationship between vascular calcification and bone homeostasis. OPG/RANK/RANKL is a cytokine network involved in osteoclast differentiation and activation as a main regulator of the critical balance between bone formation (osteoblasts) and bone resorption (osteoclasts).

The OPG cytokine, a member of the tumor necrosis factor receptor (TNFR) su-perfamily, has a name derived from two Latin words: "os" means bone and "prote-gere" to protect, which highlights its protective effects on bone. OPG is expressed by various mesenchymal-derived cells such ECs, VSMC, osteoblasts, and dendritic cells [135,136]. OPG also binds to tumor necrosis factor-related apoptosis-inducing ligand or Apo 2 ligand (TNFSF10/TRAIL) and is able to inhibit the TRAIL proapoptotic actions and promote cell survival [137]. In ECs, OPG can be localized with Weibel−Palade bodies along with von Willebrand Factor VIII in an inactive form and can be released extracellularly after proinflammatory cytokines (TNFα and Il-1 β) interaction with ECs (Fig. 3.3) [138].

RANKL is highly expressed on osteoblast and stromal cells within areas under-going bone remodeling and by T cells in lymphoid tissues. It binds to RANK and assembles into a functional homotrimeric receptor on the surface of osteoclasts as well as monocytic and dendritic cells. This process initiates multiple intracellular

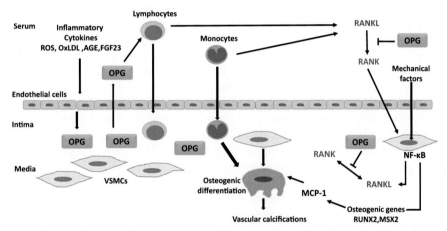

FIGURE 3.3 Role of OPG/RANK/RANKL Axis in the Pathogenesis of Vascular Calcification.

In the vasculature, RANKL—RANK—OPG signaling can be stimulated by inflammation, proinflammatory cytokines, advanced glycation end products (AGE), and reactive oxygen species (ROS). RANKL interacts with the RANK receptor and activates the NF-κB pathway causing upregulation of osteogenic genes *RUNX2, MSX2* and leading to osteogenic differentiation of VSMC. OPG inhibits osteoclastogenesis by preventing RANKL and RANK interaction as well as deposition of a mineralized matrix. In the presence of proinflammatory cytokines, FGF23 induces expression of OPG by endothelial cell (EC), lymphocytes, and VSMCs and increases transmigration of lymphocytes and monocytes into the vessel wall. The differentiation of monocytes to osteoclast-like cell is mediated by the monocyte chemotactic protein-1 (MCP-1), which is induced by osteogenic genes.

Figure adapted with permission from Rochette L, Meloux A, Rigal E, Zeller M, Cottin Y, Vergely C. The role of osteoprotegerin in the crosstalk between vessels and bone: its potential utility as a marker of cardiometabolic diseases. Pharmacol Ther 2018;182:115−32.

signaling pathways resulting in activation of the NF-κB pathway, essential in osteoclast differentiation and activity. RANKL activities can be blocked by OPG, which acts as a decoy receptor to prevent RANKL/RANK interactions. This inhibits differentiation, activation, and osteoclast survival. RANKL/RANK signaling increases bone turnover by enhancing resorption and bone loss, whereas OPG promotes bone formation by blocking RANKL/RANK interaction (Fig. 3.3) [139].

OPG has been detected in ATS plaques in mouse models as well as humans with high levels observed in unstable plaques [140]. At physiological levels, OPG production by VSMC and ECs is protective for ATS plaque calcification. In fact, studies in vitro and in animal models suggest that OPG inhibits VC and OPG-deficient mice exhibit severe alveolar bone loss with enhanced bone resorption and medial arterial calcifications of the aorta and renal arteries. However, OPG replacement induces expression of TGF-β1 and increases fibrosis in ATS plaques [141]. Apolipoprotein E-deficient ($Apoe^{-/-}$) mice, one of the most widely used murine models for atherosclerosis, express OPG adjacent to ATS foam cells and possess high levels of RANK and RANKL associated with ATS infiltrating T cells [140,142−146].

In contrast with animal models, clinical studies in large populations show that high OPG serum levels are associated with vascular calcification, advanced atherosclerosis, coronary artery disease, heart failure, and an increase in cardiovascular and cerebrovascular disease mortality. Various reported polymorphisms in the OPG gene (G1181C and T950C polymorphisms) are associated with coronary artery disease [147]. There seems to be a tight balance that regulates early-phase OPG release to protect against vascular calcification but, over the time, higher levels of OPG become detrimental by promoting inflammation and fibrosis.

In studies with TRAP-deficient mice, TRAIL molecules seem to protect against ATS, possibly by increased apoptosis of infiltrating macrophages recruited to the atherosclerotic lesion site as well as decreased VSMCs and collagen content within the plaque. Bartolo et al. have shown that TRAIL-deficient mice develop advanced atherosclerotic lesions and high tissue levels of RANKL and BMP-2 that leads to vascular calcification [148−151].

In humans, RANKL can be detected in calcific stenotic aortic valves and carotid and femoral ATZ plaque. In aortic valve myofibroblasts, RANKL stimulation increases alkaline phosphatase activity and matrix calcifications in vitro. They also appear to have increased DNA binding of Runt-related transcription factor 2 (RUNX2) or core-binding factor subunit alpha-1 (CBFA-1), a transcription factor for osteoblast differentiation and bone formation [152−154]. In patients with type 1 diabetes mellitus, microalbuminuria and age were negatively associated with RANKL levels, but no association between carotid intima media thickness and serum OPG and RANK ligand was found [155]. In general, increased RANKL serum level has been suggested as a predictor of psoriatic arthritis and cutaneous psoriasis, in which type I interferon plays a critical pathogenic role [155−157].

As such, identification of the exact mechanisms by which the RANK/RANKL/OPG system regulates cardiovascular pathophysiology still needs to be further elucidated.

Fibroblast growth factor 23

Fibroblast growth factor 23 (FGF23) is a member of the fibroblast growth factor family of proteins comprising 22 members from FGF1 to FGF23 with multiple roles from cellular proliferation, differentiation, and survival to regeneration of many tissues. All FGF family members bind fibroblast growth factor receptors (FGFRs) forming a dimer and activating several pathways including the RAS/MAPK/AKT pathway. FGF23 is secreted by osteoblasts and osteocytes and is mainly expressed by brain and bone, but can be found in salivary glands, thyroid/parathyroid gland, stomach, liver, heart, and skeletal muscle [158]. FGF23 requires posttranslational processing modifications, including proteolysis and O-glycosylation by UDP-N-acetyl-α-D-galactosamine:polypeptide N-acetylgalactosaminyl-transferase 3 (GALNT3), resulting in the mature active protein secreted in the bloodstream. FGF23 is a main regulator of calcium-phosphate-vitamin D homeostasis and part of a biological system linking bone to kidney through a complex endocrine axis that maintains bone remodeling and mineralization, important for both health and disease. An overview of FGF23 role in phosphate/calcium/vitamin D metabolism is shown in Fig. 3.4.

One of the major functions of FGF23 is to reduce serum Pi levels, decreasing its resorption and suppressing vitamin $1,25(OH)_2D3$ production. FGF23 achieves this by binding to the FGF receptor, but requires a type I transmembrane protein α-Klotho that functions as a nonenzymatic scaffold protein to promote FGF23 signaling [159]. It has been shown that FGF23 decreases the expression of NPT2a and NPT2c (sodium-dependent, inorganic phosphate (Pi) transporter type 2), the main sodium-phosphate cotransporters in the renal proximal tubule that regulates Pi resorption. Renal excretion of phosphate is an important step for total body phosphate homeostasis, and FGF23 activation increases urinary phosphate excretion by modulating these sodium-dependent Pi cotransporters (Fig. 3.4) [160].

FGF23 interactions with vitamin D have been extensively investigated and appear to establish a negative feedback loop, in which high $1,25(OH)_2$ vitamin D levels stimulate FGF23 production. Increased circulating levels of FGF23 suppress 1α-hydroxylase activity and increase 24 hydroxylase activity at the renal level, which in turn reduces circulating levels of $1,25(OH)_2$ vitamin D3 (Fig. 3.4).

FGF23 was initially identified from a gain-of-function mutation of the autosomal dominant hypophosphatemic rickets (ADHR) disorder, which is characterized by hypophosphatemia secondary to urinary phosphate wasting. In ADHR, FGF23 protein becomes resistant to proteolytic cleavage leading to increased levels of active FGF23 that result in an excessive loss of phosphate in the urine, low serum 1,25-dihydroxyvitamin D(3), and hypocalcemia causing osteomalacia or rickets [161]. Overexpression of FGF23 in transgenic mice have similar phenotype ADHR [162]. Other mutations affecting FGF23 result in hyperphosphatemic familial tumoral calcinosis (FTC) or hyperostosis-hyperphosphatemia syndrome, rare monogenic diseases that may help to elucidate the specific role of proteins and pathways that are involved and could provide insight on more common conditions.

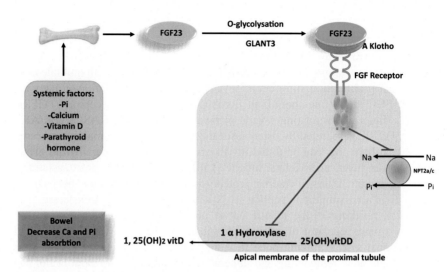

FIGURE 3.4 Role of FGF 23 in Phosphate Homeostasis.

FGF23 is secreted by osteocytes and requires O-glycosylation via UDP-N-acetyl-alpha-D-galactosamine:polypeptide N-acetylgalactosaminyl-transferase 3 (GALNT3) enzyme, which protects FGF 23 from proteolytic processing. Binding of FGF23 to the membrane α Klotho and FGFR in the apical membrane of the proximal tubule leads to inhibition of the hydroxylation of 1,25-dihydroxyvitamin-D (1,25-OH-D3), inhibition of the expression of Na phosphate cotransporter (NPT2a/2c) resulting in decreased renal tubular reabsorption of phosphate and decreased phosphate and calcium reabsorption through the small intestine. FGF23 production is under the regulation of parathyroid hormone, serum phosphate, and vitamin D levels. The overall phosphate and Vitamin D lowering effects of Klotho and FGF23 protect against vascular calcification. Most mutations in FGF23, GALNT3, and α-Klotho lead to a release of suppression of Npt2a and Npt2c, a decrease in 1,25(OH)$_2$ vitamin D levels ultimately increasing serum phosphate concentrations, which result in ectopic and vascular calcifications.

FGF23 knockout mice develop severe hyperphosphatemia, hypercalcemia, 1,25-dihydroxyvitamin D, severe growth retardation, skin atrophy, reduced bone density, soft-tissue and vascular calcifications, and decreased life span [163].O-glycosylation appears to protect FGF23 from degradation. In fact, missense mutation in GALNT3 enhances FGF23 degradation and causes familial tumoral calcinosis [164].

High FGF23 levels are strongly associated with elevated creatinine in chronic kidney disease patients and occur before any other electrolytes abnormality (such as high phosphate levels). They are also associated with progression of renal dysfunction and vascular disease in these patients. These elevated FGF23 levels have additionally been linked to increased left ventricular hypertrophy and mortality. In patients with end-stage renal disease, high FGF23 levels correlate with hyperphosphatemia and may be used as a predictor of mortality in this group of patients [165,166].

The role FGF23 in vascular calcification remains controversial. There is conflicting data both in humans and in animal models. Scialla et al. reported no correlation between FGF23 plasma level and severity of coronary artery calcium in patients with CKD [167]. In contrast, Lim et al. showed that in CKD patients, FGF23 is able to inhibit calcification of aortic smooth muscle by induction of α-Klotho and activation of the extracellular signal-regulated kinase (p-ERK), p-AKT pathway, increasing cellular proliferation [168].

Extracellular vesicles

Matrix vesicles (MVs) were first reported in the late 1960 by Bonucci and Anderson as vesicles that initiated the calcification process in the epiphyseal plate of guinea pigs, rats, and mice. Subsequent studies demonstrated that MVs are secreted from osteoblasts, odontoblasts, and chondrocytes, which are responsible for the initial step of ECM calcification. MVs possess plasma membranes and cytosolic characteristics similar to those of their originating cells [169−171].

Extracellular vesicles (EVs) are similar to MVs but are released by platelets, ECs, SMCs, erythrocytes, and leukocytes. They are composed of a plasma membrane surrounding soluble cytosolic material. EVs can be of three distinct populations from smallest to largest: exosomes formed by exocytosis of multivesicular bodies, microvesicles, or ectosomes formed directly from plasma membrane and apoptotic bodies formed during apoptosis [172].

The EV membrane contains several enzymes and membrane transporters including TNAP, ENPP1, annexin (facilitates calcium entry in EV) and ankylin protein (transports PPi into extracellular space), and type III sodium/phosphate cotransporter (NPit1, transports Pi inside of EV). The EV inner core carries calcium, phosphate, and proteins such as matrix metalloproteinase, fetuin-A, collagen fibers, and proteoglycans but also cytokines, mRNA, and noncoding RNAs. After the content of EVs are released, they can target a variety of cells functioning as intercellular signaling regulators [173−176].

An increased number of circulating EVs have been reported in several cardiovascular and metabolic disorders, including atherosclerosis, arterial medial calcification, and diabetes mellitus type 1. In menopausal women, EVs derived from circulating platelets and EC are elevated and correlate with coronary artery calcification. EVs derived from leukocyte have also been found to be elevated in patients with unstable carotid plaques, suggesting a possible role of circulating EV levels as a biomarker for these conditions.

VSMCs are also able to generate EVs, possibly as a secondary effect of high levels of intracellular calcium in the initiation of atherosclerotic plaque calcification at nucleation sites [177,178].

Conclusions

Our understanding of the molecular mechanisms and pathways involved in the vascular calcification processes has dramatically improved in the recent years because of advances in genetic testing and molecular biology techniques. Identification of specific genes involved in inherited diseases with severe forms of vascular calcification has been helpful in elucidating specific involved pathways, providing more insights as to how VC is regulated on a molecular level. VC development is an active and complex process, in which multiple signaling pathways and a variety of cells are involved in mineralization at various stages of disease and resulting in the observed heterogeneity of various forms of VC. Further studies are needed to improve our understanding of VC pathophysiology and to identify new key players that can be of therapeutic potential to halt or, even, reverse the progression of VC. The emergence of new genetic approaches will continue to allow the identification of novel common and rare variants involved in these complex processes.

Moving forward with new mechanistic insights, pharmacological strategies can be created to antagonize the pathophysiological processes that can initiate and promote VC and target to reduce calcifications and improve vascular compliance and reduce cardiovascular risk in these patients.

Abbreviations

1,25-OH-D3	1,25-dihydroxyvitamin-D
ABCC6	ATP-binding cassette transporter subfamily C member
ACDC	arterial calcification due to deficiency of CD73
ACVR1	activin A receptor, type 1
ADP	adenosine diphosphate
AGE	advanced glycation end products
ALK3	activin receptor-like kinase 3
AMP	adenosine monophosphate
AMPK	$5'$ AMP-activated protein kinase
AP-1	activator protein 1
Apoe$^{-/-}$	apolipoprotein E-deficient
ATF2	activating transcription factor 2
ATP	adenosine triphosphate
ATZ	atherosclerotic
BMP	bone morphogenetic protein
cAMP	cyclic AMP
CARDs	caspase for activation and recruitment domains
CBFA-1	core-binding factor subunit alpha-1
CCL5	C—C motif chemokine ligand 5
CD73	$5'$-nucleotidase ecto
CKD	chronic kidney disease
CTD	C terminal domain

DDX58	dead-box polypeptide 58
DMT	diabetes mellitus type 2
ECs	endothelial cells
ENPP	ectonucleotide pyrophosphatase/phosphodiesterase 1
ENPP1-asj	ectonucleotide pyrophosphatase/phosphodiesterase 1; ages with stiffened joints mice
ER	endoplasmic reticulum
ERKs	extracellular signal-regulated kinases
ESRD	end-stage renal disease
EVs	extracellular vesicles
FGF23	fibroblast growth factor 23
FGFRs	fibroblast growth factor receptors
GACI	generalized arterial calcification of infancy
GALNT3	UDP-N-acetyl-α-D-galactosamine:polypeptide N-acetylgalactosaminyl-transferase 3
GFR	glomerular filtration rate
GGCX	γ-carboxyglutamyl carboxylase
GPI	glycosyl phosphatidylinositol
GTP	guanosine-$5'$-triphosphate
HGPS	Hutchinson–Gilford progeria syndrome
IFIH1 gene	interferon induced with helicase C domain 1
IL1	interleukin 1
iMSCs	induced pluripotent stem cell-derived mesenchymal stem cells
iMSCs	pluripotent stem cell-derived mesenchymal stromal cells
INF	interferon
INF-1β, ISG15	interferon-stimulated gene 15
iPSC	induced pluripotent stem cell
IRF3	interferon regulatory factor 3
I-Smads	inhibitory Smads
JNK	c-Jun N-terminal kinase
LDLR$^{-/-}$	low-density lipoprotein receptor-deficient
LGP2	laboratory of genetics and physiology 2
LMNA	lamin A/C
MAPK	mitogen-activated protein kinases
MAVS	activating mitochondrial antiviral signaling protein
MCP-1	monocyte chemotactic protein-1
MDA5	melanoma differentiation-associated protein 5
MGP	matrix Gla protein
miR	microRNA
Msh	Homeobox 2 MSX2
mTOR1	mammalian target of rapamycin
MVs	matrix vesicles
NF-kB	nuclear factor kappa-light-chain enhancer of activated B cells
NPT2a and NPT2c	sodium-dependent inorganic phosphate transporter type 2
OPG	osteoprotegerin
P38	P38 mitogen-activated protein kinases
p-ERK	extracellular signal-regulated kinase

Pi	inorganic phosphate
PiT1/2	Na+-dependent Pi cotransporter
PPi	inorganic pyrophosphate
PXE	Pseudoxanthoma elasticum
RANK	receptor activator of nuclear factor κ B
RANKL	nuclear factor kappa-B ligand receptor activator
rh	recombinant human
RIG-I	retinoic-acid-inducible I
RLR	RIG-I-like receptor
ROS	reactive oxygen species
R-Smads	receptor-regulated-Smads
Runx2	Runt-related transcription factor 2
SLC20A2	solute carrier family 20 member 2
Smads	common mediator
SMS	Singleton-Merten syndrome
SRF	serum response factor
TCF	ternary complex factor
TGF-β	transforming growth factor-β
TGF-β2	transforming growth factor beta 2
TNAP	tissue nonspecific alkaline phosphatase
TNFR	tumor necrosis factor receptor
TNFSF10/TRAIL	tumor necrosis factor-related apoptosis-inducing ligand or Apo 2 ligand
ttw/ttw	homozygous tiptoe-walking mice
VC	vascular calcification
VSMCs	vascular smooth muscle cells

Conflict of interest

The authors declared no conflicts of interests with respect to this manuscript.

References

[1] Virchow R. Cellular pathology: as based upon physiological and pathological histology. An unabridged and unaltered republication of the English translation originally published. New York: Dover; 1863. p. 5.

[2] Mönckeberg JG. Über die reine Mediaverkalkung der Extremitätenarterien und ihr Verhalten zur Arteriosklerose. Virchows Arch Pathol Anat Physiol Klin Med 1903; 171:7.

[3] Giachelli CM. Ectopic calcification: gathering hard facts about soft tissue mineralization. Am J Pathol 1999;154(3):671−5. https://doi.org/10.1016/S0002-9440(10)65313-8.

[4] Stewart VL, Herling P, Dalinka MK. Calcification in soft tissues. J Am Med Assoc 1983;250(1):78−81.

[5] Abedin M, Tintut Y, Demer LL. Vascular calcification: mechanisms and clinical ramifications. Arterioscler Thromb Vasc Biol 2004;24(7):1161—70. https://doi.org/10.1161/01.ATV.0000133194.94939.42.

[6] Avogaro A, Fadini GP. Mechanisms of ectopic calcification: implications for diabetic vasculopathy. Cardiovasc Diagn Ther 2015;5(5):343—52. https://doi.org/10.3978/j.issn.2223-3652.2015.06.05.

[7] Amann K. Media calcification and intima calcification are distinct entities in chronic kidney disease. Clin J Am Soc Nephrol 2008;3(6):1599—605. https://doi.org/10.2215/CJN.02120508.

[8] Bryant JH, Hale-White W. A case of calcification of the arteries and obliterative endarteritis, associated with hydronephrosis, in a child aged six months. Guy's Hosp Rep 1901;55:12.

[9] Rutsch F, Ruf N, Vaingankar S, Toliat MR, Suk A, Hohne W, et al. Mutations in ENPP1 are associated with 'idiopathic' infantile arterial calcification. Nat Genet 2003;34(4):379—81. https://doi.org/10.1038/ng1221.

[10] Chong CR, Hutchins GM. Idiopathic infantile arterial calcification: the spectrum of clinical presentations. Pediatr Dev Pathol 2008;11(5):405—15. https://doi.org/10.2350/07-06-0297.1.

[11] Rutsch F, Boyer P, Nitschke Y, Ruf N, Lorenz-Depierieux B, Wittkampf T, et al. Hypophosphatemia, hyperphosphaturia, and bisphosphonate treatment are associated with survival beyond infancy in generalized arterial calcification of infancy. Circ Cardiovasc Genet 2008;1(2):133—40. https://doi.org/10.1161/CIRCGENETICS.108.797704.

[12] Nitschke Y, Baujat G, Botschen U, Wittkampf T, du Moulin M, Stella J, et al. Generalized arterial calcification of infancy and pseudoxanthoma elasticum can be caused by mutations in either ENPP1 or ABCC6. Am J Hum Genet 2012;90(1):25—39. https://doi.org/10.1016/j.ajhg.2011.11.020.

[13] Lorenz-Depiereux B, Schnabel D, Tiosano D, Hausler G, Strom TM. Loss-of-function ENPP1 mutations cause both generalized arterial calcification of infancy and autosomal-recessive hypophosphatemic rickets. Am J Hum Genet 2010;86(2):267—72. https://doi.org/10.1016/j.ajhg.2010.01.006.

[14] Hessle L, Johnson KA, Anderson HC, Narisawa S, Sali A, Goding JW, et al. Tissue-nonspecific alkaline phosphatase and plasma cell membrane glycoprotein-1 are central antagonistic regulators of bone mineralization. Proc Natl Acad Sci U S A 2002;99(14):9445—9. https://doi.org/10.1073/pnas.142063399.

[15] Krug HE, Mahowald ML, Halverson PB, Sallis JD, Cheung HS. Phosphocitrate prevents disease progression in murine progressive ankylosis. Arthritis Rheum 1993;36(11):1603—11.

[16] Drake MT, Clarke BL, Khosla S. Bisphosphonates: mechanism of action and role in clinical practice. Mayo Clin Proc 2008;83(9):1032—45. https://doi.org/10.4065/83.9.1032.

[17] Russell RG. Bisphosphonates: from bench to bedside. Ann N Y Acad Sci 2006;1068:367—401. https://doi.org/10.1196/annals.1346.041.

[18] Furusawa N, Baba H, Imura S, Fukuda M. Characteristics and mechanism of the ossification of posterior longitudinal ligament in the tip-toe walking Yoshimura (twy) mouse. Eur J Histochem 1996;40(3):199—210.

[19] Harmey D, Hessle L, Narisawa S, Johnson KA, Terkeltaub R, Millan JL. Concerted regulation of inorganic pyrophosphate and osteopontin by akp2, enpp1, and ank: an

integrated model of the pathogenesis of mineralization disorders. Am J Pathol 2004; 164(4):1199−209. https://doi.org/10.1016/S0002-9440(10)63208-7.

[20] Khan T, Sinkevicius KW, Vong S, Avakian A, Leavitt MC, Malanson H, et al. ENPP1 enzyme replacement therapy improves blood pressure and cardiovascular function in a mouse model of generalized arterial calcification of infancy. Dis Model Mech 2018; 11(10). https://doi.org/10.1242/dmm.035691.

[21] Romanelli F, Corbo A, Salehi M, Yadav MC, Salman S, Petrosian D, et al. Overexpression of tissue-nonspecific alkaline phosphatase (TNAP) in endothelial cells accelerates coronary artery disease in a mouse model of familial hypercholesterolemia. PLoS One 2017;12(10):e0186426. https://doi.org/10.1371/journal.pone.0186426.

[22] Albright RA, Stabach P, Cao W, Kavanagh D, Mullen I, Braddock AA, et al. ENPP1-Fc prevents mortality and vascular calcifications in rodent model of generalized arterial calcification of infancy. Nat Commun 2015;6:10006. https://doi.org/10.1038/ncomms10006.

[23] Nitschke Y, Yan Y, Buers I, Kintziger K, Askew K, Rutsch F. ENPP1-Fc prevents neointima formation in generalized arterial calcification of infancy through the generation of AMP. Exp Mol Med 2018;50(10):139. https://doi.org/10.1038/s12276-018-0163-5.

[24] Darier J. Pseudo-xanthome élastique. III ème congrès Intern. de Dermat de Londres 1896;23:7.

[25] Leftheriotis G, Abraham P, Le Corre Y, Le Saux O, Henrion D, Ducluzeau PH, et al. Relationship between ankle brachial index and arterial remodeling in pseudoxanthoma elasticum. J Vasc Surg 2011;54(5):1390−4. https://doi.org/10.1016/j.jvs.2011.04.041.

[26] Bergen AA, Plomp AS, Schuurman EJ, Terry S, Breuning M, Dauwerse H, et al. Mutations in ABCC6 cause pseudoxanthoma elasticum. Nat Genet 2000;25(2):228−31. https://doi.org/10.1038/76109.

[27] Le Saux O, Urban Z, Tschuch C, Csiszar K, Bacchelli B, Quaglino D, et al. Mutations in a gene encoding an ABC transporter cause pseudoxanthoma elasticum. Nat Genet 2000;25(2):223−7. https://doi.org/10.1038/76102.

[28] Ringpfeil F, Lebwohl MG, Christiano AM, Uitto J. Pseudoxanthoma elasticum: mutations in the MRP6 gene encoding a transmembrane ATP-binding cassette (ABC) transporter. Proc Natl Acad Sci U S A 2000;97(11):6001−6. https://doi.org/10.1073/pnas.100041297.

[29] Moore AF, Jablonski KA, Mason CC, McAteer JB, Arakaki RF, Goldstein BJ, et al. The association of ENPP1 K121Q with diabetes incidence is abolished by lifestyle modification in the diabetes prevention program. J Clin Endocrinol Metab 2009; 94(2):449−55. https://doi.org/10.1210/jc.2008-1583.

[30] Vasiliou V, Vasiliou K, Nebert DW. Human ATP-binding cassette (ABC) transporter family. Hum Genom 2009;3(3):281−90.

[31] Matsuzaki Y, Nakano A, Jiang QJ, Pulkkinen L, Uitto J. Tissue-specific expression of the ABCC6 gene. J Investig Dermatol 2005;125(5):900−5. https://doi.org/10.1111/j.0022-202X.2005.23897.x.

[32] Van Gils M, Nollet L, Verly E, Deianova N, Vanakker OM. Cellular signaling in pseudoxanthoma elasticum: an update. Cell Signal 2019;55:119−29. https://doi.org/10.1016/j.cellsig.2018.12.009.

[33] Borst P, Varadi A, van de Wetering K. PXE, a mysterious inborn error clarified. Trends Biochem Sci 2019;44(2):125−40. https://doi.org/10.1016/j.tibs.2018.10.005.

[34] Jiang Q, Endo M, Dibra F, Wang K, Uitto J. Pseudoxanthoma elasticum is a metabolic disease. J Investig Dermatol 2009;129(2):348−54. https://doi.org/10.1038/jid.2008.212.

[35] Hendig D, Langmann T, Kocken S, Zarbock R, Szliska C, Schmitz G, et al. Gene expression profiling of ABC transporters in dermal fibroblasts of pseudoxanthoma elasticum patients identifies new candidates involved in PXE pathogenesis. Lab Investig 2008;88(12):1303−15. https://doi.org/10.1038/labinvest.2008.96.

[36] Marchione R, Kim N, Kirsner RS. Pseudoxanthoma elasticum: new insights. J Investig Dermatol 2009;129(2):258. https://doi.org/10.1038/jid.2008.407.

[37] Ziegler SG, Ferreira CR, MacFarlane EG, Riddle RC, Tomlinson RE, Chew EY, et al. Ectopic calcification in pseudoxanthoma elasticum responds to inhibition of tissue-nonspecific alkaline phosphatase. Sci Transl Med 2017;9(393). https://doi.org/10.1126/scitranslmed.aal1669.

[38] Jansen RS, Duijst S, Mahakena S, Sommer D, Szeri F, Varadi A, et al. ABCC6-mediated ATP secretion by the liver is the main source of the mineralization inhibitor inorganic pyrophosphate in the systemic circulation-brief report. Arterioscler Thromb Vasc Biol 2014;34(9):1985−9. https://doi.org/10.1161/ATVBAHA.114.304017.

[39] Jansen RS, Kucukosmanoglu A, de Haas M, Sapthu S, Otero JA, Hegman IE, et al. ABCC6 prevents ectopic mineralization seen in pseudoxanthoma elasticum by inducing cellular nucleotide release. Proc Natl Acad Sci U S A 2013;110(50):20206−11. https://doi.org/10.1073/pnas.1319582110.

[40] Garcia-Fernandez MI, Gheduzzi D, Boraldi F, Paolinelli CD, Sanchez P, Valdivielso P, et al. Parameters of oxidative stress are present in the circulation of PXE patients. Biochim Biophys Acta 2008;1782(7−8):474−81. https://doi.org/10.1016/j.bbadis.2008.05.001.

[41] Hendig D, Zarbock R, Szliska C, Kleesiek K, Gotting C. The local calcification inhibitor matrix Gla protein in pseudoxanthoma elasticum. Clin Biochem 2008;41(6):407−12. https://doi.org/10.1016/j.clinbiochem.2007.12.023.

[42] Jiang Q, Dibra F, Lee MD, Oldenburg R, Uitto J. Overexpression of fetuin-a counteracts ectopic mineralization in a mouse model of pseudoxanthoma elasticum (abcc6(-/-)). J Investig Dermatol 2010;130(5):1288−96. https://doi.org/10.1038/jid.2009.423.

[43] St Hilaire C, Ziegler SG, Markello TC, Brusco A, Groden C, Gill F, et al. NT5E mutations and arterial calcifications. N Engl J Med 2011;364(5):432−42. https://doi.org/10.1056/NEJMoa0912923.

[44] Zhang Z, He JW, Fu WZ, Zhang CQ, Zhang ZL. Calcification of joints and arteries: second report with novel NT5E mutations and expansion of the phenotype. J Hum Genet 2015;60(10):561−4. https://doi.org/10.1038/jhg.2015.85.

[45] Antonioli L, Pacher P, Vizi ES, Hasko G. CD39 and CD73 in immunity and inflammation. Trends Mol Med 2013;19(6):355−67. https://doi.org/10.1016/j.molmed.2013.03.005.

[46] Fausther M, Lavoie EG, Goree JR, Baldini G, Dranoff JA. NT5E mutations that cause human disease are associated with intracellular mistrafficking of NT5E protein. PLoS One 2014;9(6):e98568. https://doi.org/10.1371/journal.pone.0098568.

[47] Savinov AY, Salehi M, Yadav MC, Radichev I, Millan JL, Savinova OV. Transgenic overexpression of tissue-nonspecific alkaline phosphatase (TNAP) in vascular endothelium results in generalized arterial calcification. J Am Heart Assoc 2015;4(12). https://doi.org/10.1161/JAHA.115.002499.

[48] Sheen CR, Kuss P, Narisawa S, Yadav MC, Nigro J, Wang W, et al. Pathophysiological role of vascular smooth muscle alkaline phosphatase in medial artery calcification. J Bone Miner Res 2015;30(5):824−36. https://doi.org/10.1002/jbmr.2420.

[49] Ndrepepa G, Xhepa E, Braun S, Cassese S, Fusaro M, Schunkert H, Kastrati A. Alkaline phosphatase and prognosis in patients with coronary artery disease. Eur J Clin Investig 2017;47(5):378−87. https://doi.org/10.1111/eci.12752.

[50] Panh L, Ruidavets JB, Rousseau H, Petermann A, Bongard V, Berard E, et al. Association between serum alkaline phosphatase and coronary artery calcification in a sample of primary cardiovascular prevention patients. Atherosclerosis 2017;260:81−6. https://doi.org/10.1016/j.atherosclerosis.2017.03.030.

[51] Jin H, St Hilaire C, Huang Y, Yang D, Dmitrieva NI, Negro A, et al. Increased activity of TNAP compensates for reduced adenosine production and promotes ectopic calcification in the genetic disease ACDC. Sci Signal 2016;9(458):ra121. https://doi.org/10.1126/scisignal.aaf9109.

[52] Burnstock G. Purinergic signaling in the cardiovascular system. Circ Res 2017;120(1):207−28. https://doi.org/10.1161/CIRCRESAHA.116.309726.

[53] Ralevic V, Burnstock G. Involvement of purinergic signaling in cardiovascular diseases. Drug News Perspect 2003;16(3):133−40.

[54] Tolle M, Jankowski V, Schuchardt M, Wiedon A, Huang T, Hub F, et al. Adenosine 5'-tetraphosphate is a highly potent purinergic endothelium-derived vasoconstrictor. Circ Res 2008;103(10):1100−8. https://doi.org/10.1161/CIRCRESAHA.108.177865.

[55] Burnstock G, Ralevic V. Purinergic signaling and blood vessels in health and disease. Pharmacol Rev 2014;66(1):102−92. https://doi.org/10.1124/pr.113.008029.

[56] Ferrari D, Vitiello L, Idzko M, la Sala A. Purinergic signaling in atherosclerosis. Trends Mol Med 2015;21(3):184−92. https://doi.org/10.1016/j.molmed.2014.12.008.

[57] Reiss AB, Cronstein BN. Regulation of foam cells by adenosine. Arterioscler Thromb Vasc Biol 2012;32(4):879−86. https://doi.org/10.1161/ATVBAHA.111.226878.

[58] Hale JE, Fraser JD, Price PA. The identification of matrix Gla protein in cartilage. J Biol Chem 1988;263(12):5820−4.

[59] Luo G, Ducy P, McKee MD, Pinero GJ, Loyer E, Behringer RR, Karsenty G. Spontaneous calcification of arteries and cartilage in mice lacking matrix GLA protein. Nature 1997;386(6620):78−81. https://doi.org/10.1038/386078a0.

[60] Keutel J, Jorgensen G, Gabriel P. A new autosomal-recessive hereditary syndrome. Multiple peripheral pulmonary stenosis, brachytelephalangia, inner-ear deafness, ossification or calcification of cartilages. Dtsch Med Wochenschr 1971;96(43):1676−81. https://doi.org/10.1055/s-0028-1110200.

[61] Cranenburg EC, Van Spaendonck-Zwarts KY, Bonafe L, Mittaz Crettol L, Rodiger LA, Dikkers FG, et al. Circulating matrix gamma-carboxyglutamate protein (MGP) species are refractory to vitamin K treatment in a new case of Keutel syndrome. J Thromb Haemost 2011;9(6):1225−35. https://doi.org/10.1111/j.1538-7836.2011.04263.x.

[62] Malhotra R, Burke MF, Martyn T, Shakartzi HR, Thayer TE, O'Rourke C, et al. Inhibition of bone morphogenetic protein signal transduction prevents the medial vascular calcification associated with matrix Gla protein deficiency. PLoS One 2015;10(1):e0117098. https://doi.org/10.1371/journal.pone.0117098.

[63] Strandgren C, Revechon G, Sola-Carvajal A, Eriksson M. Emerging candidate treatment strategies for Hutchinson-Gilford progeria syndrome. Biochem Soc Trans 2017;45(6):1279−93. https://doi.org/10.1042/BST20170141.

[64] Ullrich NJ, Gordon LB. Hutchinson-Gilford progeria syndrome. Handb Clin Neurol 2015;132:249−64. https://doi.org/10.1016/B978-0-444-62702-5.00018-4.

[65] Capell BC, Erdos MR, Madigan JP, Fiordalisi JJ, Varga R, Conneely KN, et al. Inhibiting farnesylation of progerin prevents the characteristic nuclear blebbing of Hutchinson-Gilford progeria syndrome. Proc Natl Acad Sci U S A 2005;102(36): 12879−84. https://doi.org/10.1073/pnas.0506001102.

[66] Eriksson M, Brown WT, Gordon LB, Glynn MW, Singer J, Scott L, et al. Recurrent de novo point mutations in lamin A cause Hutchinson-Gilford progeria syndrome. Nature 2003;423(6937):293−8. https://doi.org/10.1038/nature01629.

[67] Hanumanthappa NB, Madhusudan G, Mahimarangaiah J, Manjunath CN. Hutchinson-Gilford progeria syndrome with severe calcific aortic valve stenosis. Ann Pediatr Cardiol 2011;4(2):204−6. https://doi.org/10.4103/0974-2069.84670.

[68] Salamat M, Dhar PK, Neagu DL, Lyon JB. Aortic calcification in a patient with hutchinson-gilford progeria syndrome. Pediatr Cardiol 2010;31(6):925−6. https://doi.org/10.1007/s00246-010-9711-z.

[69] Villa-Bellosta R, Rivera-Torres J, Osorio FG, Acin-Perez R, Enriquez JA, Lopez-Otin C, Andres V. Defective extracellular pyrophosphate metabolism promotes vascular calcification in a mouse model of Hutchinson-Gilford progeria syndrome that is ameliorated on pyrophosphate treatment. Circulation 2013;127(24):2442−51. https://doi.org/10.1161/CIRCULATIONAHA.112.000571.

[70] Singleton EB, Merten DF. An unusual syndrome of widened medullary cavities of the metacarpals and phalanges, aortic calcification and abnormal dentition. Pediatr Radiol 1973;1(1):2−7.

[71] Feigenbaum A, Muller C, Yale C, Kleinheinz J, Jezewski P, Kehl HG, et al. Singleton-Merten syndrome: an autosomal dominant disorder with variable expression. Am J Med Genet 2013;161A(2):360−70. https://doi.org/10.1002/ajmg.a.35732.

[72] Rutsch F, MacDougall M, Lu C, Buers I, Mamaeva O, Nitschke Y, et al. A specific IFIH1 gain-of-function mutation causes Singleton-Merten syndrome. Am J Hum Genet 2015;96(2):275−82. https://doi.org/10.1016/j.ajhg.2014.12.014.

[73] Jang MA, Kim EK, Now H, Nguyen NT, Kim WJ, Yoo JY, et al. Mutations in DDX58, which encodes RIG-I, cause atypical Singleton-Merten syndrome. Am J Hum Genet 2015;96(2):266−74. https://doi.org/10.1016/j.ajhg.2014.11.019.

[74] Ostuni A, Infantino V, Salvia A, Miglionico R, Boraldi F, Annovi G, Bisaccia F. WITHDRAWN: epigenetic control of TNAP expression in Pseudoxanthoma elasticum fibroblasts. Cell Biol Int 2012. https://doi.org/10.1042/CBI20110314.

[75] Onomoto K, Onoguchi K, Takahasi K, Fujita T. Type I interferon production induced by RIG-I-like receptors. J Interferon Cytokine Res 2010;30(12):875−81. https://doi.org/10.1089/jir.2010.0117.

[76] Matsumiya T, Stafforini DM. Function and regulation of retinoic acid-inducible gene-I. Crit Rev Immunol 2010;30(6):489−513.

[77] Yoneyama M, Kikuchi M, Matsumoto K, Imaizumi T, Miyagishi M, Taira K, et al. Shared and unique functions of the DExD/H-box helicases RIG-I, MDA5, and LGP2 in antiviral innate immunity. J Immunol 2005;175(5):2851−8.

[78] Barral PM, Sarkar D, Su ZZ, Barber GN, DeSalle R, Racaniello VR, Fisher PB. Functions of the cytoplasmic RNA sensors RIG-I and MDA-5: key regulators of innate immunity. Pharmacol Ther 2009;124(2):219−34. https://doi.org/10.1016/j.pharmthera.2009.06.012.

[79] Wu B, Peisley A, Richards C, Yao H, Zeng X, Lin C, et al. Structural basis for dsRNA recognition, filament formation, and antiviral signal activation by MDA5. Cell 2013; 152(1−2):276−89. https://doi.org/10.1016/j.cell.2012.11.048.

[80] Heldin CH, Moustakas A. Signaling receptors for TGF-beta family members. Cold Spring Harb Perspect Biol 2016;8(8). https://doi.org/10.1101/cshperspect.a022053.

[81] Heldin CH, Moustakas A. Role of Smads in TGFbeta signaling. Cell Tissue Res 2012; 347(1):21−36. https://doi.org/10.1007/s00441-011-1190-x.

[82] Massague J. A very private TGF-beta receptor embrace. Mol Cell 2008;29(2):149−50. https://doi.org/10.1016/j.molcel.2008.01.006.

[83] Pardali E, Ten Dijke P. TGFbeta signaling and cardiovascular diseases. Int J Biol Sci 2012;8(2):195−213. https://doi.org/10.7150/ijbs.3805.

[84] Kalinina N, Agrotis A, Antropova Y, Ilyinskaya O, Smirnov V, Tararak E, Bobik A. Smad expression in human atherosclerotic lesions: evidence for impaired TGF-beta/ Smad signaling in smooth muscle cells of fibrofatty lesions. Arterioscler Thromb Vasc Biol 2004;24(8):1391−6. https://doi.org/10.1161/01.ATV.0000133605.89421.79.

[85] Dai J, Michineau S, Franck G, Desgranges P, Becquemin JP, Gervais M, Allaire E. Long term stabilization of expanding aortic aneurysms by a short course of cyclosporine A through transforming growth factor-beta induction. PLoS One 2011; 6(12):e28903. https://doi.org/10.1371/journal.pone.0028903.

[86] Feinberg MW, Jain MK. Role of transforming growth factor-beta1/Smads in regulating vascular inflammation and atherogenesis. Panminerva Med 2005;47(3):169−86.

[87] Lutgens E, Gijbels M, Smook M, Heeringa P, Gotwals P, Koteliansky VE, Daemen MJ. Transforming growth factor-beta mediates balance between inflammation and fibrosis during plaque progression. Arterioscler Thromb Vasc Biol 2002;22(6):975−82.

[88] Reifenberg K, Cheng F, Orning C, Crain J, Kupper I, Wiese E, et al. Overexpression of TGF-ss1 in macrophages reduces and stabilizes atherosclerotic plaques in ApoE-deficient mice. PLoS One 2012;7(7):e40990. https://doi.org/10.1371/journal. pone.0040990.

[89] Mallat Z, Gojova A, Marchiol-Fournigault C, Esposito B, Kamate C, Merval R, et al. Inhibition of transforming growth factor-beta signaling accelerates atherosclerosis and induces an unstable plaque phenotype in mice. Circ Res 2001;89(10):930−4.

[90] Simionescu A, Philips K, Vyavahare N. Elastin-derived peptides and TGF-beta1 induce osteogenic responses in smooth muscle cells. Biochem Biophys Res Commun 2005;334(2):524−32. https://doi.org/10.1016/j.bbrc.2005.06.119.

[91] Panutsopulos D, Papalambros E, Sigala F, Zafiropoulos A, Arvanitis DL, Spandidos DA. Protein and mRNA expression levels of VEGF-A and TGF-beta1 in different types of human coronary atherosclerotic lesions. Int J Mol Med 2005; 15(4):603−10.

[92] Redondo S, Navarro-Dorado J, Ramajo M, Medina U, Tejerina T. The complex regulation of TGF-beta in cardiovascular disease. Vasc Health Risk Manag 2012;8:533−9. https://doi.org/10.2147/VHRM.S28041.

[93] Grainger DJ. TGF-beta and atherosclerosis in man. Cardiovasc Res. 2007;74(2): 213−22. Epub; 2007 Feb 23, Review; PMID: 17382916.

[94] Wozney JM, Rosen V, Celeste AJ, Mitsock LM, Whitters MJ, Kriz RW, et al. Novel regulators of bone formation: molecular clones and activities. Science 1988; 242(4885):1528−34.

[95] Bostrom K, Watson KE, Horn S, Wortham C, Herman IM, Demer LL. Bone morphogenetic protein expression in human atherosclerotic lesions. J Clin Investig 1993; 91(4):1800−9. https://doi.org/10.1172/JCI116391.

[96] Massague J, Seoane J, Wotton D. Smad transcription factors. Genes Dev 2005;19(23): 2783−810. https://doi.org/10.1101/gad.1350705.

[97] Ross S, Hill CS. How the Smads regulate transcription. Int J Biochem Cell Biol 2008; 40(3):383−408. https://doi.org/10.1016/j.biocel.2007.09.006.

[98] Cai J, Pardali E, Sanchez-Duffhues G, ten Dijke P. BMP signaling in vascular diseases. FEBS Lett 2012;586(14):1993−2002. https://doi.org/10.1016/j.febslet.2012.04.030.

[99] Wang RN, Green J, Wang Z, Deng Y, Qiao M, Peabody M, et al. Bone Morphogenetic Protein (BMP) signaling in development and human diseases. Genes Dis 2014;1(1): 87−105. https://doi.org/10.1016/j.gendis.2014.07.005.

[100] Balderman JA, Lee HY, Mahoney CE, Handy DE, White K, Annis S, et al. Bone morphogenetic protein-2 decreases microRNA-30b and microRNA-30c to promote vascular smooth muscle cell calcification. J Am Heart Assoc 2012;1(6):e003905. https://doi.org/10.1161/JAHA.112.003905.

[101] Steitz SA, Speer MY, Curinga G, Yang HY, Haynes P, Aebersold R, et al. Smooth muscle cell phenotypic transition associated with calcification: upregulation of Cbfa1 and downregulation of smooth muscle lineage markers. Circ Res 2001;89(12):1147−54.

[102] Dyer LA, Pi X, Patterson C. The role of BMPs in endothelial cell function and dysfunction. Trends Endocrinol Metabol 2014;25(9):472−80. https://doi.org/10.1016/j.tem.2014.05.003.

[103] Garcia de Vinuesa A, Abdelilah-Seyfried S, Knaus P, Zwijsen A, Bailly S. BMP signaling in vascular biology and dysfunction. Cytokine Growth Factor Rev 2016; 27:65−79. https://doi.org/10.1016/j.cytogfr.2015.12.005.

[104] Csiszar A, Ahmad M, Smith KE, Labinskyy N, Gao Q, Kaley G, et al. Bone morphogenetic protein-2 induces proinflammatory endothelial phenotype. Am J Pathol 2006; 168(2):629−38. https://doi.org/10.2353/ajpath.2006.050284.

[105] Zhang M, Sara JD, Wang FL, Liu LP, Su LX, Zhe J, et al. Increased plasma BMP-2 levels are associated with atherosclerosis burden and coronary calcification in type 2 diabetic patients. Cardiovasc Diabetol 2015;14:64. https://doi.org/10.1186/s12933-015-0214-3.

[106] Shanahan CM, Cary NR, Salisbury JR, Proudfoot D, Weissberg PL, Edmonds ME. Medial localization of mineralization-regulating proteins in association with Monckeberg's sclerosis: evidence for smooth muscle cell-mediated vascular calcification. Circulation 1999;100(21):2168−76.

[107] Derwall M, Malhotra R, Lai CS, Beppu Y, Aikawa E, Seehra JS, et al. Inhibition of bone morphogenetic protein signaling reduces vascular calcification and atherosclerosis. Arterioscler Thromb Vasc Biol 2012;32(3):613−22. https://doi.org/10.1161/ATVBAHA.111.242594.

[108] Ankeny RF, Thourani VH, Weiss D, Vega JD, Taylor WR, Nerem RM, Jo H. Preferential activation of SMAD1/5/8 on the fibrosa endothelium in calcified human aortic valves–association with low BMP antagonists and SMAD6. PLoS One 2011;6(6): e20969. https://doi.org/10.1371/journal.pone.0020969.

[109] Galvin KM, Donovan MJ, Lynch CA, Meyer RI, Paul RJ, Lorenz JN, et al. A role for smad6 in development and homeostasis of the cardiovascular system. Nat Genet 2000; 24(2):171−4. https://doi.org/10.1038/72835.

[110] Tan HL, Glen E, Topf A, Hall D, O'Sullivan JJ, Sneddon L, et al. Nonsynonymous variants in the SMAD6 gene predispose to congenital cardiovascular malformation. Hum Mutat 2012;33(4):720–7. https://doi.org/10.1002/humu.22030.

[111] Shore EM. Fibrodysplasia ossificans progressiva: a human genetic disorder of extraskeletal bone formation, or–how does one tissue become another? Wiley Interdiscip Rev Dev Biol 2012;1(1):153–65. https://doi.org/10.1002/wdev.9.

[112] PEDERSEN K. Fetuin, a new globulin isolated from serum. Nature 1944;154:1.

[113] Schultze HE, H K, Haupt H. Charakterisierung eines niedermolekularen α2-Mukoids aus Humanserum. Naturwissenschaften 1962;49:3.

[114] Schinke T, Amendt C, Trindl A, Poschke O, Muller-Esterl W, Jahnen-Dechent W. The serum protein alpha2-HS glycoprotein/fetuin inhibits apatite formation in vitro and in mineralizing calvaria cells. A possible role in mineralization and calcium homeostasis. J Biol Chem 1996;271(34):20789–96.

[115] Chen NX, O'Neill KD, Chen X, Duan D, Wang E, Sturek MS, et al. Fetuin-A uptake in bovine vascular smooth muscle cells is calcium dependent and mediated by annexins. Am J Physiol Renal Physiol 2007;292(2):F599–606. https://doi.org/10.1152/ajprenal.00303.2006.

[116] Ketteler M, Bongartz P, Westenfeld R, Wildberger JE, Mahnken AH, Bohm R, et al. Association of low fetuin-A (AHSG) concentrations in serum with cardiovascular mortality in patients on dialysis: a cross-sectional study. Lancet 2003;361(9360):827–33. https://doi.org/10.1016/S0140-6736(03)12710-9.

[117] Oikawa O, Higuchi T, Yamazaki T, Yamamoto C, Fukuda N, Matsumoto K. Evaluation of serum fetuin-A relationships with biochemical parameters in patients on hemodialysis. Clin Exp Nephrol 2007;11(4):304–8. https://doi.org/10.1007/s10157-007-0499-y.

[118] Westenfeld R, Schafer C, Kruger T, Haarmann C, Schurgers LJ, Reutelingsperger C, et al. Fetuin-A protects against atherosclerotic calcification in CKD. J Am Soc Nephrol 2009;20(6):1264–74. https://doi.org/10.1681/ASN.2008060572.

[119] Abdel-Wahab AF, Fathy O, Al-Harizy R. Negative correlation between fetuin-A and indices of vascular disease in systemic lupus erythematosus patients with and without lupus nephritis. Arab J Nephrol Transplant 2013;6(1):11–20.

[120] Chen HY, Chiu YL, Hsu SP, Pai MF, Yang JY, Peng YS. Low serum fetuin A levels and incident stroke in patients with maintenance haemodialysis. Eur J Clin Investig 2013;43(4):387–96. https://doi.org/10.1111/eci.12057.

[121] Jung JY, Hwang YH, Lee SW, Lee H, Kim DK, Kim S, et al. Factors associated with aortic stiffness and its change over time in peritoneal dialysis patients. Nephrol Dial Transplant 2010;25(12):4041–8. https://doi.org/10.1093/ndt/gfq293.

[122] Marechal C, Schlieper G, Nguyen P, Kruger T, Coche E, Robert A, et al. Serum fetuin-A levels are associated with vascular calcifications and predict cardiovascular events in renal transplant recipients. Clin J Am Soc Nephrol 2011;6(5):974–85. https://doi.org/10.2215/CJN.06150710.

[123] Hamano T, Matsui I, Mikami S, Tomida K, Fujii N, Imai E, et al. Fetuin-mineral complex reflects extraosseous calcification stress in CKD. J Am Soc Nephrol 2010;21(11):1998–2007. https://doi.org/10.1681/ASN.2009090944.

[124] Hermans MM, Brandenburg V, Ketteler M, Kooman JP, van der Sande FM, Gladziwa U, et al. Study on the relationship of serum fetuin-A concentration with aortic stiffness in patients on dialysis. Nephrol Dial Transplant 2006;21(5):1293–9. https://doi.org/10.1093/ndt/gfk045.

[125] Jung HH, Baek HJ, Kim SW. Fetuin-A, coronary artery calcification and outcome in maintenance hemodialysis patients. Clin Nephrol 2011;75(5):391–6.

[126] Manghat P, Souleimanova I, Cheung J, Wierzbicki AS, Harrington DJ, Shearer MJ, et al. Association of bone turnover markers and arterial stiffness in pre-dialysis chronic kidney disease (CKD). Bone 2011;48(5):1127–32. https://doi.org/10.1016/j.bone.2011.01.016.

[127] Cavaco S, Viegas CS, Rafael MS, Ramos A, Magalhaes J, Blanco FJ, et al. Gla-rich protein is involved in the cross-talk between calcification and inflammation in osteoarthritis. Cell Mol Life Sci 2016;73(5):1051–65. https://doi.org/10.1007/s00018-015-2033-9.

[128] Viegas CS, Rafael MS, Enriquez JL, Teixeira A, Vitorino R, Luis IM, et al. Gla-rich protein acts as a calcification inhibitor in the human cardiovascular system. Arterioscler Thromb Vasc Biol 2015;35(2):399–408. https://doi.org/10.1161/ATVBAHA.114.304823.

[129] Khavandgar Z, Roman H, Li J, Lee S, Vali H, Brinckmann J, et al. Elastin haploinsufficiency impedes the progression of arterial calcification in MGP-deficient mice. J Bone Miner Res 2014;29(2):327–37. https://doi.org/10.1002/jbmr.2039.

[130] Koos R, Mahnken AH, Muhlenbruch G, Brandenburg V, Pflueger B, Wildberger JE, Kuhl HP. Relation of oral anticoagulation to cardiac valvular and coronary calcium assessed by multislice spiral computed tomography. Am J Cardiol 2005;96(6):747–9. https://doi.org/10.1016/j.amjcard.2005.05.014.

[131] Price PA, Faus SA, Williamson MK. Warfarin causes rapid calcification of the elastic lamellae in rat arteries and heart valves. Arterioscler Thromb Vasc Biol 1998;18(9):1400–7.

[132] Schurgers LJ, Spronk HM, Soute BA, Schiffers PM, DeMey JG, Vermeer C. Regression of warfarin-induced medial elastocalcinosis by high intake of vitamin K in rats. Blood 2007;109(7):2823–31. https://doi.org/10.1182/blood-2006-07-035345.

[133] Dalmeijer GW, van der Schouw YT, Magdeleyns EJ, Vermeer C, Verschuren WM, Boer JM, Beulens JW. Matrix Gla protein species and risk of cardiovascular events in type 2 diabetic patients. Diabetes Care 2013;36(11):3766–71. https://doi.org/10.2337/dc13-0065.

[134] Sardana M, Vasim I, Varakantam S, Kewan U, Tariq A, Koppula MR, et al. Inactive matrix Gla-protein and arterial stiffness in type 2 diabetes mellitus. Am J Hypertens 2017;30(2):196–201. https://doi.org/10.1093/ajh/hpw146.

[135] Schoppet M, Kavurma MM, Hofbauer LC, Shanahan CM. Crystallizing nanoparticles derived from vascular smooth muscle cells contain the calcification inhibitor osteoprotegerin. Biochem Biophys Res Commun 2011;407(1):103–7. https://doi.org/10.1016/j.bbrc.2011.02.117.

[136] Zannettino AC, Holding CA, Diamond P, Atkins GJ, Kostakis P, Farrugia A, et al. Osteoprotegerin (OPG) is localized to the Weibel-Palade bodies of human vascular endothelial cells and is physically associated with von Willebrand factor. J Cell Physiol 2005;204(2):714–23. https://doi.org/10.1002/jcp.20354.

[137] Bernardi S, Bossi F, Toffoli B, Fabris B. Roles and clinical applications of OPG and TRAIL as biomarkers in cardiovascular disease. BioMed Res Int 2016;2016. https://doi.org/10.1155/2016/1752854. 1752854.

[138] Baud'huin M, Duplomb L, Teletchea S, Charrier C, Maillasson M, Fouassier M, Heymann D. Factor VIII-von Willebrand factor complex inhibits osteoclastogenesis

and controls cell survival. J Biol Chem 2009;284(46):31704—13. https://doi.org/10.1074/jbc.M109.030312.

[139] Simonet WS, Lacey DL, Dunstan CR, Kelley M, Chang MS, Luthy R, et al. Osteoprotegerin: a novel secreted protein involved in the regulation of bone density. Cell 1997;89(2):309—19.

[140] Golledge J, McCann M, Mangan S, Lam A, Karan M. Osteoprotegerin and osteopontin are expressed at high concentrations within symptomatic carotid atherosclerosis. Stroke 2004;35(7):1636—41. https://doi.org/10.1161/01.STR.0000129790.00318.a3.

[141] Bucay N, Sarosi I, Dunstan CR, Morony S, Tarpley J, Capparelli C, et al. osteoprotegerin-deficient mice develop early onset osteoporosis and arterial calcification. Genes Dev 1998;12(9):1260—8.

[142] Crisafulli A, Micari A, Altavilla D, Saporito F, Sardella A, Passaniti M, et al. Serum levels of osteoprotegerin and RANKL in patients with ST elevation acute myocardial infarction. Clin Sci (Lond) 2005;109(4):389—95. https://doi.org/10.1042/CS20050058.

[143] Dhore CR, Cleutjens JP, Lutgens E, Cleutjens KB, Geusens PP, Kitslaar PJ, et al. Differential expression of bone matrix regulatory proteins in human atherosclerotic plaques. Arterioscler Thromb Vasc Biol 2001;21(12):1998—2003.

[144] Sandberg WJ, Yndestad A, Oie E, Smith C, Ueland T, Ovchinnikova O, et al. Enhanced T-cell expression of RANK ligand in acute coronary syndrome: possible role in plaque destabilization. Arterioscler Thromb Vasc Biol 2006;26(4):857—63. https://doi.org/10.1161/01.ATV.0000204334.48195.6a.

[145] Schoppet M, Al-Fakhri N, Franke FE, Katz N, Barth PJ, Maisch B, et al. Localization of osteoprotegerin, tumor necrosis factor-related apoptosis-inducing ligand, and receptor activator of nuclear factor-kappaB ligand in Monckeberg's sclerosis and atherosclerosis. J Clin Endocrinol Metab 2004;89(8):4104—12. https://doi.org/10.1210/jc.2003-031432.

[146] Schoppet M, Schaefer JR, Hofbauer LC. Low serum levels of soluble RANK ligand are associated with the presence of coronary artery disease in men. Circulation 2003;107(11):e76. author reply e76.

[147] Jia P, Wu N, Jia D, Sun Y. Association between osteoprotegerin gene polymorphisms and risk of coronary artery disease: a systematic review and meta-analysis. Balkan J Med Genet 2017;20(2):27—34. https://doi.org/10.1515/bjmg-2017-0021.

[148] Di Bartolo BA, Cartland SP, Harith HH, Bobryshev YV, Schoppet M, Kavurma MM. TRAIL-deficiency accelerates vascular calcification in atherosclerosis via modulation of RANKL. PLoS One 2013;8(9):e74211. https://doi.org/10.1371/journal.pone.0074211.

[149] Di Bartolo BA, Chan J, Bennett MR, Cartland S, Bao S, Tuch BE, Kavurma MM. TNF-related apoptosis-inducing ligand (TRAIL) protects against diabetes and atherosclerosis in Apoe (-)/(-) mice. Diabetologia 2011;54(12):3157—67. https://doi.org/10.1007/s00125-011-2308-0.

[150] Secchiero P, Candido R, Corallini F, Zacchigna S, Toffoli B, Rimondi E, et al. Systemic tumor necrosis factor-related apoptosis-inducing ligand delivery shows antiatherosclerotic activity in apolipoprotein E-null diabetic mice. Circulation 2006;114(14):1522—30. https://doi.org/10.1161/CIRCULATIONAHA.106.643841.

[151] Watt V, Chamberlain J, Steiner T, Francis S, Crossman D. TRAIL attenuates the development of atherosclerosis in apolipoprotein E deficient mice. Atherosclerosis 2011;215(2):348—54. https://doi.org/10.1016/j.atherosclerosis.2011.01.010.

[152] Heymann MF, Herisson F, Davaine JM, Charrier C, Battaglia S, Passuti N, et al. Role of the OPG/RANK/RANKL triad in calcifications of the atheromatous plaques: comparison between carotid and femoral beds. Cytokine 2012;58(2):300—6. https://doi.org/10.1016/j.cyto.2012.02.004.

[153] Kaden JJ, Bickelhaupt S, Grobholz R, Haase KK, Sarikoc A, Kilic R, et al. Receptor activator of nuclear factor kappaB ligand and osteoprotegerin regulate aortic valve calcification. J Mol Cell Cardiol 2004;36(1):57—66.

[154] Min H, Morony S, Sarosi I, Dunstan CR, Capparelli C, Scully S, et al. Osteoprotegerin reverses osteoporosis by inhibiting endosteal osteoclasts and prevents vascular calcification by blocking a process resembling osteoclastogenesis. J Exp Med 2000; 192(4):463—74.

[155] Karavanaki K, Tsouvalas E, Vakaki M, Soldatou A, Tsentidis C, Kaparos G, et al. Carotid intima media thickness and associations with serum osteoprotegerin and s-RANKL in children and adolescents with type 1 diabetes mellitus with increased risk for endothelial dysfunction. J Pediatr Endocrinol Metab 2018;31(11):1169—77. https://doi.org/10.1515/jpem-2018-0147.

[156] Montagnana M, Lippi G, Danese E, Guidi GC. The role of osteoprotegerin in cardiovascular disease. Ann Med 2013;45(3):254—64. https://doi.org/10.3109/07853890.2012.727019.

[157] Van Campenhout A, Clancy P, Golledge J. Serum osteoprotegerin as a biomarker for vascular disease. Am J Cardiol 2007;100(3):561. https://doi.org/10.1016/j.amjcard.2007.03.023.

[158] Shimada T, Mizutani S, Muto T, Yoneya T, Hino R, Takeda S, et al. Cloning and characterization of FGF23 as a causative factor of tumor-induced osteomalacia. Proc Natl Acad Sci U S A 2001;98(11):6500—5. https://doi.org/10.1073/pnas.101545198.

[159] Chen G, Liu Y, Goetz R, Fu L, Jayaraman S, Hu MC, et al. alpha-Klotho is a non-enzymatic molecular scaffold for FGF23 hormone signalling. Nature 2018; 553(7689):461—6. https://doi.org/10.1038/nature25451.

[160] Martin A, David V, Quarles LD. Regulation and function of the FGF23/klotho endocrine pathways. Physiol Rev 2012;92(1):131—55. https://doi.org/10.1152/physrev.00002.2011.

[161] White KE, Carn G, Lorenz-Depiereux B, Benet-Pages A, Strom TM, Econs MJ. Autosomal-dominant hypophosphatemic rickets (ADHR) mutations stabilize FGF-23. Kidney Int 2001;60(6):2079—86. https://doi.org/10.1046/j.1523-1755.2001.00064.x.

[162] Bai X, Miao D, Li J, Goltzman D, Karaplis AC. Transgenic mice overexpressing human fibroblast growth factor 23 (R176Q) delineate a putative role for parathyroid hormone in renal phosphate wasting disorders. Endocrinology 2004;145(11):5269—79. https://doi.org/10.1210/en.2004-0233.

[163] Shimada T, Kakitani M, Yamazaki Y, Hasegawa H, Takeuchi Y, Fujita T, et al. Targeted ablation of Fgf23 demonstrates an essential physiological role of FGF23 in phosphate and vitamin D metabolism. J Clin Investig 2004;113(4):561—8. https://doi.org/10.1172/JCI19081.

[164] Ichikawa S, Baujat G, Seyahi A, Garoufali AG, Imel EA, Padgett LR, et al. Clinical variability of familial tumoral calcinosis caused by novel GALNT3 mutations. Am J Med Genet 2010;152A(4):896—903. https://doi.org/10.1002/ajmg.a.33337.

[165] Gutierrez O, Isakova T, Rhee E, Shah A, Holmes J, Collerone G, et al. Fibroblast growth factor-23 mitigates hyperphosphatemia but accentuates calcitriol deficiency

in chronic kidney disease. J Am Soc Nephrol 2005;16(7):2205—15. https://doi.org/10.1681/ASN.2005010052.

[166] Scialla JJ, Wolf M. Roles of phosphate and fibroblast growth factor 23 in cardiovascular disease. Nat Rev Nephrol 2014;10(5):268—78. https://doi.org/10.1038/nrneph.2014.49.

[167] Scialla JJ, Lau WL, Reilly MP, Isakova T, Yang HY, Crouthamel MH, et al. Fibroblast growth factor 23 is not associated with and does not induce arterial calcification. Kidney Int 2013;83(6):1159—68. https://doi.org/10.1038/ki.2013.3.

[168] Lim K, Lu TS, Molostvov G, Lee C, Lam FT, Zehnder D, Hsiao LL. Vascular Klotho deficiency potentiates the development of human artery calcification and mediates resistance to fibroblast growth factor 23. Circulation 2012;125(18):2243—55. https://doi.org/10.1161/CIRCULATIONAHA.111.053405.

[169] Anderson HC. Vesicles associated with calcification in the matrix of epiphyseal cartilage. J Cell Biol 1969;41(1):59—72.

[170] Bonucci E. Fine structure of early cartilage calcification. J Ultrastruct Res 1967;20:18.

[171] Shapiro IM, Landis WJ, Risbud MV. Matrix vesicles: are they anchored exosomes? Bone 2015;79:29—36. https://doi.org/10.1016/j.bone.2015.05.013.

[172] Johnson SM, Dempsey C, Parker C, Mironov A, Bradley H, Saha V. Acute lymphoblastic leukaemia cells produce large extracellular vesicles containing organelles and an active cytoskeleton. J Extracell Vesicles 2017;6(1):1294339. https://doi.org/10.1080/20013078.2017.1294339.

[173] Bennett MR, Sinha S, Owens GK. Vascular smooth muscle cells in atherosclerosis. Circ Res 2016;118(4):692—702. https://doi.org/10.1161/CIRCRESAHA.115.306361.

[174] Harding CV, Heuser JE, Stahl PD. Exosomes: looking back three decades and into the future. J Cell Biol 2013;200(4):367—71. https://doi.org/10.1083/jcb.201212113.

[175] Jansen F, Nickenig G, Werner N. Extracellular vesicles in cardiovascular disease: potential applications in diagnosis, prognosis, and epidemiology. Circ Res 2017;120(10):1649—57. https://doi.org/10.1161/CIRCRESAHA.117.310752.

[176] Tanimura A, McGregor DH, Anderson HC. Matrix vesicles in atherosclerotic calcification. Proc Soc Exp Biol Med 1983;172(2):173—7.

[177] Demer LL, Tintut Y. Inflammatory, metabolic, and genetic mechanisms of vascular calcification. Arterioscler Thromb Vasc Biol 2014;34(4):715—23. https://doi.org/10.1161/ATVBAHA.113.302070.

[178] Kapustin AN, Shanahan CM. Calcium regulation of vascular smooth muscle cell-derived matrix vesicles. Trends Cardiovasc Med 2012;22(5):133—7. https://doi.org/10.1016/j.tcm.2012.07.009.

CT and calcification: understanding its role in risk prediction

4

Alexander R. van Rosendael, MD, Inge J. van den Hoogen, MD, A. Maxim Bax, Subhi J. Al'Aref, MD, Omar Al Hussein Alawamlh, MD, Daria Larine, BA, James K. Min, MD [a]

Department of Radiology and Medicine, Dalio Institute of Cardiovascular Imaging, Weill Cornell Medical College and the New York-Presbyterian Hospital, New York, New York, United States

Abbreviations

ACS	Acute coronary syndrome
ASCVD	Atherosclerotic cardiovascular disease
CAC	Coronary artery calcium
CACS	Coronary artery calcium score
CAD	Coronary artery disease
CCTA	Coronary computed tomography angiography
FFR	Fractional flow reserve
HRP	High-risk plaque
HU	Hounsfield units
ICA	Invasive coronary angiography
IVUS	Intravascular ultrasound
MACE	Major adverse cardiac events
PET	Positron emission tomography
PCI	Percutaneous coronary intervention
SPECT	Single photon emission computed tomography

Introduction

Cardiac computed tomography (CT) is an anatomical noninvasive imaging technique that provides high-resolution visualization of cardiac structures. For prevention, diagnosis, and risk stratification of cardiovascular disease, two main techniques have shown great clinical utility, namely coronary artery calcium score (CACS) and coronary computed tomography angiography (CCTA). CACS provides

---

[a] Disclosures: Dr. Min has served on the advisory boards for GE Healthcare and Arineta; has received research support from GE Healthcare, the Dalio Foundation, and the National Institutes of Health; and has equity interest in Cleerly.

Coronary Calcium. https://doi.org/10.1016/B978-0-12-816389-4.00004-9

a measure of overall coronary artery calcium burden derived from a simple, low-radiation, noncontrast chest CT scan. Absence of coronary artery calcification (CAC) has demonstrated excellent major adverse cardiac event (MACE) free survival and confers a magnitude of risk lower than the guideline-recommended threshold for statin use in almost all asymptomatic individuals [1]. CCTA provides whole coronary tree assessment of plaque burden, location, stenosis severity, and composition. In symptomatic patients with suspected atherosclerosis, CCTA serves as an excellent gatekeeper to costly invasive coronary angiography and identifies those with severe plaque burden and stenosis that may benefit from revascularization. In addition, by the identification of milder phenotypes of atherosclerosis—that can be withhold from progression by initiation of statins—CCTA demonstrated to reduce future heart attacks when added to standard care alone [2]. Improvements in automated plaque analysis have enabled whole-heart plaque burden quantification and provide detailed analysis of individual coronary plaques. Specific coronary lesions that contain quantitatively assessed high-risk plaque features (i.e., low-attenuation plaque, positive remodeling) are at increased risk to become acute coronary syndrome culprits [3]. These novel insights in atherosclerosis assessment and clinical applications of cardiac CT will be discussed in this book chapter.

Coronary artery calcium detection with noncontrast CT
Acquisition and interpretation of CACS

The development of vascular (coronary) calcification is an active pathogenic process [4,5]. Calcifying vascular cells from the arterial wall have the ability to mineralize and calcify atherosclerosis. This ossifying process occurs throughout the whole cardiovascular system and results in the formation of bone tissue (or generally called calcification or calcified plaque) in the coronary arteries or cardiac valves [6]. Using CT, CAC can be easily visualized, and earlier studies started to understand its prognostic implications and value of CAC to identify obstructive coronary artery disease (CAD) [7,8]. In 1990, Agatston et al. proposed a standardized manner to quantify CAC from ultrafast CT [9]. A prerequisite for this Agatston CACS was an electrocardiogram (ECG)-gated noncontrast CT, gated to obtain images at the specific cardiac phase with least amount of (coronary) motion. Calcium was defined as every calcified lesion with ≥ 130 Hounsfield units (HU) and an area ≥ 1 mm^2 (to exclude single high-attenuation pixels due to image noise). The area score (dependent on the area of the calcification) of each lesion is then multiplied by the density score (maximum HU per lesion) defined as $1 = 130-199$ HU, $2 = 200-299$ HU, $3 = 300-399$ HU, and $4 \geq 400$ HU. The total CACS is calculated as the summation of each of these individual plaque scores from all slices that cover the heart. Current multislice CT scanners have the ability to scan the entire heart in one single beat and have enough temporal resolution to acquire motion-free CAC scans by the acquisition of images in the diastole at low radiation dose (≈ 1 mSv).

Non-ECG-gated scans

The rapid technical development of CT scanners (including faster gantry rotations times, thinner slice thickness) has enabled CAC quantification from non-ECG-gated scans performed for other purposes, for instance screening for lung cancer. In chest CT scans, the coronary arteries are in the field of view and can be evaluated for CAC with minimal effort and without additional radiation. Budoff et al. compared CACS from 50 nongated lung CT scans with actual gated CAC scans, using the same quantification methodology as proposed by Agatston et al. [9,10]. They reported excellent correlation (intraclass correlation coefficient of 0.96) with slight overestimating of the nongated score versus the gated score especially for higher scores (potentially resulting from overestimation of calcifications in the presence of cardiac motion, as nongated scans may also be acquired during systole). Additionally, Shemesh et al. confirmed the prognostic value of a visually based CAC extent score to predict mortality [11]. Among 8782 smoking patients who underwent nongated lung CT, a three-point scale describing calcium extent (mild, moderate, or severe) per vessel (maximum score of 12) was created. Even after adjustments for age and smoking history, moderate-to-severe calcium burden was associated with an increase in cardiovascular death (HR 2.1 [95% CI 1.4−3.1]) for patients with scores of 4−12 compared to patients without calcium [11]. These results indicate that the presence of coronary calcium, even in CT scans not specifically performed for this purpose, should raise awareness of increased cardiovascular risk and may result in appropriate treatment changes. As such, the guidelines from the Society of Cardiovascular Computed Tomography (SCCT) recommend reporting of the presence and extent of CAC in all chest CT scans and physician−patient discussions about the initiation of preventive medical therapy [12].

Accuracy to detect coronary atherosclerosis

CAC development is closely associated with vascular injury and represents already a more advance stage in atherosclerotic disease. To understand the association between CAD burden and the extent of coronary calcification as measured with CACS, head-to-head comparisons with histopathology should be performed. Mautner et al. performed a calcium scan in 50 postmortem hearts and cut the coronary arteries into segments with the same section thickness as CT (3 mm) resulting in comparisons of 4325 coronary segments [13]. An excellent correlation was observed between the CACS versus histopathological calcium area ($R = 0.96$). In addition, good agreement between the modalities was observed as both CT and histopathology observed calcific deposits in 23% of the coronary segments. Calcification is considered a secondary phenomenon in atherosclerosis and is likely the result of mineralization and plaque ossification, initiated by atherosclerosis itself [5]. Although calcium amount on CT may closely correlate with the true calcium burden, atherosclerosis comprises noncalcified tissue (fibrous debris) also, which may not be detected with CACS. To this end, Rumberger et al. investigated CT detected calcium with histopathologically quantified atherosclerotic extent [14]. Similar to the prior study [13], the per-patient total histological and CT plaque areas demonstrated high correlation with each other ($R = 0.93$, $P < .001$). However, the average whole heart CT calcium areas were a factor 5 smaller than whole heart histological plaque area (22.9 vs. 133.2 mm^2). This 5-fold lower plaque area by CT was consistent on a per-vessel and per-segment level. Hence, the presence of CAC is a very specific marker for coronary atherosclerosis, but only represents one fifth of the total atherosclerotic plaque. These findings hypothesize that a certain threshold exists before plaque can be detected with CT calcium scoring (tip of the iceberg). This phenomenon is sporadically observed in patients with a CACS of 0, as can be demonstrated by contrast-enhanced CCTA. By using CCTA, the presence of (sometimes even severely stenotic) noncalcified plaque can be detected (Fig. 4.1). Simons et al. specifically looked into segments without CAC and observed that most of the histological segments demonstrated coronary plaque, but only of minimal severity (average $\approx$ 20% luminal narrowing) [15]. More specifically, segments without CAC portended a negative predictive value of 97.5 for the presence of obstructively stenotic plaque indicating that absence of CAC does not confirm the absence of atherosclerosis but generally rules out extensive CAD. Besides the value of CACS to identify coronary atherosclerosis, it has the potential to identify coronary stenosis (usually assessed with invasive coronary angiography, ICA). Observations have reported high sensitivities (85%−100%) and negative predictive value of CT detected

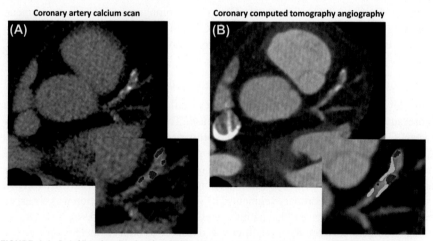

Coronary artery calcium scan   **Coronary computed tomography angiography**

FIGURE 4.1 Calcification Evaluation With CACS and Plaque Detection With CCTA.

Plaque assessment of the left anterior descending coronary artery (LAD) with the coronary artery calcium score (CACS) versus coronary computed tomography angiography (CCTA). The CACS includes only calcified lesions into its score (red colored tissue in panel A), whereas CCTA enables improved visualization of noncalcified plaque (yellow colored tissue in panel B) besides calcified plaque. It can be appreciated that calcifications represent only a minority of the total coronary plaque volume shown in CCTA.

CAC for obstructive CAD ($\geq$50% stenosis) [16−18], suggesting a clinical utility of CAC to rule out stenotic CAD. On the other hand, the presence of CAC does not necessarily relate to luminal narrowing plaque or obstructive CAD. Among 1851 patients with suspected CAD undergoing clinically indicated ICA, the specificity of any CAC for obstructive CAD was 40% that increased to 58%, 72%, and 75% for CACS thresholds of >20, >80, and >100, respectively [18]. The fact that severely calcified vessels with high CACS do not always have obstructive stenosis can be explained by the positive remodeling of coronary arteries, in which the lumen area remains preserved when plaque grows outwards (Glagov effect) [19].

Risk stratification for future cardiovascular events

Multiple large, international, prospective registries have examined the prognostic value of CACS for the prediction of MACE in several subpopulations. Given the low cost, risk, and greater ability of CACS to identify atherosclerosis than coronary stenosis, CACS is predominantly used in asymptomatic individuals for risk assessment and to help tailor preventive therapies. Important studies including the Multi-Ethnic Study of Atherosclerosis (MESA) study [20], Framingham Heart study [21], Heinz Nixdorf Recall study [22], and the Dallas Heart study [23] are population-based samples with long prospective follow-up that have consistently demonstrated a graded increase in risk for each higher CACS stratum, and most importantly, a very favorable outcome for patients without CACS. The MESA study included 6814 participants 45−84 years old from the United states with four different ethnicities (black, Hispanic, white, or Chinese) from 2000 to 2002. Individuals underwent two calcium scans, and the average CACS was used for analysis to rule out artificially wrong scores. In a recent report with >10 years of follow-up concerning the hard endpoints myocardial infarction, stroke, resuscitated cardiac arrest, or coronary death, low event rates were observed for each racial (black, Hispanic, white, or Chinese), age (45−54, 55−64, 65−74, and 75−85), and gender (male or female) subgroup: 10-year MACE rates of 1.3%−5.6% for the absence of CAC. On the other hand, event rates for CACS >300 ranged from 13.1% to 25.6% [24]. Each doubling of CAC was associated with a 14% relative increase in risk. Besides strong prognostic value on itself, CACS adds significantly to conventional cardiovascular risk scoring. Categorizing patients into low (0%−3% estimated 5-year MACE), medium (3%−10% estimated 5-year MACE), and high risk ($\geq$10% estimated 5-year MACE), the addition of CACS to a statistical model, including demographical, clinical, and laboratory results, reclassified an additional 23% of patients that experienced events into high risk and 13% of patients without events into low risk [25]. This, in combination with other reports showing improved discrimination and reclassification of events [26,27], confirms the ability of correctly downgrading risk in patients who will not experience future events and upgrading risk in patients that will in addition to risk scoring based on clinical and laboratory indices. The absence of CAC (CACS = 0) is a very powerful marker for low long-term event rates. Compared with carotid intima-media thickness below the 25th percentile, no carotid plaque, brachial flow-mediated dilation of >5% change, ankle-brachial index >0.9 and <1.3, high-sensitivity C-reactive protein <2 mg/L, homocysteine <10 µmol/L, N-terminal pro-brain natriuretic peptide <100 pg/mL, absence of microalbuminuria, no family history of CAD, no metabolic syndrome, and healthy lifestyle, a CACS of zero resulted in the greatest downward shift in cardiovascular risk [28]. The Heinz Nixdorf Recall registry had a similar design and aim as the MESA study and included 4129 healthy individuals from Germany who were followed for cardiac death and myocardial infarction [29]. At 5 years of follow-up and 93 events, CACS added significantly to the Framingham risk score or National Cholesterol Education Panel Adult Treatment Panel (ATP) III guidelines as demonstrated by an increase in area under the receiver operation characteristic curve (AUC) by 0.05−0.10 points [22]. Strong prognostic value of CACS has been reproduced in other cohorts such as populations referred for suspected CAD [30], diabetics [31], hypertension [32], smoking, and familial history [23]. Among 25,253 patients referred by their primary care physician, risk-adjusted relative risks were 2.2, 4.5, 6.4, 9.2, 10.4, and 12.5 for CACS of 11−100, 101−299, 300−399, 400−699, 700−999, and >1000, respectively, when compared with a score of 0 [30]. To further enable integration of CACS into current risk assessment, a risk score was created using 10-year follow-up MESA data with external validation in other population based cohorts (Heinz Nixdorf Recall study and Dallas Heart study) [33]. The addition of CAC to a clinical model increased the AUC significantly (0.80 vs. 0.75, $P < .001$), with similarly good performance in the external validation cohorts (AUC 0.779 and 0.816) indicating the robustness of CACS.

Importance of extracoronary calcifications

Besides the evaluation of CAC, coronary artery calcium scans also allow for the assessment of calcifications in extracoronary structures. Similar to the relation between CAC and future MACE [20], prognostic value of extracoronary calcium for the prediction of MACE has been demonstrated [34].

Calcifications on mammograms—so-called breast arterial calcifications (BACs)—have been shown to relate to CAC and given the high number of performed scans are of recent interest. BACs are circularly shaped calcified lesions in the media layer of arteries in the breast (e.g., Mönckeberg's medial calcific sclerosis) and can be found on mammography intended for breast cancer screening [35]. The prevalence of BAC is estimated at 12.7% (95% CI 10.4%−15.1%) and increases significantly with age [36]. Furthermore, risk factors for BAC are parity, as well as postmenopausal status [37,38]. Although BAC is different from CAC—which usually develops in the intima layer of the coronary artery—both are the consequence of inflammatory interactions with the smooth arterial muscle cells, thereby stiffening the arterial wall [39]. BAC demonstrated to be significantly associated with CAC; the sensitivity of BAC for the presence of CAC was 63%, whereas the specificity was 76% [12,40]. On the other hand, research concerning the relationship between traditional cardiovascular risk factors and the presence of BAC shows mixed results;

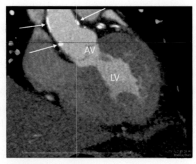

FIGURE 4.2 Aortic Root Calcification on CCTA.

The figure shows calcifications of the aortic root (white arrows) on coronary computed tomography angiography (CCTA).

women with diabetes were more prone to have BAC (OR 1.88; 95% CI 1.36−2.59), whereas women with hypertension were not (OR 1.08; 95% CI 0.98−1.19). Unexpectedly, the prevalence of BAC was significantly lower among women who smoked (OR 0.48; 95% CI 0.39−0.60) [36]. Although these discrepancies might be caused by different outcome definitions, further research should improve understanding of the potential role of BAC in CAD evaluation.

Besides CAC, calcifications can be present in the aortic root, arch, thoracic, and abdominal part and can be quantified with the Agatston method as well (Fig. 4.2) [9]. Aortic calcification leads to stiffening of the aorta, results in increase of wall stress, a final common pathway toward rupture of plaque [41]. Several substudies from the MESA study have been conducted to explore the correlation between the amount of thoracic aortic calcifications and MACE. Unadjusted survival analyses mainly show significant predictive value for MACE. However, after adjusting for clinical risk factors and CACS, these effects were significantly modified [21]. Budoff et al. reported predictive value of aortic thoracic calcification only in women (HR 2.15; 95% CI 1.10−4.17) and no longer in men after adjustments for traditional risk factors and CACS (HR 0.92; 95% CI 0.64−1.32) [42]. Similarly, as demonstrated by the Early Identification of Subclinical Atherosclerosis by Noninvasive Imaging Research (EISNER) study, thoracic aortic calcifications did not independently attribute to better risk assessment for cardiac events. Aortic arch calcifications have been topic of research as well. The Rotterdam study found an independent association of this calcification and cardiovascular mortality (HR 2.72; 95% CI 1.85−4.02) [44]. In addition, calcifications of the abdominal aorta were associated with cardiac events in a meta-analysis including 11,250 patients (RR 1.81; 95% CI 1.54−2.14) [45].

CACS for prediction noncardiac events

CAC scores increase chronologically with age, but significant heterogeneity of CACS exists within age groups. This potential enables to compare CACS of patients with their age counterparts, which places individuals at relatively increased or decreased cardiovascular risk [46]. Besides age, CAC is associated with traditional cardiovascular risk factors and likely (partially) represent the cumulative burden of risk factor exposure. However, CACS is not merely a summation of risk factors, as patients without them can have CAC, which provides similar prognostic information as in patients with risk factors [47]. CACS also "captures" increased risk because of genetic or environmental factors that are generally difficult to assess [48]. The characteristic of CACS to quantify overall coronary arterial calcification has posed CACS as an accurate estimate of vascular age or overall patient health. Indeed, studies have shown that CACS independently predicted all-cause mortality, cancer, chronic kidney disease, chronic obstructive pulmonary disease, or hip fractures. Using nonreferral, population-based data from the MESA cohort, 20% of the first occurrences of a noncardiovascular event occurred in individuals with CACS >400 (this group represented 10% of the total cohort) and 70% of events occurred in patients with CACS >0 [46]. In addition, low CACS levels have been associated with low rates of dementia and heart failure in the (very) elderly [49]. Individuals with CACS = 0 seem to be protected from chronic diseases and elderly without CACS are at low risk for mortality and morbidity representing a group of healthy agers. The identification of this subpopulation is of importance for public health officials or for distribution of healthcare resources as these individuals will not likely need healthcare resources. It is plausible that these individuals have lived a life without cardiovascular risk factors, abundance of stress and eaten healthy. However, future research should focus on genetic factors of these individuals as traditional cardiovascular disease factors perform poorly in predicting persistent CACS = 0 [50].

Clinical use of CACS
CACS to guide preventive medical therapy

Risk assessment and treatment decisions for preventive medical therapy (aspirin or statin) are guided by risk scores. The pooled cohort atherosclerotic cardiovascular disease (ASCVD) risk equation as used in the United States incorporates age, gender, race, total cholesterol, high-density lipoprotein cholesterol, systolic blood pressure, diabetes, and current smoking status for individuals aged 40–79 [1]. In Europe, risk assessment is based on the systematic coronary risk evaluation (SCORE) chart that is based on age, gender, systolic blood pressure, and cholesterol levels; and specific conditions (chronic kidney disease, diabetes, or markedly elevated single risk factors) at higher risk levels. The American College of Cardiology (ACC) and American Heart Association (AHA) guidelines recommend the initiation of statin therapy when the estimated 10-year ASCVD event rates exceed 7.5% and recommend statin use regardless of ASCVD score for specific subgroups (very high LDL-cholesterol, diabetes with intermediate LDL-cholesterol levels). The European Society of Cardiology (ESC) guidelines recommend lowering the serum LDL-cholesterol below specific levels for certain risk groups (low, moderate, high, and very high risk) that are based on the SCORE calculation and distinct high-risk subpopulations [51]. Inclusion of variables in the scores is generally based using state-of-the-art statistical methods which resulted that (potentially important) risk factors such as family history, C-reactive protein, or diet were not incorporated. Furthermore, the ACC/AHA pooled cohort equations may be subject to significant overestimation of risk among several age and race populations [52]. These scores are designed to provide risk estimations on population level based on parsimonious models but may therefore be inaccurate on individual basis. CACS can be added to traditional risk assessment to improve the certainty of risk for individuals. Especially among intermediate risk patients, where risk scoring and CACS are often discordant, CACS can up- or down-classify risk and subsequent intensity of medial therapy. Of special interest are patients with CACS = 0, who have long-term low event rates, even in the presence of cardiovascular risk factors. Nasir et al. included 4758 individuals from the MESA cohort with available ASCVD risk scores, without baseline statin therapy and complete 10-year follow-up for ASCVD consisting of myocardial infarction, resuscitated cardiac arrest, coronary heart disease death, and fatal and nonfatal strokes [50]. According to the current guidelines, participants were subdivided into "statin recommended" ($\geq$7.5% 10-year ASCVD risk), "statin considered" (5%$-$7.5% 10-year ASCVD risk), or "statin not recommended" ($<$5% 10-year ASCVD risk). Of the included participants, 50% were recommended, 12% were considered, and 38% were not recommended for statin therapy; 58% had a CACS = 0, and 44% of the statin candidates (recommended or considered) had CACS = 0. The observed 10-year ASCVD event rates among all participants with CACS = 0 (regardless of their risk score) was 5.2% and for CACS >0 the event rate was 12.0%. Importantly, even among individuals who received a statin recommendation according to the ASCVD score, the addition of a CACS = 0 was associated with an observed ASCVD rate below the statin recommendation threshold of 7.5%. The 10-year ASCVD rates for CACS = 0 among participants with no recommendation for statin were 1.3% and 1.5% for those with an indication to consider statins. Hence, CACS = 0 has the ability to reclassify individuals to below the recommendation for statins in all statin indication classes. These findings have important implications given the large number of individuals with CACS = 0 among statin consideration (57%) and statin recommendation (41%) and may warrant that statins do not need to be started yet. The statin recommendations are based on 10-year events and do not mandate prescription of statins as a dichotomous decision. Over time, CAC can develop, ASCVD risk scores can change, and a risk estimation for a longer follow-up period may be more important for certain individuals. Therefore, these risks should inform patients and physicians in their discussion when considering medication initiation, lifestyle changes, side effects, costs, and burden of use, and treatment decisions should be based on these considerations [50]. For instance, a patient with an elevated serum cholesterol and a recommendation for statin may prefer lifestyle and diet changes to reduce risk in first place before direct prescription of statins. On the other hand, in the extremes of ASCVD risk scores, the addition of CACS has limited impact on the reclassification of patients regarding their statin indication. The observed 10-year ASCVD rates for individuals with an ASCVD risk score <5% was below the statin recommendation of 7.5% when CACS was either zero or above zero. For patients with ASCVD risk score >20%, the observed ASCVD rate was above the statin recommendation threshold for both CACS = 0 and CACS >0. Although CACS can increase risk stratification in these extremes of risk, the presence or absence of CAC will not have enough impact to reduce or increase risk below and above the recommended risk threshold to recommend statin according to the preventive guidelines of the United States. In addition, only 21% of patients with ASCVD risk score <5% had CACS >0. On the other hand, intermediate risk individuals with ASCVD scores ranging from 5% to 20%, a CACS = 0 was associated with <5% observed ASCVD and CACS >0 reclassified participants to ≥7.4% risk. When adding CACS to participants with the ASCVD risk score of 5%−20%, 49% could be reclassified to below the risk threshold of 7.5% for statin recommendation. Fig. 4.3 represents an algorithm how CACS will reclassify risk and can help to tailor statin therapy to the right

Using 10-year ASCVD risk estimate plus coronary artery calcium (CAC) score to guide statin therapy				
Patient's 10-year atherosclerotic cardiovascular disease (ASCVD) risk estimate:	**<5%**	**5-7.5%**	**>7.5-20%**	**>20%**
Consulting ASCVD risk estimate alone	Statin not recommended	Consider for statin	Recommend statin	Recommend statin
Consulting ASCVD risk estimate + CAC **If CAC score =0**	Statin not recommended	Statin not recommended	Statin not recommended	Recommend statin
If CAC score >0	Statin not recommended	Consider for statin	Recommend statin	Recommend statin
Does CAC score modify treatment plan?	✕ **CAC not effective for this population**	✓ **CAC can reclassify risk up or down**	✓ **CAC can reclassify risk up or down**	✕ **CAC not effective for this population**

FIGURE 4.3 Proposal for Adding CACS to ASCVD Risk Score in Selected Patients to Guide Statin Initiation.

The figure shows an approach where coronary artery calcium score (CACS) is added to the traditional atherosclerotic cardiovascular disease (ASCVD) score in selected patients to reclassify patients with regards to the prescription of statins. In patients at very low (<5% 10-year ASCVD predicted risk) or high risk (>20% predicted 10-year ASCVD predicted risk), the addition of CACS will not significantly reclassify the recommendation for statins but most value of CACS is in patients at intermediate risk (ASCVD score between 5% and 20%).

Derived from: Greenland P, Blaha MJ, Budoff MJ, Erbel R, Watson KE. Coronary Calcium Score and Cardio-vascular Risk. Journal of the American College of Cardiology 2018;72:434−47.

patients. Similar findings were reproduced using statin recommendations as advocated by the ESC guidelines. Of the 3575 without lipid-lowering medication at baseline and complete 10-year follow-up for ASCVD, 1288 (34.4%) fulfilled the criteria for statin recommendation when applying the ESC criteria, which appeared to be less strict than the ACC/AHA guidelines that recommended statin use in 56.1% [53]. Participants with CACS = 0 had low ASCVD rates when statins were not recommended (1.5%) and when they were recommended (5.7%). Interestingly, CACS >100 was associated with high event rates also when statins were not indicated (8.7%) and when they were indicated (17.4%). This study reinforces the value of CACS to better select patients who are at such low risk that they are less likely to receive benefit from statins and in addition, high CACS—regardless of clinical risk—warrants statin therapy. In addition, high CACS identifies a subgroup of patients at the very high risk (17.4% event rate when statins were also indicated), which may help to convince patients that intensive preventive therapies and lifestyle changes are definitely needed.

CACS, medication adherence and lifestyle change

Visualization of coronary calcium on a CT scan may provide a more understandable and distinct measure of risk than the presence of (usually asymptomatic) risk factors such as hypertension or dyslipidemia. This may help patients to understand their risk to help keep or switch to a healthy lifestyle and increase medication adherence. In addition, physicians may be encouraged to increase prescription of preventive medications when a high burden of coronary calcium is detected. The EISNER study randomized 2137 volunteers to a group that did and did not undergo CACS before risk factor counseling and the primary endpoint was change in CAD risk profiles at 4 years [54]. Compared with the no scan group, individuals undergoing CACS had significantly greater reduction in systolic blood pressure, serum LDL-cholesterol levels, and reduction of waist circumference in those with increased abdominal girth at baseline. Moreover, the Framingham risk score increased in the no scan group but remained unchanged in the scan group, while individuals were 4 years older at follow-up. For medical resource utilization, more scan participant had initiation of new antihypertensive medications and a tendency was observed toward greater use of lipid-lowering medication. Importantly, no increase in subsequent overall cost of management, noninvasive or invasive coronary testing, or revascularization procedures were observed. Among the individuals undergoing scanning, increasing baseline CACS was associated with proportionally greater improvement in most CAD risk factors, greater weight loss was observed among overweight individuals with CACS ≥100, and a trend was observed toward more exercise with increasing CACS. Similarly, a progressive increase in new preventive medication prescription occurred with increasing CACS. For instance, 35% of patients with CACS 1−99 were started on lipid-lowering drugs, while 65% of individuals with CACS ≥400 were initiated. These observations are important as patients with the highest CACS are at highest risk for future events and will therefore receive most benefit from risk reduction therapies. Similar results were observed among participants from the MESA study where initiation of new preventive therapy was assessed approximately 2 years after baseline for those not on medications. In addition, the continuation of preventive therapy at 3 years for those on medication at baseline was assessed [55]. After extensive adjustment for confounding, patients with CACS >400 were 1.5 times more likely to be initiated on preventive therapies and also continuation of medication was significantly higher in patients with elevated CACS. These observations of coronary calcium and preventive therapies may be explained by the conviction of physicians to increase medication prescriptions after the observation of CAC; and on the other hand, patients to continue medication and live a healthier lifestyle. Although these changes are important to reduce risk and increase overall health of individuals, there are no studies that demonstrate a reduction in hard events with the use of CACS. However, it is unlikely that such a randomized trial will ever be performed given the very large sample size needed with long-term follow-up to detect differences in clinical events [56].

CACS in symptomatic patients with suspected CAD

A CACS = 0 demonstrates an excellent long-term outcome because it is almost always associated with no to little CAD. Calcified plaques are ruled out but the presence of noncalcified plaque, which is more prevalent than calcified plaque, is not detected with the Agatston method. Besides risk stratification in asymptomatic patients, CACS can be used in symptomatic patients for diagnostic purposes. Anatomical imaging in symptomatic patient with suspected CAD aims to detect obstructive coronary lesions ($\geq$50% stenosis) that cause ischemia (hemodynamically significant CAD), which can be treated with medical therapy or revascularization to alleviate symptoms. Although CACS = 0 is associated with little CAD, it does not completely rule out significant noncalcified plaque, as exemplified in Fig. 4.4 [57]. From the Coronary CT Angiography Evaluation for Clinical Outcomes: An International Multicenter (CONFIRM) registry, maximal stenosis severity of the three coronary arteries was assessed in 10,037 symptomatic patients undergoing CACS and CCTA for evaluation of CAD. Albeit very low, 3.5% of the patients with CACS = 0 demonstrated $\geq$50% stenosis and 1.4% demonstrated $\geq$70% stenosis [58]. Symptoms status, and clinical risk profile are important aspects to consider for the decision-making process after the performance of CACS. Although the prevalence of obstructive stenosis is likely to be even lower than 3.5% in asymptomatic patients (where CACS is usually performed for risk stratification), other cohorts observed an increasing rate of CAD among patients with CACS = 0 with increasing clinical risk profile or typical angina [59]. As such, another report including 40 patients with suspected acute coronary syndrome (ACS) showed a CACS = 0 in 13 (13%) of the patients and in 12 (85%) of them coronary plaque was observed on CCTA of which 5 (39%) were obstructive [60]. Although CACS = 0 among low-to-intermediate clinical risk patients does not completely rule out obstructive stenosis, the CAD extent is generally not severe. Of 3501 patients undergoing CACS and

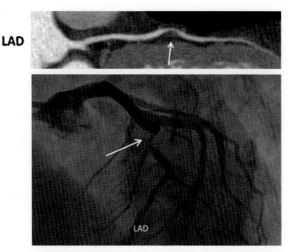

FIGURE 4.4 Noncalcified Plaque on CCTA.

This symptomatic high-risk (diabetes, hypertension) patient had a coronary artery calcium score of zero, but the subsequently performed coronary computed tomography angiography showed an obstructive stenosis (arrow) in the mid left anterior descending coronary artery (LAD), which was confirmed on invasive coronary angiography and subsequently stented.

Derived from: van Rosendael AR, Dimitriu-Leen AC, Bax JJ, Kroft LJ, Scholte AJ. One-stop-shop cardiac CT: Calcium score, angiography, and myocardial perfusion. J Nucl Cardiol 2015.

single photon emission computed tomography (SPECT) myocardial perfusion imaging, 868 (25%) patients had a CACS = 0, and none demonstrated myocardial ischemia. Given the low rate of obstructive CAD following CACS = 0 in low-to-intermediate risk patients, little additional value of further invasive or invasive coronary testing can be expected to improve certainty about the diagnosis. Therefore, the absence of CAC can clinically serve as an excellent gatekeeper to further testing in selected, symptomatic patients. For high-risk patients and CACS = 0, the likelihood of obstructive CAD is higher and additional testing may be needed.

Atherosclerosis detection with CCTA
Acquisition and interpretation of CCTA

CCTA differs from a calcium scan by the use of contrast agent and scan acquisition that enables (in addition to calcium burden as quantified with CACS) assessment of stenosis severity, noncalcified plaque burden, coronary anatomy (anomalies, congenital heart disease) and more specific features like positive lesion remodeling or low-attenuation plaque. To maximize image quality, the administration of beta-blocker to reduce cardiac motion during acquisition and (sublingual) nitroglyc-erin to induce coronary vasodilation are recommended. Furthermore, ECG gating is required to align CT data acquisition with the cardiac cycle. After placement of an intravenous line for contrast media injection, an acquisition protocol needs to be selected based on the patient's body mass index, heart rate, and whether additional bypass grafts or aorta need to be included. Then, acquisition of the images needs to be correctly timed during maximal contrast enhancement of the coronary arteries to ensure good artery visualization. A retrospective ECG-gated acquisition mode col-lects data during the entire heart beat (systole and diastole) and allows for functional cardiac assessment as left ventricular ejection fraction but exposes the patient to more radiation. This mode enables data reconstruction at series from 0% to 100% of the cardiac cycle (R—R interval) that provides a lot of data facilitating identifica-tion of the optimal cardiac phase with no or as few motion artifacts as possible. A prospective (step and shoot) ECG-gated mode applies radiation dose only during the diastole where least motion of the coronary arteries can be expected. Finally, a prospective ECG-gated high-pitch mode acquires data while the patient table is moved at a high pitch so that the entire heart scan be covered within one single heart-beat, decreasing radiation exposure to as low as 1 mSv [61]. A concern of CCTA used to be the high radiation exposure; but due to rapid improvements in technology, shorter gantry rotation times, and the use of radiation reduction techniques, the dosage has decreased dramatically the last decade. A radiation dose survey from 64 hospitals from 32 countries enrolling 4502 patients reported a 78% reduction in exposure in 2017 compared with 2007 without an increase in nondiagnostic scans (1.9% in total) [62]. The mean effective dose estimate was 5.1 mSv with potential to reduce even further given the large heterogeneity between study sites with the lowest and highest exposure (37-fold variability in median dose length product). Postpro-cessing of data should be performed with dedicated software. First, the raw data should be reconstructed using the thinnest possible slice thickness (i.e., 0.5 or 0.6 mm depending on the CT system). Then, an overview of the coronary artery anatomy should be appreciated, and an image quality check helps to foresee possible difficulties caused by severe calcium or motion artifacts. In addition, multiplanar reformations that enable visualization of the whole coronary artery in one view, which helps in stenosis severity assessment and overall extent of CAD can be created (Fig. 4.5).

CAD reporting with CCTA

Atherosclerosis reporting from CCTA should be performed in accordance with the Coronary Artery Disease Reporting and Data System (CAD-RADS) to communi-cate and standardize CCTA findings and to facilitate decision-making for further patient management [63]. The CAD-RADS classification should be applied on a per-patient basis and represents the highest grade stenosis from the coronary tree. Stenosis severity is subdivided into 5 categories (0%, 1%—24%, 25%—49%, 50% —69%, 70%—99%, and 100% or occlusion). In addition to this, specific

MPR of LAD

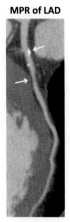

FIGURE 4.5 Multiplanar Reformation of the LAD.

This figure shows a multiplanar reformation of the left anterior descending coronary artery (LAD). In the proximal part, a calcified plaque is observed with mild stenosis (<30%). In the mid segment, a noncalcified plaque is observed with 30%–50% stenosis.

recommendations for further patient management (initiation of preventive medical therapies, symptom-guided antiischemic therapy, referral for ICA or revascularization) are given. Fig. 4.6 shows an overview of the CAD-RADS categories including the specific management recommendations. Specific modifiers complement the

TABLE 2 CAD-RADS Reporting and Data System for Patients Presenting With Stable Chest Pain

	Degree of Maximal Coronary Stenosis	Interpretation	Further Cardiac Investigation	Management
CAD-RADS 0	0% (No plaque or stenosis)	Documented absence of CAD*	None	Reassurance. Consider non-atherosclerotic causes of chest pain
CAD-RADS 1	1-24% - Minimal stenosis or plaque with no stenosis**	Minimal non-obstructive CAD	None	Consider non-atherosclerotic causes of chest pain Consider preventive therapy and risk factor modification
CAD-RADS 2	25-49% - Mild stenosis	Mild non-obstructive CAD	None	Consider non-atherosclerotic causes of chest pain Consider preventive therapy and risk factor modification, particularly for patients with non-obstructive plaque in multiple segments.
CAD-RADS 3	50-69% stenosis	Moderate stenosis	Consider functional assessment	Consider symptom-guided anti-ischemic and preventive pharmacotherapy as well as risk factor modification per guideline-directed care*** Other treatments should be considered per guideline-directed care***
CAD-RADS 4	A - 70-99% stenosis or B - Left main >50% or 3-vessel obstructive (≥70%) disease	Severe stenosis	A: Consider ICA**** or functional assessment B: ICA is recommended	Consider symptom-guided anti-ischemic and preventive pharmacotherapy as well as risk factor modification per guideline-directed care*** Other treatments (including options of revascularization) should be considered per guideline-directed care***
CAD-RADS 5	100% (total occlusion)	Total coronary occlusion	Consider ICA and/or viability assessment	Consider symptom-guided anti-ischemic and preventive pharmacotherapy as well as risk factors modification per guideline-directed care*** Other treatments (including options of revascularization) should be considered per guideline-directed care***
CAD-RADS N	Non-diagnostic study	Obstructive CAD cannot be excluded	Additional or alternative evaluation may be needed	

The CAD-RADS classification should be applied on a per-patient basis for the clinically most relevant (usually highest-grade) stenosis. All vessels greater than 1.5 mm in diameter should be graded for stenosis severity. CAD-RADS will not apply for smaller vessels (<1.5 mm in diameter). MODIFIERS: If more than one modifier is present, the symbol "/" (slash) should follow each modifier in the following order: First: modifier N (non-diagnostic). Second: modifier S (stent). Third: modifier G (graft). Fourth: modifier V (vulnerability). *CAD = coronary artery disease. **CAD-RADS 1 - This category should also include the presence of plaque with positive remodeling and no evidence of stenosis. ***Guideline-directed care per ACC Stable Ischemic Heart Disease Guidelines (Fihn et al. JACC 2012) (26). ****ICA = invasive coronary angiography.

FIGURE 4.6 CAD-RADS for Patients Presenting With Stable Chest Pain

This figure shows the Coronary Artery Disease Reporting and Data System (CAD-RADS) classifications to report coronary computed tomography angiography (CCTA) findings according to the maximal stenosis per patient. Furthermore, the CAD-RADS categories are clinically interpreted, recommendations are given for further diagnostic testing, and individualized patient management is advised.

Derived from: Cury RC, Abbara S, Achenbach S et al. Coronary Artery Disease - Reporting and Data System (CAD-RADS): An Expert Consensus Document of SCCT, ACR and NASCI: Endorsed by the ACC. JACC Cardiovasc Imaging 2016;9:1099–113.

previously mentioned CAD-RADS categories to provide more clinically relevant information: N, the study is not fully evaluable or nondiagnostic; S, to indicate the presence of coronary stents; G, to indicate the presence of grafts; and V, to indicate the presence of vulnerable plaque. Vulnerable plaque has high-risk plaque features that are independently associated with future ACS [3]. These features include positive remodeling, low-attenuation plaque, spotty calcification, and the napkin ring sign (Fig. 4.7). If a coronary lesion demonstrates two or more of those features by CCTA, the modifier V (vulnerability) should be added to the CAD-RADS classification. Such a lesion is called a high-risk plaque (HRP). Although not proven yet, it is likely that these patients deserve more aggressive therapy to modify their increased risk than their counterparts without vulnerable plaque (Figs. 4.8 and 4.9). In a study using data from the CONFIRM registry, the performance of CAD-RADS to risk stratify patients was established. Among 5039 patients without known CAD referred for CCTA due to suspected CAD, the 5-year cumulative major cardiac event-free survival rates were 95.2% for CAD-RADS = 0, 87.2% for CAD-RADS = 1, 84.2% for CAD-RADS = 2, 83.1% for CAD-RADS = 3, 80.1% for CAD-RADS = 4a, 72.7% for CAD-RADS = 4b, and 69.3% for CAD-RADS = 5 (Fig. 4.10).

CCTA and histopathological atherosclerosis

With the rapid advancements in the CCTA technology, ex vivo histological comparisons have declined as a shift has been made in the direction of correlating CT with intravascular ultrasound (IVUS), as the latter modality is far more easily applicable and has established itself as a reliable alternative for plaque analysis [64−68]. Atherosclerotic plaques are visualized on CCTA as structures located outside the lumen that are either calcified or possess an attenuation value that is lower than the vessel lumen filled with contrast medium. Those structures can be visually or qualitatively be classified as noncalcified, mixed/partially calcified, or calcified [69−71]. On histopathology, atherosclerotic lesions are often classified based on the Stary classification where a plaque can be categorized in one of eight types based on its composition and structure [72]. Appearing bright with a high-attenuation signal, calcium is the most commonly recognized component in a plaque on CCTA. This is in contrast to lipid-rich or fibrous-rich plaque that appears hypoattenuated [73,74]. Fig. 4.11 displays a cross-sectional correlation of histopathology with CCTA of a calcified plaque (7a), lipid-rich plaque (7b), and a fibrous-rich plaque (7c). Although CCTA is considered the main noninvasive diagnostic tool for the detection of calcified lesions, its performance however can be reduced by the presence of two common artifacts, beam-hardening and blooming, especially in highly calcified plaques, which could lead to over or underestimation of stenosis severity [75,76]. In addition, a limitation exists in the detection of early stage lesions that are calcium free. In a head-to-head comparison of CCTA and histopathology among 322 histologically determined plaques, 79% of all plaques were detected by CCTA. The "missed" lesions were generally early lesions according to the Stary

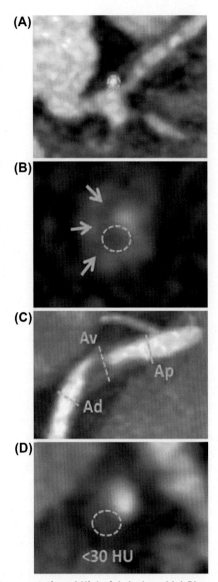

FIGURE 4.7 Pictorial Representation of High-risk (vulnerable) Plaque Features.

These include (A) spotty calcium, defined as punctate calcium within a plaque; (B) napkin ring sign, defined as central low-attenuation plaque with a peripheral rim of higher CT attenuation (arrows); (C) positive remodeling, defined as the ratio of outer vessel diameter at the site of plaque divided by the average outer diameter of the proximal and distal vessel greater than 1.1, or Av/[(Ap þ Ad)/2] >1.1; and (D) low-attenuation plaque, defined as noncalcified plaque with internal attenuation less than 30 Hounsfield units (HU). Please note that a combination of two or more high-risk features is necessary to designate the plaque as high-risk for Coronary Artery Disease Reporting and Data System (CAD-RADS).

Derived from: Cury RC, Abbara S, Achenbach S et al. Coronary Artery Disease - Reporting and Data System (CAD-RADS): An Expert Consensus Document of SCCT, ACR and NASCI: Endorsed by the ACC. JACC Cardiovasc Imaging 2016;9:1099–113.

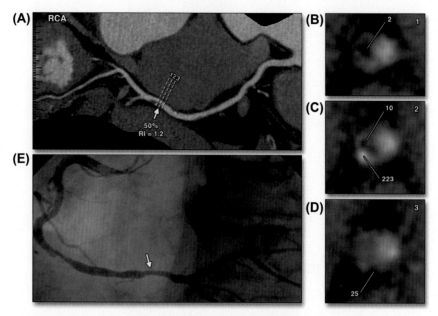

FIGURE 4.8 High-risk Plaque Features Associated With Mild Diameter Stenosis.

These figures display the images of a 66-year old man with atypical exertional chest pain on statin therapy; atherosclerotic cardiovascular disease (ASCVD) score of 9.0% with low-density lipoprotein of 75 mg/dL. After an ambiguous nuclear stress test, he underwent coronary computed tomography angiography (CCTA) followed by invasive coronary angiography (ICA) due to persistent symptoms. (A) CCTA: Multiplanar reconstruction (MPR) of the right coronary artery (RCA) with a ≤50% distal stenosis (white arrow). (B–D) CCTA: cross-sectional analysis with HRP features: a low-attenuation plaque adjacent to the lumen (2–25 Hounsfield units (HU); B), spotty calcification (223 HU; C), and positive remodeling (D). (E) ICA: stenosis <50% (white arrow).

Derived from: Hecht HS, Achenbach S, Kondo T, Narula J. High-Risk Plaque Features on Coronary CT Angiog-
raphy. JACC Cardiovasc Imaging 2015;8:1336–9.

classification; CCTA detected 29% of early lesions (Stary I–III) and 100% of advanced (Stary IV–VII) plaques [77]. The authors also observed that CCTA detected lesions are most likely histologically more advanced plaques when they appear mixed or calcified, while other plaques that are detected but lack a calcium attenuation may be considered as early stage. The distinction between those plaque types mainly relies on density measurements (HU), where recorded attenuation signals tend to significantly vary among different plaque compositions.

Two notable studies that have tested the ability of CCTA in the detection of plaque composition against histopathology have clearly shown that a quantitative classification of plaque can be established on the basis of measuring densities. One of those studies by Becker et al. has investigated this in the early stages of

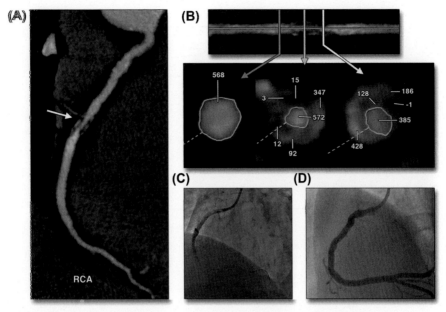

FIGURE 4.9 High-Risk Plaque Features Associated With Severe Diameter Stenosis.

These figures display the images of a 62-year-old woman with continued exertional chest pain on multiple medications. She underwent coronary computed tomography angiography (CCTA) followed by invasive coronary angiography (ICA). (A-B) CCTA: Multiplanar reconstruction (MPR) (A) and straightened view (B) of the right coronary artery (RCA) with a >70% mid-RCA stenosis (white arrow). Cross-sectional images show HRP features: a low-attenuation plaque adjacent to the lumen (−1, 3 and 12 Hounsfield units (HU)) and positive remodeling. (C–D) ICA: no reflow phenomenon during stent placement (C) and intracoronary nicorandil restored flow (D).

Derived from: Hecht HS, Achenbach S, Kondo T, Narula J. High-Risk Plaque Features on Coronary CT Angiography. JACC Cardiovasc Imaging 2015;8:1336–9.

utilization of CCTA technology, where they have shown that lipid-rich plaques had a significantly lower attenuation (47 ± 9 HU) as compared to fibrous-rich plaques (104 ± 28 HU) after those lesions had been histologically classified [70]. Those results were in agreement with the second study published later by Nikolaou et al. (47 ± 13 HU in lipid-rich plaques and 87 ± 29 in fibrous-rich plaques) [78]. Carrascosa et al. have also demonstrated similar results in living subjects when they assessed CCTA to compare densities in different plaques that were defined by IVUS (71.5 ± 32.1 in lipid-rich plaques and 116.3 ± 35.7 in fibrous-rich plaques) [79].

Although measuring plaque density has proven accuracy in distinguishing the different plaque types, nevertheless, caution should be taken when classifying plaques solely on attenuation as intraluminal attenuation, due to the contrast agent,

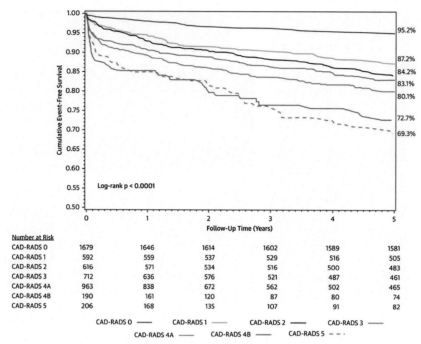

FIGURE 4.10 Cumulative 5-year Event Free Survival Rates According to the CAD-RADS Categories.

This figure demonstrates the cumulative event-free survival from death or myocardial infarction and the number at risk through follow-up displayed across Coronary Artery Disease Reporting and Data System (CAD-RADS) scores.

Derived from: Xie JX, Cury RC, Leipsic J et al. The Coronary Artery Disease-Reporting and Data System (CAD-RADS): Prognostic and clinical implications associated with standardized coronary computed tomography angiography reporting. JACC Cardiovasc Imaging 2018;11:78–89.

can affect plaque attenuation [80]. Similarly, Halliburton et al. have noted this issue as a potential limitation in their study [81]. Additionally, the latter group also pointed out to the unclear potential role that postmortem changes may play in the disruption of the results. Furthermore, ex vivo conditions may meddle with the vascular structure [82].

Accuracy to detect stenosis versus ICA

ICA has traditionally been considered the gold standard for coronary stenosis assessment and comparisons with CCTA have been extensively performed, showing good agreement. In a multicenter study including 291 patients, coronary segments $\geq$1.5 mm diameter were analyzed by CCTA and conventional ICA by independent core laboratories. The patient-based diagnostic accuracy of quantitative CCTA to identify $\geq$50 stenosis by ICA revealed an AUC of 0.93 (95% CI

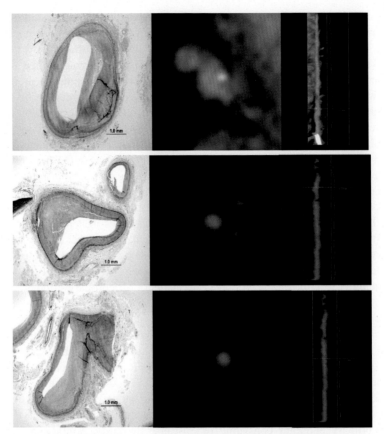

FIGURE 4.11 Cross-sectional Correlations of Histology Specimens With CCTA.

These figures demonstrate the histology specimen (left), oblique cross-section on coronary computed tomography angiography (CCTA) (middle), and a multiplanar reconstruction of the total coronary artery (right): (A) calcium predominant plaque; (B) lipid-rich plaque; (C) fibrous-rich plaque.

0.90—0.96) with a sensitivity of 85% (95% CI 79%—90%), a specificity of 90% (95% CI 83%—94%), a positive predictive value of 91% (95% CI 86%—95%), and a negative predictive value of 83% (95% CI 75%—89%). In addition, accuracy measures were good on a per-vessel basis with an AUC of 0.91 (95% CI 0.88—0.93) to identify ≥50%. These findings were observed in a high-risk population of symptomatic patients referred for ICA due to suspected CAD with a high prevalence of ≥50% stenosis by ICA (56% of the patients and 31% of the coronary segments). These promising results were replicated in another multicenter trial including 230 patients showing a sensitivity of 95%, specificity of 83%, positive predictive value of 64%, and negative predictive value of 99% for ≥50% stenosis [83]. The almost absence of false negatives makes CCTA an excellent test to rule out obstructive

CAD and therefore gatekeeper to ICA. This is reflected in recommendations from the European guidelines to perform CCTA in a low-intermediate risk population to rule out significant CAD [84]. Although $\geq$50 or $\geq$70% stenosis by ICA has been considered relevant to validate CCTA stenosis assessment against because these lesions were targets for revascularization, more recent literature demonstrated that few of these plaques actually are related to reduced myocardial perfusion and ischemia, so that revascularization will likely not result in symptoms relief. Schuijf et al. demonstrated that only 50% of patients with $\geq$50% on CCTA had myocardial ischemia on SPECT imaging [85]. The poor correlation between stenosis severity and coronary flow was also observed with fractional flow reserve (FFR) as reference standard: an invasive index of physiological significance of a coronary stenosis that can be easily measured during ICA. In short, the coronary pressure distal to a lesion is divided by the pressure proximal to the lesion (aortic pressure) during pharmacologically induced hyperemia and a value $\leq$ 0.80 identifies ischemia-causing coronary stenosis with high accuracy [86]. Using data from the Fractional Flow Reserve versus Angiography in Multivessel Evaluation (FAME) study, quantitative ICA stenosis severity was compared with FFR among 509 patients (1414 lesions) undergoing ICA with routine FFR measurements of all coronary arteries [87]. Of lesions categorized as 50%−70% stenosis, 35% demonstrated FFR $\leq$0.80, 80% of lesions between 71% and 90% stenosis, and 96% among lesions categorized 91% −99% stenosis were ischemic. Two important studies from the FAME data demonstrated the clinical importance of FFR identified hemodynamically significant lesions. Among 1220 patients with stable CAD, FFR was assessed in all visible coronary stenoses on ICA, and patients who had at least one vessel with FFR $\leq$0.80 were randomly assigned to undergo FFR-guided percutaneous coronary intervention (PCI) plus medical therapy or to receive medical therapy alone. The rate of the primary endpoint consisting of all-cause mortality, nonfatal myocardial infarction of urgent revascularization within 2 years was significantly lower in the PCI group than in the medial therapy-guided group (8.1% vs. 19.5%, $P < .001$) driven by a lower rate of urgent revascularization [88]. In another study, Tonino et al. demonstrated lower rates of MACE for FFR-guided revascularization (for lesions with FFR $\leq$0.80) compared with angiography-guided revascularization (stenting of all indicated lesions based on the stenosis severity) [89]. Currently, measurement of FFR is recommended of intermediate coronary lesions to assess the hemodynamically significance and revascularization only needs to be performed in vessels with FFR $\leq$0.80 [90]. In a recent study of 208 patients undergoing CCTA and FFR of all three coronary arteries, a sensitivity of 90% (95% CI 82%−95%), specificity of 60% (95% CI 51%−69%), positive predictive value of 64% (95% CI 55%−73%), and a negative predictive value of 89% (95% CI 80%−95%) were observed [91]. All patients underwent also SPECT and positron emission tomography (PET) imaging that revealed a sensitivity of 57% (95% CI 46%−6%) and a specificity of 94% (95% CI 88%−98%) for SPECT and a sensitivity of 87% (95% CI 78%−93%) and a specificity of 84% (95% CI 75%−89%) for PET. Hence, in comparison with stenosis severity thresholds as reference standard, CCTA still holds the excellent rule-out ability but has a higher rate of false positive findings when FFR ≤ 0.80 is the reference standard. A meta-analysis reported similar results with a sensitivity for CCTA detected obstructive CAD of 90% (95% CI 86% -93%) and a specificity of 39% (95% CI 34%-44%) when compared with FFR [92]. The lower specificity and related false positive rate of CCTA for functionally significant CAD reflects the poor relation between higher grade stenosis (between 50% and 95%) and flow reduction and are inherent to anatomical coronary stenosis imaging. On the other hand, milder stenosis ($<40\%$) do not often cause ischemia. To this end, FFR is usually added when lesions exceed 40% diameter stenosis in ICA. In the field of CCTA, an integrated approach with the addition of functional noninvasive testing (PET, SPECT, stress cardiac magnetic resonance, stress CT myocardial perfusion, or stress echocardiography) in case of detected CAD (also called hybrid imaging) may be the optimal approach. Selective functional imaging for patients with $\geq 50\%$ stenosis utilizes the high negative predictive value of CCTA for vessels with $<50\%$ stenosis and aims to improved diagnosis of ischemic vessels for those with $\geq 50\%$ stenosis. Among 384 patients with suspected CAD, stress CT myocardial perfusion was consecutively performed to CCTA when this test showed obstructive CAD and patients were followed for referral for ICA and revascularization during 1 year. When obstructive CAD was followed by a normal stress CT perfusion, 15% of the patients were eventually revascularized and in case of abnormal stress CT perfusion with $\geq 50\%$ stenosis, 76% of the patients received revascularization indicating the clinical utility of this approach [93]. Cardiac hybrid imaging is a novel field, and future research should investigate whether it is cost effective, is accurate to detect ischemic vessels, and reduces unnecessary ICA referral (improves rates of ICA that show lesions that need revascularization), and adds significantly in these fields compared with stand-alone imaging.

Clinical use of CCTA

Diagnostic role of CCTA in patients with suspected stable CAD

Stable CAD involves a reversible imbalance between myocardial blood supply and metabolic demand generally induced by emotional, physical, or other stress-triggers with different underlying pathophysiology (e.g., plaque-related obstruction of epicardial arteries, focal or diffuse spasms of normal or diseased arteries, microvascular dysfunction, left ventricular dysfunction due to prior acute myocardial necrosis or hibernation). Its clinical presentation can be either symptomatic as chest pain (e.g., typical, atypical, nonanginal) [94], or asymptomatic (e.g., silent) as often is observed in certain patient populations as diabetics [95]. To date, the golden standard for the diagnosis of stable CAD is ICA, despite its nonnegligible risks and significant costs [84]. If necessary, ICA will be combined with appropriate confirmation through FFR measurements. CCTA is proposed as a noninvasive modality for the precise anatomic evaluation of the coronary arteries and henceforward the presence of coronary atherosclerosis [96,97]. It has proven to accurately identify the presence and severity of CAD in symptomatic individuals suspected of CAD [98]. For instance, CCTA has a sensitivity of 95%−99% and a specificity of 64%−83% for the detection of CAD in populations with low-to-medium prevalence of this condition [83]. Additionally, its negative predictive value is excellent, while its positive predictive remains limited due to the tendency of CCTA to overestimate the severity of coronary stenoses [99]. Hence as a rule, pretest probability (PTP) always needs to be considered when choosing the "right" diagnostic test for the "right" individual because of its interdependence with test performance. Therefore, the ESC guidelines mainly restrict CCTA to individuals with low-intermediate PTP for CAD or to individuals in which noninvasive stress testing was inconclusive [84]. Thus, CCTA may be specifically useful in the rule out of CAD in this specific patient population. So far, two large randomized clinical trials have been performed regarding the benefit of CCTA for the diagnosis of CAD. First, the Scottish Computed Tomography of the Heart (SCOT-HEART) Trial—a large multicenter RCT at 12 sites across Scotland—randomly assigned 4146 patients with suspicion of CAD (age 57 ± 10 years, 56% males, intermediate-high PTP of CAD) in a 1:1 manner to either standard care or to standard care plus CCTA to investigate the primary endpoint of diagnostic certainty of angina pectoris secondary to CAD at 6 weeks [100]. By comparing 2073 patients assigned to standard care with 2073 patients assigned to standard care plus CCTA, they demonstrated that addition of CCTA to standard care improved the diagnostic certainty of chest pain due to CAD. For example, at 6 weeks CCTA reclassified the diagnosis of CAD in 27% patients and the diagnosis of angina pectoris due to CAD in 23% (vs. both 1% in standard care, $P < .001$). More importantly, they showed that the use of CCTA lowered the need for downstream stress-testing, increased the use of ICA and altered treatment strategies that were associated with a 38% reduction in myocardial infarction (e.g., fatal, nonfatal). Thus, in symptomatic patients, the diagnosis of CAD is important for further clinical management [101,102].

Second, the Computed Tomography versus Exercise Testing in Suspected Coronary Artery Disease (CRESCENT) Trial—a pragmatic multicenter randomized trial at four sites across the Netherlands—randomly assigned 350 patients with stable chest pain (age 55 ± 8 years, 45% males, intermediate PTP of CAD) to either CAC-scoring followed by CCTA (if CAC was 1−400) or functional testing (e.g., exercise ECG, myocardial perfusion imaging, stress echocardiography) [103]. Primary endpoint of the study was clinical effectiveness, defined as absence of chest pain symptoms after 1 year. Secondary outcome was safety, described as the occurrence of adverse events over time. The following adverse events were collected: all-cause death, nonfatal myocardial infarction, major stroke, unstable angina with objective ischemia and/or requiring revascularization, unplanned cardiac evaluations and late revascularization >90 days. By comparing 242 patients assigned to CCTA with 108 patients assigned to functional testing, this study showed that CCTA is an effective and safe alternative to functional testing. For instance, at 1 year, 39% patients randomized to CCTA reported relief of anginal complaints (vs. 25% in functional testing, $P = .012$). Besides, after mean follow-up of 1.2 years, the event-free survival was significantly better in CCTA patients compared to functional testing patients (96.7% vs. 89.8%, $P = .011$).

Diagnostic role of CCTA in patients with suspected unstable CAD

Standard care of patients presenting with acute chest pain to the emergency department classically involves the use of serial ECG, serial cardiac biomarkers (e.g., high-sensitive cardiac troponin as a marker for cardiomyocyte necrosis), and if appropriate additional stress testing for further risk stratification [104]. The primary goal of this approach is to effectively exclude ACS—unstable angina pectoris, non-ST-elevation myocardial infarction or ST-elevation myocardial infarction—and other serious conditions rather than the detection of CAD [105]. In patients without prior cardiovascular events with a negative ECG and negative cardiac biomarkers, the incidence of ACS is low (1%—8%). Yet, missing the diagnosis of ACS can have crucial consequences regarding the morbidity and mortality of patients [106]. Given the congestion of emergency departments and the limited number of specialized chest pain observation units nowadays, triage strategies must therefore not only be greatly accurate but also highly efficient in the rule out of ACS [107].

In this context, evidence has been published that an early CCTA strategy in low-to-intermediate risk patients presenting with acute chest pain to the emergency department can accurately, safely, and efficiently rule out ACS [108]. On the other hand, in patients with a higher PTP for CAD or patients with known CAD, this CCTA approach is not useful [104]. Similarly, CCTA in the acute setting in patients with prior PCI or coronary artery bypass grafting has not been established. However, an important consideration for using CCTA in the acute setting is that it can also exclude other life-threatening causes of chest pain such as pulmonary embolism, aortic dissection, or tension pneumothorax [105].

To date, four randomized controlled trials compared an early CCTA approach with a standard care approach in low-to-intermediate risk patients suspected of ACS with a nonischemic ECG and inconclusive cardiac biomarkers [109—112]. By comparing both triage approaches, no differences in adverse outcomes (e.g., incidence of MI, postdischarge visits to the emergency department, postdischarge rehospitalizations) were observed [113]. However, the CCTA approach showed a reduction in costs and a decreased length of stay. Consequently, current guidelines propose that CCTA can be considered as an alternative to ICA to exclude CAD in low-to-intermediate patients once ECG and/or if cardiac biomarkers are inconclusive [104].

Prognostic role of CCTA for future MACE

CCTA provides accurate whole heart visualization of total 17-coronary segment atherosclerosis, and multiple reports have shown the relationship between increased extent of CAD and heightened risk for all-cause mortality and coronary events

[114−116]. One of the first studies demonstrated that measures of angiographic disease extent, location, and distribution detected by three distinct CCTA grading systems significantly predicted death in patients presenting with chest pain symptoms during an approximately 1-year follow-up [114]. Classifying patients according to no, nonobstructive, obstructive 1-vessel, obstructive 2-vessel, or obstructive 3-vessel/left main CAD resulted in a proportionally elevated event rate with each increasing disease severity category. Similar worsening survival rates were observed when categorizing patients according to stenosis severity of the proximal left anterior descending artery or number of diseased segments, and higher risk was also associated with more proximal disease. Importantly, CCTA identified CAD has prognostic value independent from clinical risk profile and the investigated subpopulation. Among 1884 patients with suspected CAD without clinical risk factors, obstructive CAD resulted in a 6 times heightened risk for MACE (death, nonfatal MI, unstable angina, and late revascularization) during 5.6 ± 1.3 years of follow-up compared with a normal CCTA [47]. Because of the absence of risk factors, 92% of the patients were categorized as either low or intermediate risk according to their pretest likelihood that is largely at odds with the actual observed MACE rates for patients with obstructive CAD (36.3%). Strong risk stratification of events with CCTA is also true for patients with diabetes, a population assumed to be at high risk and sometimes called as CAD-equivalent [117]. In addition, in patients with diabetes and normal CCTA, the 5-year mortality rates were comparable to those of nondiabetics patients, highlighting that normal CCTA identifies a subgroup of patients with favorable prognosis. Besides severe CAD, CCTA can detect milder, nonobstructive CAD that is prognostically important, especially if present in multiple vessels or segments. This specific question has been addressed in a study by Bittencourt et al. including 3242 patients undergoing CCTA with follow-up for myocardial infarction or cardiovascular death. Compared with absence of CAD, extensive nonobstructive CAD (5 or more coronary segments with plaque) conferred a similar increase in risk compared to nonextensive obstructive CAD (adjusted HR 3.1 vs. 3.0) [118]. Hence, these results suggest that regardless of whether obstructive or nonobstructive disease is present, the extent of plaque enhances risk assessment. This has important implications for the choice of noninvasive tests. Functional noninvasive tests aim to identify the consequence of CAD, namely myocardial ischemia, which is usually caused by severely stenotic or extensive CAD. As nonobstructive CAD is not often related to ischemia, CCTA may detect CAD at an earlier disease stage than functional testing, enabling earlier onset of therapy. In the Prospective Multicenter Imaging Study for Evaluation of Chest Pain (PROMISE) trial, 4500 stable chest pain patients were randomly assigned to CCTA and 4602 randomly assigned to functional testing and followed for a median of 26 months. MACE rates were lower in patients with a normal CCTA (0.9%) than a normal functional test (2.1%) and most events (54%, 74/137) occurred in patients with nonobstructive CAD. Overall, anatomical assessment with CAD provided significantly better prognostic information than functional testing (AUC: 0.72 vs. 0.64, $P = .04$). The lower event rate for patients with a normal CCTA can be explained by the absence of CAD, while a normal functional test does not rule nonflow limiting CAD. Although this milder CAD does not produce myocardial ischemia, it does contribute to the development of acute coronary events. Given the strong risk stratification of CCTA detected measures of atherosclerosis, the next step is to reduce this risk associated with coronary atherosclerosis by prescription of preventive medical therapies (i.e., statins, aspirin), lifestyle modifications and revascularization of significant stenosis. In the previously described SCOT-HEART trial investigating the clinical utility of the addition of CCTA to usual care, the treating physicians were encouraged to prescribe preventive therapies for patients with nonobstructive or obstructive CAD in the CCTA arm, and in case of high clinical risk profile (ASSIGN score >20) for patients in the standard care group. During a follow-up of 4.8 (range 3—7) years, patients assigned to CCTA were more likely than patients assigned to the standard care groups to have commenced preventive therapies (19.4% vs. 14.7%) and antianginal therapies (13.2% vs. 10.7%). The previously seen higher rates of coronary revascularization during shorter follow-up were not present anymore at 5 years. Importantly, the rate of the primary long-term endpoint consisting of death from coronary heart disease or nonfatal myocardial infarction was lower in the CCTA group than the standard care group (2.3% vs. 3.9%, $P < .001$). The authors suggest that the use of CCTA resulted in more correct diagnoses of coronary heart disease than standard care alone, which led to the use of appropriate subsequent therapy. In addition, patients who receive a correct diagnosis may have greater motivation to implement healthy lifestyle changes. These hypotheses are further reinforced by the fact that the event rates in the two randomized arms were similar after approximately 7 weeks, a time in which diagnoses were confirmed and alterations in treatment were made but did not affect outcomes yet.

CCTA for primary prevention

No data are available that support the screening of asymptomatic individuals without known CAD by using CCTA, therefore routine screening of this population is discouraged by global guidelines [119,120]. However, some papers described to potential of CAD screening with CCTA in specific high-risk subpopulations with the aim to detect especially those patients at low risk for future events [121]. Moreover, in asymptomatic patients with diabetes—in which atherosclerosis manifests in a more accelerated and progressive manner, the risk for developing CAD is doubled and threshold for symptoms is higher [122—124]—specific subjects need to be identified with subclinical CAD in whom more aggressive lifestyle or treatment changes would allow prevention of progression of the disease and reduce future adverse events [125]. Potentially, CCTA can provide a pivotal role in tailored therapy in diabetics and hence be part of the solution of this quest [126].

CCTA and plaque characterization, relation with ischemia and clinical outcomes

Besides the presence, severity, extent, and location of CAD, CCTA allows for assessment of plaque composition and more detailed plaque characterization. By visual assessment, plaques are categorized as noncalcified, partially calcified and calcified. Over time plaques become usually more calcified and especially in the elderly CCTA shows great extent of calcium. Even by this relatively crude manner of plaque composition differentiation, observational studies recognized different magnitudes of risk for different composition types. However, caution is advised while investigating this topic to fully adjust for overall plaque burden and location, as these represent much stronger predictive value and are closely correlated with chronological age, a known strong predictor of MACE. In a multivariable Cox regression model including measures of total plaque burden (i.e., number of diseased coronary segments and significant CAD), the number of segments with calcified plaque demonstrated a hazard ratio of 1.1, whereas the hazard ratio for number of noncalcified plaques was 1.2 and 1.3 for mixed plaques to predict all-cause mortality, nonfatal myocardial infarction and unstable angina requiring revascularization [127]. The higher risk for noncalcified plaque tissue is in line with observations of CCTA at time of ACS. Motoyama et al. imaged 38 patients with CCTA at time of ACS and compared plaque features with 33 patients who underwent CCTA because of stable angina. They observed that disrupted atherosclerotic plaques causing ACS demonstrated significantly more frequent low-attenuation plaque (79% vs. 9%, $P < .001$)—defined as plaque tissue <30 HU as this corresponds with lipid pool on IVUS—less frequent large calcifications (22% vs. 55%, $P = .004$), more frequent positive remodeling—defined as an increase in diameter at the plaque of 10% compared with the "healthy" reference segment—and more frequent spotty calcifications (63% vs. 21%, $P < .001$). Moreover, a lesion with positive remodeling, low-attenuation plaque, and spotty calcification was a culprit lesion in 95% of all lesions. In addition, the value of these high-risk plaque features in prospective cohort studies has been demonstrated. In a registry of 3158 patients undergoing CCTA, 88 patients developed ACS and the predictive value of significant stenosis (defined as $\geq 70\%$ luminal narrowing) and HRP was investigated. ACS occurred in 16.3% of HRP + patients and in 1.4% of HRP - patients; ACS occurred in 5.5% of patients with significant stenosis and in 2.1% of those without. Furthermore, if HRP and significant stenosis were present, 19% of those developed ACS and if both features were absent, only 1.1% developed ACS during a mean follow-up duration of 3.9 ± 2.4 years. High-risk plaque features in these studies have been measured by visual inspection that is subject to observer variability and requires substantial reading experience. Currently, several semiautomated plaque quantification software packages are available that enable fast plaque, coronary lumen, and vessel wall quantification for objective assessment of advanced plaque features as volumes of noncalcified plaque, remodeling index, volume of the coronary lumen, maximal cross-sectional area stenosis or lesion length, in good agreement with IVUS [128].

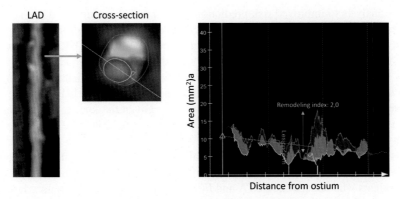

FIGURE 4.12 Quantitative Plaque Assessment of a Lesion in the LAD.

The figure on the left panel demonstrates a multiplanar reconstruction of a left anterior descending coronary artery (LAD) with a substantial amount of noncalcified plaque with some calcified spots. A cross-sectional view with quantitative plaque composition, lumen a vessel wall evaluation at the level of the blue arrow is displayed in the middle panel. The yellow circle represents the lumen, the orange circle represents the vessel wall, and the color-coded tissue represents plaque. Based on the Hounsfield units, the white color indicated calcified plaque, dark green fibrous tissue, light green fibrofatty tissue, and red tissue necrotic core. This cross-sectional view shows spotty calcification, necrotic core, and positive remodeling (remodeling index of 2.1), which indicates that this lesion is a high-risk plaque. The right panel gives an overview of the quantified plaque volume of the total vessel that enables quantification of features such as lesion length, remodeling index, or maximal stenosis severity (at cross section indicated with the yellow line).

Fig. 4.12 demonstrates an example of lesion in the left anterior descending artery with quantitative plaque analysis. Although not possible yet due to considerable time for manual adjustment of lumen and vessel wall border annotation, further improvements in quantitative software have the potential for rapid whole coronary tree analysis to receive quantitative estimates of total plaque volume, maximal luminal narrowing, and high-risk plaque features within seconds. The true independent value of quantified plaque characteristics above and beyond total plaque burden has been demonstrated by the Incident Coronary Syndromes Identified by Computed Tomography (ICONIC) study [3]. In this study, 234 patients were identified with ACS occurrence after previously having been evaluated with CCTA for suspected CAD. These patients were propensity score matched with non-ACS patients based on clinical risk factors, and CCTA evaluated CAD severity and extent. One core laboratory performed blinded adjudication of ACS and culprit lesions, and another lab performed quantification of CCTAs for several plaque characteristics. Independent predictors for ACS were diameter stenosis, fibrofatty plaque volume, necrotic core volume, maximal cress-sectional plaque burden, and the presence of HRP. Importantly, only one fourth of the actual lesions that were precursors for ACS were obstructively stenotic. These results highlight the importance of detailed

CCTA analysis to improved risk prediction and support the addition of the vulnerable plaque modifier (V) in the CAD-RADS guidelines. Another study investigated the effect of statin use on CCTA-defined plaque composition and HRP presence. In total, CCTAs from 1255 patients who underwent their follow-up scan at least 2 years from baseline were quantitatively analyzed. Statin use was significantly associated with slower progression of overall plaque volume, reduction in HRP presence but showed higher calcified plaque progression [129]. Given the known risk reductive effects of statins, calcification of coronary plaque over time may be protective for events and serve as treatment targets.

Besides helpful in identifying patients at risk for future events, detailed plaque characterization improves identification of vessels that cause ischemia. Among 252 patients who underwent CCTA and FFR measurement of 407 coronary lesions, spotty calcification, low-attenuation plaque, positive remodeling, and plaque burden all added independently associated to the diagnosis of ischemia beyond obstructive stenosis with CCTA [130]. In addition, a comparable study demonstrated that several plaque features, positive remodeling, and noncalcified plaque increased diagnosis of vessel specific ischemia beyond stenosis severity (AUC 0.86 for stenosis and AUC 0.90 for stenosis + positive remodeling + noncalcified plaque volume, $P < .001$) [131]. The pathophysiological explanation for the associations of high-risk plaque features (especially positive remodeling and low-attenuation plaque) are not fully understood but several mechanisms have been hypothesized. Positive remodeling represents atheroma development with associated outward remodeling of the vessel wall, while the lumen is preserved from narrowing. However, this present plaque results in increased resistance of the coronary arteries that can cause myocardial ischemia. As the diameter stenosis can be mild, but the plaque burden large with a high remodeling index, positive remodeling will be associated with ischemia independently from stenosis severity. The independent association of low-attenuation plaque and ischemia can be caused by more endothelial dysfunction [132]. The interesting observation that high-risk plaque features are able to predict ACS and ischemia reflect a comparable pathophysiology of both clinical outcomes. For instance, revascularization of hemodynamically significant lesions results in reduction in future ACS [87]. On the other hand, stenosis severity of culprit precursors is usually mild (nonobstructive) in contrast to ischemia causing lesions. Importantly, atherosclerosis is dynamic process and imaging tests only provide a snapshot of the disease at a certain time point. Several studies have documented the clinical importance of detailed plaque characterization that will be studied further and should find its way into clinical CCTA reports.

CCTA in chronic total occlusions

Beyond the aforementioned uses of CCTA, a novel application has been in the setting of chronic total occlusions, CTOs. A chronic total occlusion is defined as an angiographically documented or clinically suspected complete interruption of anterograde coronary flow (thrombolysis in myocardial infarction—TIMI—0 flow) of greater than 3 months standing [133]. While accounting for 10%–20% of lesions identified in CAD patients, CTOs are generally regarded "the final frontier" and challenging for PCI [134]. Untreated CTOs are associated with poor prognosis, and many providers choose medical management over a high-risk PCI procedure [135]. In a report from 2012 including 1697 patients presenting with CTOs, recanalization with PCI was attempted only 10% of cases with a 70% success rate [136]. The decision to attempt PCI instead of medical therapy or coronary artery bypass grafting should be done on individual basis and guided by clinical, angiographic, and technical considerations. Due to its technical difficulty, PCI on CTO requires specific operator expertise. With ICA, there is an incomplete visualization of vessel trajectory, morphology, occlusion sites, and most importantly, calcification that may be encountered by the PCI wire, and most of these factors are predictors for recanalization success [137]. With the guidance of CCTA, there is potential to improve selection of lesions to increase the currently moderate success rates. Because CCTA provides whole heart anatomy—in contrast to luminography—those features that determine success rates for PCI can be visualized. By assessing calcified plaque via CCTA, several studies have shown that calcification involving ≥50% of the vessel cross-sectional area predicts PCI failure [138–140]. Tortuosity or bending defined as >45°angle at occlusion site or proximal to the occlusion site is the most prominent predictor of wiring failure [141]. Occlusion length can often be underestimated by ICA and has been shown to be more accurately measured by CCTA. Multiple occlusion sites, which occur in 25% of CTOs, are also more easily identified on CCTA and have been shown to have a negative effect on the success of recanalization [139].

A summation of predictive factors from CT-derived visualizations was developed to predict recanalization success in CTO. Named the CT-RECTOR score, this noninvasive tool consists of six variables weighted one point each: presence of multiple occlusions, severe calcification, bending, CTO age ≥12 months, blunt stump morphology, and previously failed PCI. As the score increases from 0 to 6, the likelihood of successful guidewire crossing decreases from 95% to 22% for a score greater than ≥3 [139].

CCTA can be highly valuable for the purposes of procedure planning and case selection [142]. Understanding of the calcium distribution in the region of interest can guide the selection of wires or debulking devices that can be used to negotiate the calcified vessel crossing. To facilitate this, some centers have imported CCTA datasets directly into the catheterization laboratory to assist in real-time vessel navigation. As CCTA imaging is increasingly used in the setting of CTOs, the success of the PCI can be enhanced and employed for a larger number of appropriately identified cases. Given the potential benefit of revascularization of CTO's on angina severity, left ventricular function, and survival, closing the success rate gap on PCI performance will be an exciting advancement for the treatment of patients with this condition.

Conclusion

Cardiac CT has emerged as a clinically relevant technique for the evaluation of asymptomatic and symptomatic patients with suspected CAD and has found its way into several clinical practice guidelines. Assessment of calcium burden with a simple CACS provides individualized risk estimation for MACE and can tailor risk reduction therapies including lifelong statin therapy beyond clinical risk scores. In patients with chest pain, a CACS = 0 is associated with a very low rate of stenotic, ischemic lesions in low-to-intermediate risk patients and these patients do not need further coronary testing. CCTA has emerged as a valid first test for the evaluation symptomatic patients and can reliably rule out obstructive and hemodynamically significant CAD due to its high negative predictive value. The presence of obstructive CAD does not necessarily relate to myocardial ischemia (which remains the hallmark of revascularization), and future research should study whether the addition of functional tests will generate an ideal noninvasive approach for the selection of patients that may benefit from coronary stenting or bypass grafting. The unique ability of CCTA to detect CAD at an early stage provides the potential to start treatment in the beginning of the disease process that relates to significant reductions of future heart attacks compared with standard care alone. Finally, detailed plaque analysis with quantitative software identifies individual coronary plaques that are at higher risk to for disruption independent from their stenosis or overall plaque volume.

References

[1] Goff Jr DC, Lloyd-Jones DM, Bennett G, et al. 2013 ACC/AHA guideline on the assessment of cardiovascular risk: a report of the American college of cardiology/American heart association task force on practice guidelines. J Am Coll Cardiol 2014;63:2935−59.

[2] Newby DE, Adamson PD, Berry C, et al. Coronary CT angiography and 5-year risk of myocardial infarction. N Engl J Med 2018;379:924−33.

[3] Chang HJ, Lin FY, Lee SE, et al. Coronary atherosclerotic precursors of acute coronary syndromes. J Am Coll Cardiol 2018;71:2511−22.

[4] Tintut Y, Alfonso Z, Saini T, et al. Multilineage potential of cells from the artery wall. Circulation 2003;108:2505−10.

[5] Demer LL, Watson KE, Bostrom K. Mechanism of calcification in atherosclerosis. Trends Cardiovasc Med 1994;4:45−9.

[6] Buerger L, Oppenheimer A. Bone formation in sclerotic arteries. J Exp Med 1908;10:354−67.

[7] Hamby RI, Tabrah F, Wisoff BG, Hartstein ML. Coronary artery calcification: clinical implications and angiographic correlates. Am Heart J 1974;87:565−70.

[8] Detrano R, Markovic D, Simpfendorfer C, et al. Digital subtraction fluoroscopy: a new method of detecting coronary calcifications with improved sensitivity for the prediction of coronary disease. Circulation 1985;71:725−32.

[9]   Agatston AS, Janowitz WR, Hildner FJ, Zusmer NR, Viamonte Jr M, Detrano R. Quantification of coronary artery calcium using ultrafast computed tomography. J Am Coll Cardiol 1990;15:827—32.

[10]   Budoff MJ, Nasir K, Kinney GL, et al. Coronary artery and thoracic calcium on non-contrast thoracic CT scans: comparison of ungated and gated examinations in patients from the COPD Gene cohort. J Cardiovasc Comput Tomogr 2011;5:113—8.

[11]   Shemesh J, Henschke CI, Shaham D, et al. Ordinal scoring of coronary artery calcifications on low-dose CT scans of the chest is predictive of death from cardiovascular disease. Radiology 2010;257:541—8.

[12]   Hecht H, Blaha MJ, Berman DS, et al. Clinical indications for coronary artery calcium scoring in asymptomatic patients: Expert consensus statement from the Society of Cardiovascular Computed Tomography. J Cardiovasc Comput Tomogr 2017;11:157—68.

[13]   Mautner GC, Mautner SL, Froehlich J, et al. Coronary artery calcification: assessment with electron beam CT and histomorphometric correlation. Radiology 1994;192: 619—23.

[14]   Rumberger JA, Simons DB, Fitzpatrick LA, Sheedy PF, Schwartz RS. Coronary artery calcium area by electron-beam computed tomography and coronary atherosclerotic plaque area. A histopathologic correlative study. Circulation 1995;92:2157—62.

[15]   Simons DB, Schwartz RS, Edwards WD, Sheedy PF, Breen JF, Rumberger JA. Noninvasive definition of anatomic coronary artery disease by ultrafast computed tomographic scanning: a quantitative pathologic comparison study. J Am Coll Cardiol 1992;20:1118—26.

[16]   Tanenbaum SR, Kondos GT, Veselik KE, Prendergast MR, Brundage BH, Chomka EV. Detection of calcific deposits in coronary arteries by ultrafast computed tomography and correlation with angiography. Am J Cardiol 1989;63:870—2.

[17]   Baumgart D, Schmermund A, Goerge G, et al. Comparison of electron beam computed tomography with intracoronary ultrasound and coronary angiography for detection of coronary atherosclerosis. J Am Coll Cardiol 1997;30:57—64.

[18]   Budoff MJ, Diamond GA, Raggi P, et al. Continuous probabilistic prediction of angiographically significant coronary artery disease using electron beam tomography. Circulation 2002;105:1791—6.

[19]   Glagov S, Weisenberg E, Zarins CK, Stankunavicius R, Kolettis GJ. Compensatory enlargement of human atherosclerotic coronary arteries. N Engl J Med 1987;316: 1371—5.

[20]   Detrano R, Guerci AD, Carr JJ, et al. Coronary calcium as a predictor of coronary events in four racial or ethnic groups. N Engl J Med 2008;358:1336—45.

[21]   Hoffmann U, Massaro JM, D'Agostino SrSr, Kathiresan S, Fox CS, O'Donnell CJ. Cardiovascular event prediction and risk reclassification by coronary, aortic, and valvular calcification in the Framingham heart study. J Am Heart Assoc 2016;5.

[22]   Erbel R, Mohlenkamp S, Moebus S, et al. Coronary risk stratification, discrimination, and reclassification improvement based on quantification of subclinical coronary atherosclerosis: the Heinz Nixdorf Recall study. J Am Coll Cardiol 2010;56: 1397—406.

[23]   Paixao AR, Ayers CR, El Sabbagh A, et al. Coronary artery calcium improves risk classification in younger populations. JACC Cardiovasc Imaging 2015;8:1285—93.

[24]   Budoff MJ, Young R, Burke G, et al. Ten-year association of coronary artery calcium with atherosclerotic cardiovascular disease (ASCVD) events: the multi-ethnic study of atherosclerosis (MESA). Eur Heart J 2018;39:2401—8.

[25] Polonsky TS, McClelland RL, Jorgensen NW, et al. Coronary artery calcium score and risk classification for coronary heart disease prediction. JAMA 2010;303:1610−6.

[26] Yeboah J, McClelland RL, Polonsky TS, et al. Comparison of novel risk markers for improvement in cardiovascular risk assessment in intermediate-risk individuals. JAMA 2012;308:788−95.

[27] Shaw LJ, Giambrone AE, Blaha MJ, et al. Long-term prognosis after coronary artery calcification testing in asymptomatic patients: a cohort study. Ann Intern Med 2015; 163:14−21.

[28] Blaha MJ, Cainzos-Achirica M, Greenland P, et al. Role of coronary artery calcium score of zero and other negative risk markers for cardiovascular disease: the multi-ethnic study of atherosclerosis (MESA). Circulation 2016;133:849−58.

[29] Schmermund A, Mohlenkamp S, Stang A, et al. Assessment of clinically silent athero-sclerotic disease and established and novel risk factors for predicting myocardial infarction and cardiac death in healthy middle-aged subjects: rationale and design of the Heinz Nixdorf RECALL Study. Risk Factors, Evaluation of Coronary Calcium and Lifestyle. Am Heart J 2002;144:212−8.

[30] Budoff MJ, Shaw LJ, Liu ST, et al. Long-term prognosis associated with coronary calcification: observations from a registry of 25,253 patients. J Am Coll Cardiol 2007;49:1860−70.

[31] Raggi P, Shaw LJ, Berman DS, Callister TQ. Prognostic value of coronary artery cal-cium screening in subjects with and without diabetes. J Am Coll Cardiol 2004;43: 1663−9.

[32] Erbel R, Lehmann N, Mohlenkamp S, et al. Subclinical coronary atherosclerosis pre-dicts cardiovascular risk in different stages of hypertension: result of the Heinz Nixdorf Recall Study. Hypertension 2012;59:44−53.

[33] McClelland RL, Jorgensen NW, Budoff M, et al. 10-Year coronary heart disease risk prediction using coronary artery calcium and traditional risk factors: derivation in the MESA (Multi-Ethnic study of atherosclerosis) with validation in the HNR (Heinz Nix-dorf Recall) study and the DHS (Dallas heart study). J Am Coll Cardiol 2015;66: 1643−53.

[34] Tison GH, Guo M, Blaha MJ, et al. Multisite extracoronary calcification indicates increased risk of coronary heart disease and all-cause mortality: the Multi-Ethnic Study of Atherosclerosis. J Cardiovasc Comput Tomogr 2015;9:406−14.

[35] Duhn V, D'Orsi ET, Johnson S, D'Orsi CJ, Adams AL, O'Neill WC. Breast arterial calcification: a marker of medial vascular calcification in chronic kidney disease. Clin J Am Soc Nephrol 2011;6:377−82.

[36] Hendriks EJ, de Jong PA, van der Graaf Y, Mali WP, van der Schouw YT, Beulens JW. Breast arterial calcifications: a systematic review and meta-analysis of their determi-nants and their association with cardiovascular events. Atherosclerosis 2015;239: 11−20.

[37] Maas AH, van der Schouw YT, Mali WP, van der Graaf Y. Prevalence and determi-nants of breast arterial calcium in women at high risk of cardiovascular disease. Am J Cardiol 2004;94:655−9.

[38] Sedighi N, Radmard AR, Radmehr A, Hashemi P, Hajizadeh A, Taheri AP. Breast arte-rial calcification and risk of carotid atherosclerosis: focusing on the preferentially affected layer of the vessel wall. Eur J Radiol 2011;79:250−6.

[39] Lanzer P, Boehm M, Sorribas V, et al. Medial vascular calcification revisited: review and perspectives. Eur Heart J 2014;35:1515−25.

[40] Margolies L, Salvatore M, Hecht HS, et al. Digital mammography and screening for coronary artery disease. JACC Cardiovasc Imaging 2016;9:350–60.

[41] Hoshino T, Chow LA, Hsu JJ, et al. Mechanical stress analysis of a rigid inclusion in distensible material: a model of atherosclerotic calcification and plaque vulnerability. Am J Physiol Heart Circ Physiol 2009;297:H802–10.

[42] Budoff MJ, Nasir K, Katz R, et al. Thoracic aortic calcification and coronary heart disease events: the multi-ethnic study of atherosclerosis (MESA). Atherosclerosis 2011; 215:196–202.

[43] Desai MY, Cremer PC, Schoenhagen P. Thoracic aortic calcification: diagnostic, prognostic, and management considerations. JACC Cardiovasc Imaging 2018;11:1012–26.

[44] Bos D, Leening MJ, Kavousi M, et al. Comparison of atherosclerotic calcification in major vessel beds on the risk of all-cause and cause-specific mortality: the rotterdam study. Circulation Cardiovasc Imaging 2015;8.

[45] Bastos Goncalves F, Voute MT, Hoeks SE, et al. Calcification of the abdominal aorta as an independent predictor of cardiovascular events: a meta-analysis. Heart 2012;98: 988–94.

[46] Handy CE, Desai CS, Dardari ZA, et al. The association of coronary artery calcium with noncardiovascular disease: the multi-ethnic study of atherosclerosis. JACC Cardiovasc Imaging 2016;9:568–76.

[47] Cheruvu C, Precious B, Naoum C, et al. Long term prognostic utility of coronary CT angiography in patients with no modifiable coronary artery disease risk factors: results from the 5 year follow-up of the CONFIRM International Multicenter Registry. J Cardiovasc Comput Tomogr 2016;10:22–7.

[48] Greenland P, Blaha MJ, Budoff MJ, Erbel R, Watson KE. Coronary calcium score and cardiovascular risk. J Am Coll Cardiol 2018;72:434–47.

[49] Kuller LH, Lopez OL, Gottdiener JS, et al. Subclinical atherosclerosis, cardiac and kidney function, heart failure, and dementia in the very elderly. J Am Heart Assoc 2017;6.

[50] Whelton SP, Silverman MG, McEvoy JW, et al. Predictors of long-term healthy arterial aging: coronary artery calcium nondevelopment in the MESA study. JACC Cardiovasc Imaging 2015;8:1393–400.

[51] Gulizia MM, Colivicchi F, Abrignani MG, et al. Consensus document ANMCO/ ANCE/ARCA/GICR-IACPR/GISE/SICOA: long-term antiplatelet therapy in patients with coronary artery disease. Eur Heart J Suppl 2018;20:F1–f74.

[52] DeFilippis AP, Young R, McEvoy JW, et al. Risk score overestimation: the impact of individual cardiovascular risk factors and preventive therapies on the performance of the American Heart Association-American College of Cardiology-Atherosclerotic Cardiovascular Disease risk score in a modern multi-ethnic cohort. Eur Heart J 2017;38:598–608.

[53] Mahabadi AA, Mohlenkamp S, Lehmann N, et al. CAC score improves coronary and CV risk assessment above statin indication by ESC and AHA/ACC primary prevention guidelines. JACC Cardiovascular imaging 2017;10:143–53.

[54] Rozanski A, Gransar H, Shaw LJ, et al. Impact of coronary artery calcium scanning on coronary risk factors and downstream testing the EISNER (Early Identification of Subclinical Atherosclerosis by Noninvasive Imaging Research) prospective randomized trial. J Am Coll Cardiol 2011;57:1622–32.

[55] Nasir K, McClelland RL, Blumenthal RS, et al. Coronary artery calcium in relation to initiation and continuation of cardiovascular preventive medications: the Multi-Ethnic Study of Atherosclerosis (MESA). Circ Cardiovasc Qual Outcomes 2010;3:228−35.

[56] Polonsky TS, Greenland P. Viewing the value of coronary artery calcium testing from different perspectives. JAMA Cardiology 2018;3(10):908−10.

[57] van Rosendael AR, Dimitriu-Leen AC, Bax JJ, Kroft LJ, Scholte AJ. One-stop-shop cardiac CT: calcium score, angiography, and myocardial perfusion. J Nucl Cardiol 2015;23(5):1176−9.

[58] Villines TC, Hulten EA, Shaw LJ, et al. Prevalence and severity of coronary artery disease and adverse events among symptomatic patients with coronary artery calcification scores of zero undergoing coronary computed tomography angiography: results from the CONFIRM (Coronary CT Angiography Evaluation for Clinical Outcomes: an International Multicenter) registry. J Am Coll Cardiol 2011;58:2533−40.

[59] van Werkhoven JM, de Boer SM, Schuijf JD, et al. Impact of clinical presentation and pretest likelihood on the relation between calcium score and computed tomographic coronary angiography. Am J Cardiol 2010;106:1675−9.

[60] Henneman MM, Schuijf JD, Pundziute G, et al. Noninvasive evaluation with multislice computed tomography in suspected acute coronary syndrome: plaque morphology on multislice computed tomography versus coronary calcium score. J Am Coll Cardiol 2008;52:216−22.

[61] Karlo CA, Leschka S, Stolzmann P, Glaser-Gallion N, Wildermuth S, Alkadhi H. A systematic approach for analysis, interpretation, and reporting of coronary CTA studies. Insights into imaging 2012;3:215−28.

[62] Stocker TJ, Deseive S, Leipsic J, et al. Reduction in radiation exposure in cardiovascular computed tomography imaging: results from the prospective multicenter registry on RadiaTion dose estimates of cardiac CT AngIOgraphy IN daily practice in 2017 (PROTECTION VI). Eur Heart J 2018;39(41):3715−23.

[63] Cury RC, Abbara S, Achenbach S, et al. Coronary artery disease − reporting and data system (CAD-RADS): an Expert consensus document of SCCT, ACR and NASCI: endorsed by the ACC. JACC Cardiovasc Imaging 2016;9:1099−113.

[64] Kwan AC, Cater G, Vargas J, Bluemke DA. Beyond coronary stenosis: coronary computed tomographic angiography for the assessment of atherosclerotic plaque burden. Current Cardiovasc Imaging Rep 2013;6:89−101.

[65] Voros S, Rinehart S, Qian Z, et al. Coronary atherosclerosis imaging by coronary CT angiography: current status, correlation with intravascular interrogation and meta-analysis. JACC Cardiovasc Imaging 2011;4:537−48.

[66] van Velzen JE, de Graaf FR, de Graaf MA, et al. Comprehensive assessment of spotty calcifications on computed tomography angiography: comparison to plaque characteristics on intravascular ultrasound with radiofrequency backscatter analysis. J Nucl Cardiol 2011;18:893−903.

[67] Obaid DR, Calvert PA, Brown A, et al. Coronary CT angiography features of ruptured and high-risk atherosclerotic plaques: correlation with intra-vascular ultrasound. J Cardiovasc Comput Tomogr 2017;11:455−61.

[68] Pundziute G, Schuijf JD, Jukema JW, et al. Evaluation of plaque characteristics in acute coronary syndromes: non-invasive assessment with multi-slice computed tomography and invasive evaluation with intravascular ultrasound radiofrequency data analysis. Eur Heart J 2008;29:2373−81.

[69] Munnur RK, Cameron JD, Ko BS, Meredith IT, Wong DTL. Cardiac CT: atherosclerosis to acute coronary syndrome. Cardiovasc Diagn Ther 2014;4:430−48.

[70] Becker CR, Nikolaou K, Muders M, et al. Ex vivo coronary atherosclerotic plaque characterization with multi-detector-row CT. Eur Radiol 2003;13:2094−8.

[71] Raff GL, Chair, Abidov A, et al. SCCT guidelines for the interpretation and reporting of coronary computed tomographic angiography. J Cardiovasc Comput Tomogr 2009; 3:122−36.

[72] Stary HC. Natural history and histological classification of atherosclerotic lesions: an update. Arterioscler Thromb Vasc Biol 2000;20:1177−8.

[73] Owen DR, Lindsay AC, Choudhury RP, Fayad ZA. Imaging of atherosclerosis. Annu Rev Med 2011;62:25−40.

[74] Ibanez B, Badimon JJ, Garcia MJ. Diagnosis of atherosclerosis by imaging. Am J Med 2009;122:S15−25.

[75] Liu W, Zhang Y, Yu CM, et al. Current understanding of coronary artery calcification. J Geriatr Cardiol 2015;12:668−75.

[76] Qi L, Tang LJ, Xu Y, et al. The diagnostic performance of coronary CT angiography for the assessment of coronary stenosis in calcified plaque. PLoS One 2016;11:e0154852.

[77] Leschka S, Seitun S, Dettmer M, et al. Ex vivo evaluation of coronary atherosclerotic plaques: characterization with dual-source CT in comparison with histopathology. J Cardiovasc Comput Tomogr 2010;4:301−8.

[78] Nikolaou K, Becker CR, Muders M, et al. Multidetector-row computed tomography and magnetic resonance imaging of atherosclerotic lesions in human ex vivo coronary arteries. Atherosclerosis 2004;174:243−52.

[79] Carrascosa PM, Capunay CM, Garcia-Merletti P, Carrascosa J, Garcia MF. Characterization of coronary atherosclerotic plaques by multidetector computed tomography. Am J Cardiol 2006;97:598−602.

[80] Cademartiri F, Mollet NR, Runza G, et al. Influence of intracoronary attenuation on coronary plaque measurements using multislice computed tomography: observations in an ex vivo model of coronary computed tomography angiography. Eur Radiol 2005;15:1426−31.

[81] Halliburton SS, Schoenhagen P, Nair A, et al. Contrast enhancement of coronary atherosclerotic plaque: a high-resolution, multidetector-row computed tomography study of pressure-perfused, human ex-vivo coronary arteries. Coron Artery Dis 2006;17:553−60.

[82] Barreto M, Schoenhagen P, Nair A, et al. Potential of dual-energy computed tomography to characterize atherosclerotic plaque: ex vivo assessment of human coronary arteries in comparison to histology. J Cardiovasc Comput Tomogr 2008;2:234−42.

[83] Budoff MJ, Dowe D, Jollis JG, et al. Diagnostic performance of 64-multidetector row coronary computed tomographic angiography for evaluation of coronary artery stenosis in individuals without known coronary artery disease: results from the prospective multicenter ACCURACY (Assessment by Coronary Computed Tomographic Angiography of Individuals Undergoing Invasive Coronary Angiography) trial. J Am Coll Cardiol 2008;52:1724−32.

[84] Task Force M, Montalescot G, Sechtem U, et al. 2013 ESC guidelines on the management of stable coronary artery disease: the Task Force on the management of stable coronary artery disease of the European Society of Cardiology. Eur Heart J 2013; 34:2949−3003.

[85] Schuijf JD, Wijns W, Jukema JW, et al. Relationship between noninvasive coronary angiography with multi-slice computed tomography and myocardial perfusion imaging. J Am Coll Cardiol 2006;48:2508−14.

[86] Pijls NH, Van Gelder B, Van der Voort P, et al. Fractional flow reserve. A useful index to evaluate the influence of an epicardial coronary stenosis on myocardial blood flow. Circulation 1995;92:3183−93.

[87] Tonino PA, Fearon WF, De Bruyne B, et al. Angiographic versus functional severity of coronary artery stenoses in the FAME study fractional flow reserve versus angiography in multivessel evaluation. J Am Coll Cardiol 2010;55:2816−21.

[88] De Bruyne B, Pijls NH, Kalesan B, et al. Fractional flow reserve-guided PCI versus medical therapy in stable coronary disease. N Engl J Med 2012;367:991−1001.

[89] Tonino PA, De Bruyne B, Pijls NH, et al. Fractional flow reserve versus angiography for guiding percutaneous coronary intervention. N Engl J Med 2009;360:213−24.

[90] Sousa-Uva M, Neumann FJ, Ahlsson A, et al. 2018 ESC/EACTS Guidelines on myocardial revascularization. Eur J Cardiothorac Surg 2018;55(1):4−90.

[91] Danad I, Raijmakers PG, Driessen RS, et al. Comparison of coronary CT angiography, SPECT, PET, and hybrid imaging for diagnosis of ischemic heart disease determined by fractional flow reserve. JAMA Cardiol 2017;2:1100−7.

[92] Danad I, Szymonifka J, Twisk JW, et al. Diagnostic performance of cardiac imaging methods to diagnose ischaemia-causing coronary artery disease when directly compared with fractional flow reserve as a reference standard: a meta-analysis. Eur Heart J 2016;38(13):991−8.

[93] van Rosendael AR, Dimitriu-Leen AC, de Graaf MA, et al. Impact of computed tomography myocardial perfusion following computed tomography coronary angiography on downstream referral for invasive coronary angiography, revascularization and, outcome at 12 months. Eur Heart J Cardiovasc Imaging 2017;18(9):969−77.

[94] Padley SPG, Roditi G, Nicol ED. Bsci/Bscct. Chest pain of recent onset: assessment and diagnosis (CG95). A step change in the requirement for cardiovascular CT. Clin Radiol 2017;72:751−3.

[95] Zellweger MJ, Hachamovitch R, Kang X, et al. Prognostic relevance of symptoms versus objective evidence of coronary artery disease in diabetic patients. Eur Heart J 2004;25:543−50.

[96] Achenbach S, Ulzheimer S, Baum U, et al. Noninvasive coronary angiography by retrospectively ECG-gated multislice spiral CT. Circulation 2000;102:2823−8.

[97] Nieman K, Oudkerk M, Rensing BJ, et al. Coronary angiography with multi-slice computed tomography. Lancet 2001;357:599−603.

[98] Miller JM, Rochitte CE, Dewey M, et al. Diagnostic performance of coronary angiography by 64-row CT. N Engl J Med 2008;359:2324−36.

[99] Raff GL, Gallagher MJ, O'Neill WW, Goldstein JA. Diagnostic accuracy of noninvasive coronary angiography using 64-slice spiral computed tomography. J Am Coll Cardiol 2005;46:552−7.

[100] investigators S-H. CT coronary angiography in patients with suspected angina due to coronary heart disease (SCOT-HEART): an open-label, parallel-group, multicentre trial. Lancet 2015;385:2383−91.

[101] Mark DB, Nelson CL, Califf RM, et al. Continuing evolution of therapy for coronary artery disease. Initial results from the era of coronary angioplasty. Circulation 1994;89:2015−25.

[102] Yusuf S, Zucker D, Peduzzi P, et al. Effect of coronary artery bypass graft surgery on survival: overview of 10-year results from randomised trials by the Coronary Artery Bypass Graft Surgery Trialists Collaboration. Lancet 1994;344:563–70.

[103] Lubbers M, Dedic A, Coenen A, et al. Calcium imaging and selective computed tomography angiography in comparison to functional testing for suspected coronary artery disease: the multicentre, randomized CRESCENT trial. Eur Heart J 2016;37: 1232–43.

[104] Roffi M, Patrono C, Collet JP, et al. 2015 ESC guidelines for the management of acute coronary syndromes in patients presenting without persistent ST-segment elevation: task force for the management of acute coronary syndromes in patients presenting without persistent ST-segment elevation of the european Society of cardiology (ESC). Eur Heart J 2016;37:267–315.

[105] Amsterdam EA, Kirk JD, Bluemke DA, et al. Testing of low-risk patients presenting to the emergency department with chest pain: a scientific statement from the American Heart Association. Circulation 2010;122:1756–76.

[106] Pope JH, Aufderheide TP, Ruthazer R, et al. Missed diagnoses of acute cardiac ischemia in the emergency department. N Engl J Med 2000;342:1163–70.

[107] Raff GL, Chinnaiyan KM, Cury RC, et al. SCCT guidelines on the use of coronary computed tomographic angiography for patients presenting with acute chest pain to the emergency department: a report of the Society of Cardiovascular Computed Tomography Guidelines Committee. J Cardiovasc Comput 2014;8:254–71.

[108] Samad Z, Hakeem A, Mahmood SS, et al. A meta-analysis and systematic review of computed tomography angiography as a diagnostic triage tool for patients with chest pain presenting to the emergency department. J Nucl Cardiol 2012;19:364–76.

[109] Goldstein JA, Chinnaiyan KM, Abidov A, et al. The CT-STAT (coronary computed tomographic angiography for systematic triage of acute chest pain patients to treatment) trial. J Am Coll Cardiol 2011;58:1414–22.

[110] Goldstein JA, Gallagher MJ, O'Neill WW, Ross MA, O'Neil BJ, Raff GL. A randomized controlled trial of multi-slice coronary computed tomography for evaluation of acute chest pain. J Am Coll Cardiol 2007;49:863–71.

[111] Hoffmann U, Truong QA, Schoenfeld DA, et al. Coronary CT angiography versus standard evaluation in acute chest pain. N Engl J Med 2012;367:299–308.

[112] Litt HI, Gatsonis C, Snyder B, et al. CT angiography for safe discharge of patients with possible acute coronary syndromes. N Engl J Med 2012;366:1393–403.

[113] Hulten E, Pickett C, Bittencourt MS, et al. Outcomes after coronary computed tomography angiography in the emergency department: a systematic review and meta-analysis of randomized, controlled trials. J Am Coll Cardiol 2013;61:880–92.

[114] Min JK, Shaw LJ, Devereux RB, et al. Prognostic value of multidetector coronary computed tomographic angiography for prediction of all-cause mortality. J Am Coll Cardiol 2007;50:1161–70.

[115] Cho I, Chang HJ, OH B, et al. Incremental prognostic utility of coronary CT angiography for asymptomatic patients based upon extent and severity of coronary artery calcium: results from the COronary CT Angiography EvaluatioN for Clinical Outcomes InteRnational Multicenter (CONFIRM) study. Eur Heart J 2015;36:501–8.

[116] Nielsen LH, Botker HE, Sorensen HT, et al. Prognostic assessment of stable coronary artery disease as determined by coronary computed tomography angiography: a Danish multicentre cohort study. Eur Heart J 2017;38:413–21.

[117] Blanke P, Naoum C, Ahmadi A, et al. Long-term prognostic utility of coronary CT angiography in stable patients with diabetes mellitus. JACC Cardiovasc Imaging 2016;9:1280–8.

[118] Bittencourt MS, Hulten E, Ghoshhajra B, et al. Prognostic value of nonobstructive and obstructive coronary artery disease detected by coronary computed tomography angiography to identify cardiovascular events. Circulation Cardiovasc Imaging 2014;7: 282–91.

[119] Greenland P, Alpert JS, Beller GA, et al. 2010 ACCF/AHA guideline for assessment of cardiovascular risk in asymptomatic adults: a report of the American College of cardiology foundation/American heart association task force on practice guidelines. J Am Coll Cardiol 2010;56. e50-103.

[120] Piepoli MF, Hoes AW, Agewall S, et al. 2016 European guidelines on cardiovascular disease prevention in clinical practice: the sixth joint task force of the European society of cardiology and other societies on cardiovascular disease prevention in clinical practice (constituted by representatives of 10 societies and by invited experts) Developed with the special contribution of the European Association for Cardiovascular Prevention & Rehabilitation (EACPR). Eur Heart J 2016;37:2315–81.

[121] van den Hoogen IJ, de Graaf MA, Roos CJ, et al. Prognostic value of coronary computed tomography angiography in diabetic patients without chest pain syndrome. J Nucl Cardiol 2016;23:24–36.

[122] Emerging Risk Factors C, Sarwar N, Gao P, et al. Diabetes mellitus, fasting blood glucose concentration, and risk of vascular disease: a collaborative meta-analysis of 102 prospective studies. Lancet 2010;375:2215–22.

[123] Kannel WB, McGee DL. Diabetes and cardiovascular disease. The Framingham study. J Am Med Assoc 1979;241:2035–8.

[124] Ambepityia G, Kopelman PG, Ingram D, Swash M, Mills PG, Timmis AD. Exertional myocardial ischemia in diabetes: a quantitative analysis of anginal perceptual threshold and the influence of autonomic function. J Am Coll Cardiol 1990;15:72–7.

[125] Fox CS, Golden SH, Anderson C, et al. Update on prevention of cardiovascular disease in adults with type 2 diabetes mellitus in light of recent evidence: a scientific statement from the American heart association and the American diabetes association. Diabetes Care 2015;38:1777–803.

[126] Celeng C, Maurovich-Horvat P, Ghoshhajra BB, Merkely B, Leiner T, Takx RA. Prognostic value of coronary computed tomography angiography in patients with diabetes: a meta-analysis. Diabetes Care 2016;39:1274–80.

[127] van Werkhoven JM, Schuijf JD, Gaemperli O, et al. Incremental prognostic value of multi-slice computed tomography coronary angiography over coronary artery calcium scoring in patients with suspected coronary artery disease. Eur Heart J 2009;30: 2622–9.

[128] Boogers MJ, Broersen A, van Velzen JE, et al. Automated quantification of coronary plaque with computed tomography: comparison with intravascular ultrasound using a dedicated registration algorithm for fusion-based quantification. Eur Heart J 2012;33: 1007–16.

[129] Lee SE, Chang HJ, Sung JM, et al. Effects of statins on coronary atherosclerotic plaques: the PARADIGM (progression of AtheRosclerotic PlAque DetermIned by computed TomoGraphic angiography imaging) study. JACC Cardiovascular imaging 2018;11(10):1475–84.

[130] Park HB, Heo R, o Hartaigh B, et al. Atherosclerotic plaque characteristics by CT angiography identify coronary lesions that cause ischemia: a direct comparison to fractional flow reserve. JACC Cardiovasc Imaging 2015;8:1–10.

[131] Driessen RS, Stuijfzand WJ, Raijmakers PG, et al. Effect of plaque burden and morphology on myocardial blood flow and fractional flow reserve. J Am Coll Cardiol 2018;71:499–509.

[132] Lavi S, Bae JH, Rihal CS, et al. Segmental coronary endothelial dysfunction in patients with minimal atherosclerosis is associated with necrotic core plaques. Heart 2009;95:1525–30.

[133] Sianos G, Werner GS, Galassi AR, et al. Recanalisation of chronic total coronary occlusions: 2012 consensus document from the EuroCTO club. EuroIntervention 2012;8:139–45.

[134] Mitomo S, Demir OM, Colombo A, Nakamura S, Chieffo A. What the surgeon needs to know about percutaneous coronary intervention treatment of chronic total occlusions. Ann Cardiothorac Surg 2018;7:533–45.

[135] Stone GW, Reifart NJ, Moussa I, et al. Percutaneous recanalization of chronically occluded coronary arteries: a consensus document: part II. Circulation 2005;112:2530–7.

[136] Fefer P, Knudtson ML, Cheema AN, et al. Current perspectives on coronary chronic total occlusions: the Canadian multicenter chronic total occlusions registry. J Am Coll Cardiol 2012;59:991–7.

[137] Magro M, Schultz C, Simsek C, et al. Computed tomography as a tool for percutaneous coronary intervention of chronic total occlusions. EuroIntervention 2010;6(Suppl. G):G123–31.

[138] Soon KH, Cox N, Wong A, et al. CT coronary angiography predicts the outcome of percutaneous coronary intervention of chronic total occlusion. J Interv Cardiol 2007;20:359–66.

[139] Opolski MP, Achenbach S, Schuhback A, et al. Coronary computed tomographic prediction rule for time-efficient guidewire crossing through chronic total occlusion: insights from the CT-RECTOR multicenter registry (Computed Tomography Registry of Chronic Total Occlusion Revascularization). JACC Cardiovasc Interv 2015;8:257–67.

[140] Mollet NR, Hoye A, Lemos PA, et al. Value of preprocedure multislice computed tomographic coronary angiography to predict the outcome of percutaneous recanalization of chronic total occlusions. Am J Cardiol 2005;95:240–3.

[141] Ehara M, Terashima M, Kawai M, et al. Impact of multislice computed tomography to estimate difficulty in wire crossing in percutaneous coronary intervention for chronic total occlusion. J Invasive Cardiol 2009;21:575–82.

[142] Rolf A, Werner GS, Schuhback A, et al. Preprocedural coronary CT angiography significantly improves success rates of PCI for chronic total occlusion. Int J Cardiovasc Imaging 2013;29:1819–27.

Intravascular imaging and coronary calcification

Akiko Maehara, MD [1,2], **Mitsuaki Matsumura, BS** [3], **Ziad A. Ali, MD, DPhil** [4,5,6], **Gary S. Mintz, MD** [7]

[1]*Professor, Center for Interventional Vascular Therapy, New York-Presbyterian Hospital/ Columbia University Medical Center, New York, NY, United States;* [2]*Director of Intravascular Imaging Core Laboratory, Cardiovascular Research Foundation, New York, NY, United States;* [3]*Assistant Director of Intravascular Imaging and Physiology Core Laboratories, Cardiovascular Research Foundation, New York, NY, United States;* [4]*Director of Intravascular Imaging and Physiology, Center for Interventional Vascular Therapy, New York-Presbyterian Hospital/ Columbia University Medical Center, New York, NY, United States;* [5]*Director of Angiographic Core Laboratory, Cardiovascular Research Foundation, New York, NY, United States;* [6]*Department of Cardiology, St. Francis Hospital, Roslyn, NY, United States;* [7]*Senior Medical Advisor, Cardiovascular Research Foundation, New York, NY, United States*

Abbreviations

AUC	Area under the curve
CABG	Coronary artery bypass grafting
CrCl	Creatinine clearance
CTO	Chronic total occlusion
DES	Drug-eluting stents
ELCA	Excimer laser coronary angiography
IQR	Interquartile range
ISR	In-stent restenosis
IVUS	Intravascular ultrasound
LMCA	Left main coronary artery
MI	Myocardial infarction
MSA	Minimum stent area
OA	Orbital atherectomy
OCT	Optical coherence tomography
OR	Odds ratio
PCI	Percutaneous coronary intervention
RA	Rotational atherectomy
STEMI	ST-segment elevation myocardial infarction
VH	Virtual histology

Identification of coronary calcium by intravascular imaging
Validation of intravascular imaging calcium detection versus histology

The accuracy of calcium detection by intravascular imaging depends on the total amount of calcium. Because ultrasound does not penetrate calcium and all ultrasound is reflected at the surface of calcified plaque, when using intravascular ultrasound (IVUS), calcified plaque appears as "bright echoes with acoustic shadowing" (more sensitive), sometimes with reverberations (more specific) [1]. IVUS can detect only the leading edge and not the trailing edge of calcium and cannot determine the thickness of calcium. By IVUS, calcified plaques are categorized according to the location. If the leading edge of the calcium appears within the shallowest 50% of the plaque plus media thickness, calcium is categorized as superficial; if it appears within the deepest 50%, calcium is categorized as deep. The angle of calcium can be measured in degrees using the center of the lumen. Using known pullback speed, the length of calcium can be measured. Calcium can produce reverberations, which are multiple equidistant reflections that result from oscillation of ultrasound between transducer and calcium and cause concentric arcs at equidistance. This indicates the smooth surface of calcium that allows reflections rather than scattering (Fig. 5.1).

Using histology as a gold standard, Kostamaa et al. [2] reported high sensitivity (89%) and specificity (97%) of detecting calcium deposits using 25 MHz IVUS (Table 5.1).

Friedrich et al. [3] reported that 30 MHz IVUS was able to detect dense calcified deposits with high sensitivity (89%) and specificity (100%), but it was limited in visualizing accumulation of microcalcification (≤ 0.05 mm in size, sensitivity 17%, specificity 100%). When the surface area of calcium was measured using 30 MHz IVUS (multiplying calcium circumference and length), the correlation of calcium surface area between IVUS and histology ($r = 0.84$) was better than the correlation between IVUS calcium angle and histology calcium surface area ($r = 0.41$) [4]; however, to standardize the severity of IVUS calcium independent of vessel size, calcium angle is most often used clinically.

By optical coherence tomography (OCT), calcified plaque is defined as a low-intensity signal with sharply delineated borders [5]. By OCT, both the leading edge and the trailing edge of calcium are visible; the angle, thickness, and area can be measured along with the length of calcium (Fig. 5.2).

Kume et al. [6] examined 32 superficial calcified plaques by OCT and 40 MHz IVUS using histology as a gold standard and showed that the measurements of calcium angle were similarly good by both IVUS and OCT; however, different from IVUS, calcium area measurement can be done by OCT with good correlation compared with histology ($r = 0.84$). Similarly, Mehanna et al. [7] evaluated 55 calcified plaques using OCT and cryo-imaging, calculated calcium volume using Simpson's rule, and compared calcium volume by OCT versus cryo-imaging.

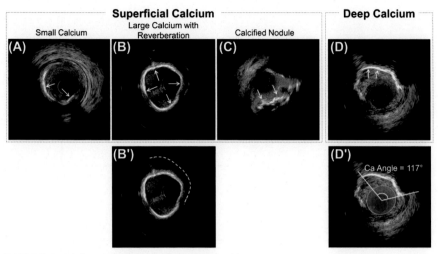

FIGURE 5.1 Calcium Type and Measurement by IVUS.

IVUS defines calcium as "bright echoes with acoustic shadowing" and further categorized as "superficial" (white arrows in A, B, C) if the leading edge of the calcium appears within the most shallow 50% of the plaque plus media thickness, or "deep" (white arrows in D) if the leading edge of the calcium appears within the deepest 50%. "Reverberations" appear as equidistant concentric arcs (blue and yellow dotted lines in B'). "Calcified nodule" is defined as a convex and irregular luminal surface with signal attenuation (C). Calcium angle can be measured in degrees using the center of the lumen. In (D') the blue area indicates the lumen area, and the calcium angle was measured as 117 degrees shown in yellow.

Although the correlation was good for calcium angle ($r = 0.90$), it was modest for calcium volume ($r = 0.76$), especially when calcified plaque with an ambiguous trailing edge was included. Saita et al. [8] categorized four types of coronary calcium by histology and compared with OCT: (1) superficial dense calcification; (2) deep intimal calcification; (3) scattered microcalcification; and (4) calcified nodule. The leading edge was visible in 100% of superficial dense calcium and calcified nodule, and the leading edge was visible in 90% (27/30) of scattered microcalcification, but the leading edge was not visible in any deep intimal calcification due to superficial lipidic plaque and/or the limited penetration of OCT. Both the leading and trailing edges of calcification were clearly identified in 55% of superficial dense calcifications in which there was a good correlation of the maximum thickness of calcification between the OCT and histology ($r^2 = 0.97$). In the rest of the 45% superficial dense calcifications, the trailing edge was not visible due to attenuation from residual lipid or necrotic core associated with the calcified plaque. All three calcified nodules contained fibrin, which caused attenuation, and no trailing edge was visible.

Table 5.1 Comparison between IVUS or OCT and histology.

Intravascular imaging modality	Author (reference)	Number of calcium segments	Calcium type	Sensitivity	Specificity	Regression coefficient
IVUS, 25 MHz	Kostamaa [2]	38	Any calcium	89%	97%	—
IVUS, 25 MHz	Kostamaa	38	Length of calcium	—	—	$r = 0.79$
IVUS, 30 MHz	Friedrich [3]	33	Any calcium	64%	100%	—
IVUS, 30 MHz	Friedrich	18	Dense calcium (>0.05 mm)	89%	100%	—
IVUS, 30 MHz	Friedrich	12	Microcalcification (≤0.05 mm)	17%	100%	—
IVUS, 30 MHz	Scott [4]	253	Surface area of any calcium	—	—	$r = 0.82$
IVUS, 30 MHz	Scott	253	Angle of any calcium	—	—	$r = 0.41$
OCT	Kume [6]	32	Area of superficial calcium	—	—	$r = 0.84$
OCT	Mehanna [7]	55	Angle of any calcium	—	—	$r = 0.90$
OCT	Mehanna	55	Volume of any calcium	—	—	$r = 0.76$
OCT	Saita [8]	105	Superficial dense calcium	100%	—	—
OCT	Saita	105	Calcium thickness of superficial dense calcium with visible trailing edge	—	—	$r^2 = 0.97$
OCT	Saita	30	Microcalcification	90%	—	—
OCT	Saita	20	Deep intimal calcium	0%	—	—
OCT	Saita	3	Calcified nodule	100%	—	—

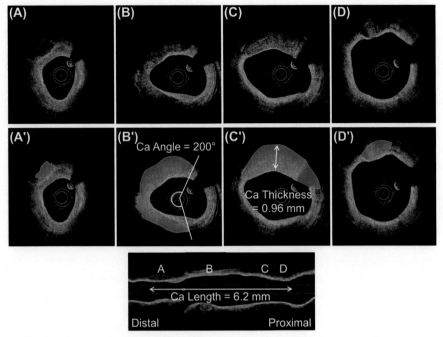

FIGURE 5.2 Calcium Measurement by OCT.

(A′−D′) are the same as (A−D), but with annotation. In each frame, (A′−D′), calcium is shown in gray. Calcium length (bottom image) can be calculated as 6.2 mm in length by multiplying the number of frames (31 frames in this case) with calcium by the frame interval (0.2 mm in this case). Calcium angle (B′) is 200 degrees (white angle) using the center of the lumen. Calcium thickness (C′) is 0.96 mm (white double-headed arrows) from the leading edge to the trailing edge.

Calcium detection comparing intravascular imaging and coronary angiography

In 1995, Mintz et al. [9] showed that in 1117 patients with stable angina, the prevalence of coronary artery target lesion calcium was 73% by IVUS (25 MHz or 30 MHz) and 38% by angiography. Angiographically, moderate calcification was defined as radio-opacities noted only during the cardiac cycle before contrast injection, whereas severe calcification was defined as radio-opacities observed without cardiac motion, usually affecting both sides of the arterial lumen [10]. The rest were considered as none/mild angiographic calcification. Using IVUS calcium as the gold standard, angiographic calcium detection showed poor sensitivity (48%) and good specificity (89%), but with an 11% false-positive rate. In this report, 8% had no angiographic calcium, but there was an IVUS calcium arc >180 degrees.

In 2017, we revisited this issue including 440 patients with stable coronary artery disease studied with coronary angiography, IVUS, and OCT [11]. The prevalence of culprit lesion calcium was 83% by IVUS (40 MHz), 77% by OCT, and 40% by angiography. The sensitivity and specificity of angiography to detect any IVUS calcium were 48% and 99%, respectively, and the sensitivity and specificity to detect any OCT calcium were 51% and 95%, respectively. Table 5.2 summarizes the OCT and IVUS findings stratified by angiographic calcium severity.

Each measurement of OCT or IVUS calcium severity increased with greater amounts of angiographic calcium. By receiver operating characteristic analysis, the IVUS maximum calcium angle that predicted angiographically visible calcium (moderate or severe) was a cutoff of 110 degrees (area under the curve [AUC] = 0.80). The OCT maximum calcium angle, maximum calcium thickness, and length that predicted angiographically visible calcium was a cutoff of 101 degrees (AUC = 0.78), 0.57 mm (AUC = 0.80), and 4.0 mm (AUC = 0.81), respectively. Among 74 lesions with IVUS calcium angle >180 degrees, 21.6% (16/74) did not show angiographically visible calcium. Sixteen lesions without angiographically visible calcium had thinner calcium by OCT (maximum calcium thickness 0.71 mm [interquartile range (IQR) 0.52−0.89] vs. 0.95 mm [0.75−1.15], $P = .004$); most angiographically invisible calcium were <0.5 mm thick and had a shorter calcium length (calcium length 11 mm [6−18] vs. 16 mm [11−23], $P = .01$), although a similar maximum calcium angle (190 degrees [146−300] vs. 250 degrees [170−320], $P = .047$). Of note, lesions with angiographically invisible

Table 5.2 OCT and IVUS findings stratified by angiographic calcium severity.

	Angiographic calcium severity			
	None/mild (n = 236)	Moderate (n = 133)	Severe (n = 44)	P Value
Pre-PCI OCT findings				
Presence of calcium	63.1%	96.2%	100%	<.001
Maximum calcium angle, °	82 (52, 128)	111 (69, 171)	235 (150, 331)	<.001
Maximum calcium thickness, mm	0.55 (0.41, 0.81)	0.79 (0.58, 1.04)	0.98 (0.76, 1.18)	<.001
Calcium length, mm	4.0 (2.0, 8.0)	8.0 (4.0, 13.8)	15.5 (12.0, 23.3)	<.001
Pre-PCI IVUS findings				
Presence of calcium	71.5%	99.2%	100%	<.001
Maximum calcium angle, °	78 (52, 120)	123 (76, 174)	252 (167, 325)	<.001

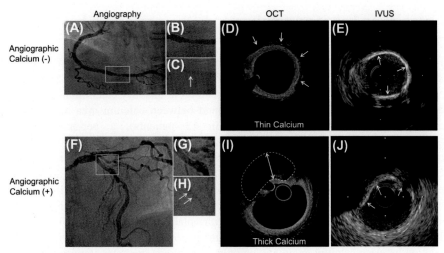

FIGURE 5.3 Representative Cases Showing Discordance or Concordance for the Detection of Calcium Between Angiography and OCT or IVUS.

Upper panels (A—E) show an example of discordance and lower panels (F—J) show an example of concordance. Zoomed images (white square in A or F) are shown with (B or G) or without contrast (C or H). Although there is no angiographic calcium (C), OCT shows a thin large angle of calcium (white arrows in D), and IVUS shows a large angle of calcium (E). Without cardiac motion, angiographic calcium (severe calcium) is observed (H). OCT shows thick calcium (I, the yellow dotted area shows calcium area and the yellow double-headed arrows indicate calcium thickness), and IVUS shows the modest angle of calcium (J).

calcium had a greater final minimum stent area (MSA) (8.1 mm^2 [6.6—9.3]) versus 58 lesions with angiographically visible calcium (5.9 mm^2 [4.6—7.3], $P = .001$). Representative cases are shown in Fig. 5.3.

Among 102 lesions without any OCT calcium, 4.9% (5/102) of lesions had angiographic calcium, and among 76 lesions without any IVUS calcium (also not visible by OCT), 1.3% (1/76) of lesions had angiographic calcium because IVUS superficial attenuated plaque and OCT superficial lipidic plaque masked the underlying calcium.

Thus, the earlier and more recent data were consistent, even though they were acquired 20 years apart and even though there have been significant improvements in coronary angiography; however, it appears that angiographically invisible calcium (only detectable by IVUS or OCT) indicates smaller amounts of calcium (thin calcium if it is large in area or circumference) and not necessarily due to the poor resolution of angiography. Although there can be tremendous clinical and individual patient variability, in general angiographically invisible calcium did not inhibit stent expansion.

Association between calcium and atherosclerotic burden or plaque vulnerability

IVUS and histology analysis showed that the amount of coronary calcification correlated with plaque burden, not with degree of lumen compromise [12−14]. Sangiorgi et al. [12] used a nondecalcified histology method to quantify calcium area in coronary arteries. A good correlation was found between calcium area and plaque area on a per-heart basis ($r = 0.87$ in 13 patients), a per-artery basis ($r = 0.7−0.89$ in 37 coronary arteries), and a per-segment basis ($r = 0.52$ in 723 segments), but there was no correlation between calcium area and lumen dimensions. Using IVUS, Mintz et al. [13] showed the association between calcium angle and plaque burden in 1442 patients (calcium angle was 17 ± 33 degrees in lesions with plaque burden $\leq 50\%$; 78 ± 84 degrees in lesions with plaque burden 51% −75%; 98 ± 100 degrees in lesions with plaque burden 76%−90%; and 121 ± 118 degrees in lesions with plaque burden >90%), whereas there was no correlation between calcium angle and angiographic lumen diameter stenosis. Finally, the Agatston calcium score showed a fair correlation with total atheroma volume on a per-patient basis using 3-vessel IVUS ($r = 0.31$) [15].

Ehara et al. [16] showed that "spotty calcium," defined as calcium angle <90 degrees, was more common in 61 patients with acute myocardial infarction (MI) compared with 47 patients with stable angina (number of spotty calcium: 1.4 ± 1.3 vs. 0.5 ± 0.8, $P < .01$), although total culprit calcium length was shorter in patients with MI (2.2 ± 1.6 mm vs. 4.3 ± 3.2 mm, $P < .01$). Using OCT with a similar cohort (53 ST-segment elevation MI [STEMI] versus 55 stable patients) and methodology found no difference in the number of spotty calcium deposits, although there was a greater overall amount of calcium in patients with stable angina compared with STEMI [17]. Using histology as the gold standard, we evaluated 2294 segments from 151 coronary arteries using grayscale IVUS [18]. Spotty calcium (calcium angle <90 degrees) was found in 14% of segments, and larger calcium (≥90 degrees) was found in 18% of segments. Plaques with spotty calcium had more fibroatheromas than plaques with larger calcium deposits (62% vs. 22%, $P < .001$).

In 676 patients with prepercutaneous coronary intervention (PCI) virtual histology (VH)-IVUS evaluation of the culprit lesion, we reported that the calcium angle at the minimum lumen area sites were smaller in patients with STEMI compared with patients with NSTEMI or stable coronary artery disease (35 ± 61 degrees in STEMI vs. 46 ± 65 degrees in NSTEMI vs. 55 ± 75 degrees in stable coronary artery disease, $P = .001$) [19]. We also showed that a greater amount of total calcium in the culprit lesion was associated with more fibroatheromas and negative remodeling compared to lesions with less calcium [20]. When the total calcium volume was quantified by OCT, total calcium volume of the culprit lesions correlated with a smaller lipid volume index and thicker fibrous caps [21].

Including all observations and noting the inconsistencies, perhaps spotty calcification or small calcium at the culprit lesion is a marker, but not a cause, of a fibroatheroma that can cause an acute thrombotic event, and the amount of calcium detected either by IVUS or OCT is a marker of advanced atherosclerosis but is not correlated with luminal stenosis or plaque vulnerability.

Different types of calcium
Calcified nodule

In 1996, an IVUS-diagnosed calcified nodule was reported for the first time in three cases in which coronary angiography showed only a filling defect suggesting a thrombus [22]. Using histology as a gold standard, an IVUS calcified nodule was subsequently defined as a convex and irregular luminal surface with signal attenuation [23] (Fig. 5.4).

In the Providing Regional Observations to Study Predictors of Events in the Coronary Tree (PROSPECT) study, we reported that the prevalence of an IVUS-identified calcified nodule was 17% per vessel and 30% per patient based on 3-vessel VH-IVUS in 623 patients [24]. Patients with calcified nodules were significantly older and had a greater plaque volume and more thick-cap fibroatheromas but fewer nonculprit lesion major adverse cardiac events at 3-year follow-up.

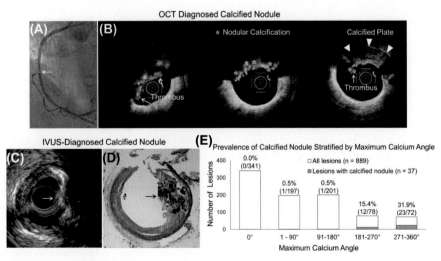

FIGURE 5.4 Calcified Nodule.

The OCT images (B) correspond to the white arrow in the angiogram (A). By OCT, a calcified nodule is defined as an accumulation of small calcium fragments (white stars) with disruption of the fibrous cap on the calcified plate (white triangles). Platelet-rich thrombus is indicated by white arrows. By IVUS (C), a calcified nodule (white arrow) is defined as a convex and irregular luminal surface with signal attenuation. The corresponding histology image (D) indicates a calcified nodule (black arrow). Prevalence of calcified nodule increases with greater maximum calcium angle (E).

By OCT, a calcified nodule was defined as an accumulation of small calcium fragments with disruption of the fibrous cap on the calcified plate. Typically, a calcified nodule was accompanied by strong OCT attenuation mimicking that of a red blood cell—rich thrombus [8]. High attenuation can be explained by the multiple direction of scattering due to the accumulation of small calcium fragments and the accompanying fibrin.

We reported that the prevalence of a calcified nodule in de novo coronary artery disease was 4.2% (37/889) and that it was more frequent in the ostium or mid right coronary artery having a hinge motion [25]. Among 37 calcified nodules, 17 that presented as an acute coronary syndrome more often had an accompanied thrombus compared to a stable presentation (82% vs. 20%, $P < .001$). The clinical and anatomical predictors of a calcified nodule were an underlying greater amount of calcium, hinge motion of the coronary artery, and hemodialysis. When we included only lesions with a maximum calcium angle >270 degrees, the prevalence of a calcified nodule was 31.9% (23/72), and if we included only acute coronary syndrome lesions and lesions with a maximum calcium angle >270 degrees, the prevalence was 42.9% (9/21). Thus, looking only at severely calcified lesions, the prevalence of a calcified nodule causing an acute coronary syndrome was not rare, although pathological reports showed that the prevalence of a calcified nodule among "all" patients who suddenly died due to a coronary thrombotic event was only 3%—5% [26].

Medial calcification and thin intimal calcium in renal insufficiency

Gruberg et al. [27] showed IVUS evaluation among patients with estimated creatinine clearance (CrCl) >70 mL/min ($n = 39$), CrCl 50—69 mL/min ($n = 41$), and CrCl <50 mL/min ($n = 37$) and hemodialysis-dependent patients ($n = 25$). Although the maximum calcium angle was similar in nonhemodialysis patients of differing CrCl (164 ± 84 degrees, 179 ± 99 degrees, and 189 ± 118 degrees, respectively), it was significantly larger in hemodialysis patients (243 ± 107 degrees, $P = .03$). In a large scale VH-IVUS analysis (898 culprit lesions and 752 nonculprit lesions in 762 patients), diminishing renal function without hemodialysis was associated with increased coronary calcification and decreased coronary vessel and lumen size, with a graded response according to the reduction in CrCl in both culprit and nonculprit lesions [28].

Mönckeberg medial calcification is known to be associated with hemodialysis [29]. We analyzed 64 hemodialysis patients who underwent preintervention OCT compared to 64 controls without renal insufficiency by matching age, diabetes mellitus, sex, and culprit vessel location [30]. The hemodialysis-dependent group had a greater calcium angle in both culprit and nonculprit lesions and a greater maximum calcium angle in the distal vessel segment indicating diffuse calcium distribution. Hemodialysis patients also had a higher prevalence of nonatherosclerotic thin intimal calcium (arc of calcium >30 degrees within intima <0.5 mm thick) in addition to medial calcification (Fig. 5.5).

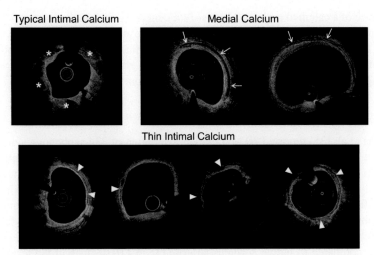

FIGURE 5.5 Calcium Related to Renal Insufficiency.

Typical intimal calcium accompanies significant amounts of plaque (white stars). Medial calcium without plaque can be diagnosed by OCT by confirming the continuity with the adjacent media layer (white arrows). Thin intimal calcium is defined as thin calcium within an intima <0.5 mm thick (white triangles) more frequently in the dialysis patient.

Reprinted from Chin CY, Matsumura M, Maehara A, Zhang W, Lee CT, Yamamoto MH, Song L, Parviz Y, Jhalani NB, Mohan S, Ratner LE, Cohen DJ, Ben-Yehuda O, Stone GW, Shlofmitz RA, Kakuta T, Mintz GS, Ali ZA. Coronary plaque characteristics in hemodialysis-dependent patients as assessed by optical coherence tomography. Am J Cardiol. 2017; 119:1313–1319. Copyright (2017), with permission from Elsevier Inc.

This thin intimal calcium or medial calcium may cause an overestimation of calcium burden by IVUS and may contribute to the lack of correlation between increased coronary artery calcification scores versus long-term outcomes in patients with renal insufficiency.

Acceleration of calcium postcoronary artery bypass grafting

In long-term serial angiographic studies after coronary artery bypass grafting (CABG), there is progressive narrowing in the native coronary arteries proximal to the distal graft anastomosis [31,32]. By IVUS, we compared 41 patients with patent grafts to the left coronary artery (mean duration from CABG of 8.2 ± 6.1 years) and showed that left main coronary artery (LMCA) disease proximal to the patent graft had greater calcium with negative remodeling compared to LMCA disease in non-CABG patients [33]. In a second study using OCT, we compared LMCA disease in 76 CABG patients with a patent graft to the left coronary artery versus LMCA disease in 146 non-CABG patients by matching clinical factors [34]. CABG patients had greater amounts of calcium (thicker calcium, larger calcium angle, and longer calcium length) compared with those without CABG, a greater prevalence of a calcified nodule (37% vs. 14%), more calcium at the carina (30% vs. 14%, unusual in typical atherosclerosis), and more thin calcium (26% vs. 17%, similar to what was seen in dialysis patients). These findings suggested that reduced flow within the LMCA in the setting of a patent graft likely resulted in vessel shrinkage and nonatherosclerotic calcium progression.

Calcification in saphenous vein grafts

Castagna et al. [35] reported 30 MHz IVUS evaluation of 334 saphenous vein graft lesions. Calcium was observed in 40%, and independent predictors of calcium were graft age, insulin-treated diabetes mellitus, and smoking. Importantly, calcium was uniformly distributed, and when the segment was divided into lesion and proximal or distal reference, the mean angle and total length of calcium of each segment were remarkably similar (mean angle: 151 ± 107 degrees in the lesion, 175 ± 121 degrees in the proximal reference, and 177 ± 121 degrees in the distal reference), quite different compared to native coronary artery (115 ± 110 degrees in the lesion and 42 ± 80 degrees in the reference in 1117 stable angina patients studied during the same period at the same hospital) [9]. In addition, graft calcification occurred more frequently within the wall (65%) and less frequently within the plaque (35%), also quite different compared with calcification in native coronary arteries (superficial in 72% and only deep location in 28%). Roleder et al. [36] reported 32 vein graft lesions (11.6 ± 4.9 years after surgery) evaluated by OCT; calcification was found in 44% (87 ± 54 degrees) and increased in relation to graft age. Aorto-ostial vein graft lesions were often negatively remodeled and sometimes either calcified and/or densely fibrotic causing poor stent expansion [37].

PCI outcomes in relation to lesion calcium
Stent expansion

Despite general agreement that stent expansion is limited by target lesion calcium and that the severity of stent underexpansion is related to the amount and location of the calcium, limited data confirm this. Vavuranakis et al. [38] evaluated 27 moderate-to-severely calcified de novo lesions (angle of calcium of 181 ± 60 degrees) with pre- and post-PCI IVUS. The calcium angle was inversely correlated with stent expansion ($r = -0.8$), and adequate stent expansion (defined as MSA >90% compared to the reference lumen area) was achieved in only 59% of lesions after high-pressure ballooning (20 atm).

Kobayashi et al. [39] extended this concept using OCT. They enrolled 51 de novo lesions treated with second-generation drug-eluting stents (DES) and confirmed that stent expansion was inversely related to the calcium angle as well as calcium area. We developed an OCT calcium score for predicting stent expansion [40] (Fig. 5.6).

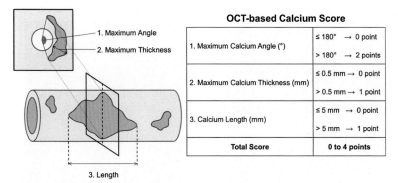

OCT-based Calcium Score

1. Maximum Calcium Angle (°)	≤ 180° → 0 point
	> 180° → 2 points
2. Maximum Calcium Thickness (mm)	≤ 0.5 mm → 0 point
	> 0.5 mm → 1 point
3. Calcium Length (mm)	≤ 5 mm → 0 point
	> 5 mm → 1 point
Total Score	**0 to 4 points**

FIGURE 5.6 Calcium Score to Predict Stent Expansion.

OCT-based calcium score is defined as 2 points for maximum calcium angle >180 degrees, 1 point for maximum calcium thickness >0.5 mm, and 1 point for calcium length >5 mm; the total calcium score ranges from 0 to 4 points.

Reprinted from Fujino A, Mintz GS, Matsumura M, Lee T, Kim SY, Hoshino M, Usui E, Yonetsu T, Haag ES, Shlofmitz RA, Kakuta T, Maehara A. A new optical coherence tomography-based calcium scoring system to predict stent underexpansion. EuroIntervention. 2018;13:e2182–e2189, Copyright (2018), with permission from Europa Digital & Publishing.

In 128 calcified lesions, stent expansion at the maximum calcium site was independently associated with the maximum calcium angle (per 180 degrees: regression coefficient [95% confidence interval (CI) −7.4 [−12.6 to −2.2]), maximum calcium thickness (per 0.5 mm: regression coefficient −3.4 [−6.4 to −0.5]), and total calcium length (per 5 mm: regression coefficient −2.3 [−4.1 to −0.6]). An OCT-based calcium score was then defined as 2 points for maximum calcium angle >180 degrees, 1 point for maximum calcium thickness >0.5 mm, and 1 point for calcium length >5 mm for a total calcium score of 0–4 points. This calcium score was confirmed as predicting stent expansion in a separate validation cohort (stent expansion was 99% [IQR 93–108] in score 0; 85% [78–93] in score 1; 86% [77–100] in score 2; 80% [73–85] in score 3; and 78% [70–86] in score 4, $P < .01$).

Calcium fracture during PCI has been associated with better stent expansion. In 261 calcified lesions treated only by ballooning and stent implantation, calcium fracture was observed in 10.7% (28 of 261) of the lesions [41] (Fig. 5.7).

Quantitative calcium characteristics predicting calcium fracture were minimum calcium thickness (cutoff 0.24 mm, AUC = 0.75) and maximum calcium angle (cutoff 225 degrees, AUC = 0.92). The presence of calcium fracture was independently associated with greater stent expansion (regression coefficient 5.1 [95% CI, 2.8–10.7]). Because most fractures (27 of 28) occurred in calcium with a minimum calcium thickness <0.5 mm, calcium modification before stenting should be considered if the minimum calcium thickness is > 0.5 mm. Although less sensitive, IVUS can also detect calcium fracture by disconnection of calcified plaque along with newly visible vascular deep tissue at the site of a previously solid arc of calcium (Fig. 5.8).

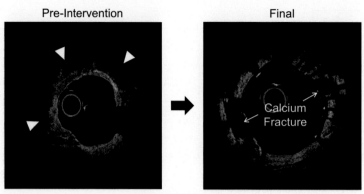

FIGURE 5.7 Calcium Fracture Evaluated by OCT.

OCT images pre- and post-PCI were matched. The pre-PCI OCT image shows thick calcium (white triangles) that were completely separated in the post-PCI OCT image (white arrows).

Stent edge dissections

In the balloon angioplasty era IVUS showed dissections in 76% (31/41) of lesions; 74% (23/31) of dissections were related to calcium deposits, mostly located adjacent to the dissection, and the size of dissection was larger in the setting of calcium compared to the absence of calcium [42]. In a large IVUS registry, we reported the prevalence of untreated stent edge dissections in 7.7% of patients (159/2062) [43]. The presence of calcified plaque was an independent predictor of stent edge dissection (relative risk 1.72, $P = .04$) in addition to the presence of attenuated

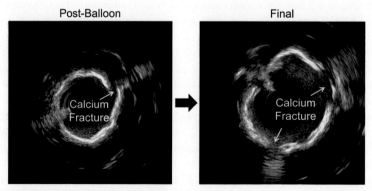

FIGURE 5.8 Calcium Fracture Evaluated by IVUS.

IVUS images after conventional and after high-pressure ballooning were matched. Calcium fracture can be observed as a separation of calcium (yellow arrow), which was further separated in the final IVUS image (yellow arrow at 2 o'clock). Another new calcium fracture also appeared (yellow arrow at seven o'clock).

plaque (presumably indicating fibroatheroma), larger plaque burden, and greater stretch of the vessel. A residual large stent edge dissection with a smaller lumen area was associated with target lesion revascularization during 1-year follow-up. By OCT, the morphological predictors of stent edge dissection were larger angle of calcified plaque in addition to the presence of a thin-cap fibroatheroma and greater vessel stretch, consistent with the IVUS findings [44]. After stenting, the stent edge often had a "hazy" angiographic appearance but without a clearly defined intimal tear, dissection, thrombus, or stenosis. In 13 such patients studied with IVUS, Grewal et al. [45] concluded that the haziness was often associated with calcified plaque.

Acute stent deformation or chronic stent fracture

In patients with planned follow-up angiography, the prevalence of stent fracture has been reported as 1.7%–2.9% [46–48], and angiographic calcium was one of the morphologies associated with stent fracture [47]. IVUS can diagnose stent fracture as (1) absence of stent struts; (2) malalignment of stent struts; or (3) overlapped stent struts within a single stent where there should be no overlapping struts [49]. We reported 14 sirolimus-eluting and 6 paclitaxel-eluting stents, in which IVUS-detected calcium was observed in 70% (14/20). In our IVUS-guided PCI cohort treated with second-generation DES, the prevalence of IVUS-diagnosed acute stent deformation (defined as intra-stent wrinkling and overlapping with stent elongation or shortening) was 1.1% (17 deformations in 1489 lesions) in which angiographic calcium was observed in 59% (10/17) [50]. In summary, the presence of calcium could be a cause or a coexisting complex morphology associated with stent fracture or deformation.

Acute stent malapposition

When the same lesions were evaluated by both IVUS and OCT in two studies, stent strut malapposition was observed by OCT more than twice as frequently as with IVUS (39% vs. 14% [51] and 38.5% vs. 19.3% [52]). Similarly, in large IVUS and OCT cross-sectional studies, the prevalence of acute stent strut malapposition was 8%–15% by IVUS [53–55] and 36%–62% by OCT [56–58] with approximately half of the acute malappositions resolving at 9-month follow-up. Calcified plaque can cause acute stent strut malapposition [54,55] because rigid calcified plaque cannot be stretched by stenting. In a subset of 812 pre-PCI IVUS lesions from the Assessment of Dual AntiPlatelet Therapy With Drug Eluting Stents (ADAPT-DES) study, the maximum superficial calcium angle was larger (138 degrees [IQR 121–154] vs. 108 degrees [102–114], $P = .001$) and calcified nodules more frequent (12% vs. 4.8%, $P = .003$) in lesions with acute malapposition compared to those without malapposition [55]; however, in the absence of stent underexpansion, neither OCT- nor IVUS-detected acute malapposition has been associated with adverse early or long-term outcomes regardless of the amount of malapposition [53–58].

Right coronary ostial lesions

Angiographic severe calcification was found in 8%−24% of unselected patients [59−62], whereas the prevalence of severe calcification was higher in lesions at the ostial right coronary artery (30%−50%) [63,64]. IVUS confirmed that the maximum calcium angle of a target lesion was significantly greater in the aorto-ostial location compared to non-aorto-ostial lesions (157 ± 127 degrees vs. 114 ± 110 degrees, $P < .01$) [13] and negative vessel remodeling was very frequent (84%) [65]. As described earlier, a calcified nodule appears more frequently in the right coronary ostium. Tsunoda et al. [64] showed in serial IVUS observations of 11 ostial right coronary artery in-stent restenosis (ISR) cases (immediately after stent implantation and at the time of ISR) that MSA decreased significantly (recoil rate, median of 25% [range, 0−44]) and angiographic calcium of target lesion was associated with ISR.

Left main coronary artery lesion

LMCA lesion morphology is different between aorto-ostial location versus the distal bifurcation. Maximum calcium angle is larger in distal LMCA lesions compared to aorto-ostial lesions (195 ± 101 degrees vs. 78 ± 65 degrees, $P < .01$), with a larger plaque burden, more concentric plaque, and positive remodeling [66]. Similar to non-LMCA lesions, maximum calcium angle is associated with stent underexpansion.

Chronic total occlusion

Angiographic severe calcification in a chronic total occlusion (CTO) is a strong predictor of procedural failure regardless of the antegrade or retrograde approach [67]. To cross a severely calcified CTO segment using IVUS guidance, the location without severe calcium is chosen to make reentry from the subintimal space to the true lumen after wire tracking (i.e., intended medial dissection) [68]. In 67 relatively benign CTO lesions with IVUS evaluation after guidewire crossing (all were treated by the antegrade approach with 51% having a tapered type of occlusion), IVUS-detected calcium was found in 96% of CTO lesions, at the proximal end of the CTO in 78%, and at the distal end of CTO in 59% [69]. Intralesional maximum calcium was greater in the CTO with a blunt stump versus a tapered stump (163 ± 82 degrees vs. 127 ± 84 degrees, $P = .04$), and when each calcium deposit was analyzed every 1 mm, 68% of calcium deposits were <90 degrees.

Calcium modification

PCI outcome stratified by angiographic calcification

In DES studies, angiographic severe calcification has been reported in 8%−24% of unselected patients and has been associated with more comorbidities and multivessel disease [59–62]. Even after adjusting for baseline clinical and anatomical factors, long-term outcomes including cardiac death, MI, and target vessel revascularization were worse compared to patients without severe calcification.

Cutting balloon or scoring balloon

The cutting balloon has blades to make radially directed, longitudinal cuts to expand the vessel by limiting uncontrolled dissection [70]. The scoring balloon has nitinol wires on the surface of the balloon to focus forces to induce plaque disruption [71]. Tang et al. [70] randomized 92 patients with IVUS-defined severe calcium (maximum calcium angle $\geq$180 degrees) to conventional balloon versus cutting balloon before sirolimus-eluting stent implantation. After crossover of seven patients from conventional balloon to cutting balloon due to poor balloon dilatation, minimum stent area was larger in lesions treated by cutting balloon (6.3 ± 0.4 mm^2) compared with conventional balloon (5.0 ± 0.3 mm^2), $P = .03$. Sugawara et al. [71] showed OCT findings in 32 angiographic severely calcified lesions treated by a scoring balloon. Complete calcium fracture was observed in 68% of cross sections (117/172) and 81% (26/32) of lesions. The cutoff value of calcium thickness to predict complete calcium fracture was determined to be 0.57 mm (AUC $= 0.93$). In the Comparison of Strategies to PREPARE Severely CALCified Coronary Lesions (PREPARE-CALC) Trial, 278 severely calcified lesions in 200 patients were randomized to lesion preparation either by balloon lesion modification (3% cutting or 97% scoring balloon) versus rotational atherectomy (RA) following sirolimus-eluting stent (Orsiro) implantation for both groups [72]. Strategy success as the primary endpoint to test superiority (defined as successful stent delivery and expansion with <20% in-stent diameter stenosis) was significantly better after RA pretreatment compared to modified balloon predilatation (98% vs. 81%, $P = .0001$), and 9-month in-stent late lumen loss as the coprimary endpoint to test noninferiority was met (P for noninferiority $= .02$).

Excimer laser coronary angioplasty

Excimer laser coronary angioplasty (ELCA) rapidly converts water into exploding vapor bubbles that then modify the plaque, especially when lasing in a contrast-filled lumen [73,74]. In 48 patients with sequential 30 MHz IVUS evaluation (pre, post-ELCA, after adjunctive therapy), Mintz et al. [73] showed an increase of the lumen area from 1.4 ± 0.5 mm^2 (pre) to 2.7 ± 0.8 mm^2 (post-ELCA), $P < .01$, without a change of superficial calcium angle and without calcium removal; however, 16 lesions showed fragmented appearance of the superficial calcium, presumably calcium fracture. Twenty-three lesions underwent adjunctive balloon angioplasty; there was further lumen increase by stretching the vessel area (2.7 ± 0.9 mm^2 to 4.4 ± 1.3 mm^2, $P < .001$). Later, Ambrosini et al. [74] treated 100 lesions in 80 patients (57 angiographic calcified lesions, 32 balloon resistant lesions, and 11 CTOs) with standard laser therapy (60 fluence and 40 Hz) in 41 patients and with increased laser therapy (60 fluence and 80 Hz or 80 fluence and 80 Hz) in 39 patients. Procedural success was achieved in 94% of cases with acceptable stent expansion (mean diameter stenosis = $13 \pm 19\%$).

Atherectomy

RA and orbital atherectomy (OA) are designed to ablate calcified plaque. The concept of "differential cutting" is the ability to selectively ablate calcified plaque while maintaining the integrity of normal vessel or soft, noncalcified plaque [75]. The prospective, randomized ROTAXUS (Rotational Atherectomy Prior to Taxus Stent Treatment for Complex Native Coronary Artery Disease) trial was performed to determine whether lesion preparation with RA before paclitaxel-eluting stent implantation provided benefits compared with paclitaxel-eluting stent with balloon predilatation alone in angiographic moderate-to-severe calcified lesions [75]. Among 240 patients, strategy success was higher with RA pretreatment (92.5% vs. 83.3%, $P = .03$); however, despite greater acute lumen gain with RA, 9-month angiographic follow-up revealed higher late lumen loss in the RA group. Rates of restenosis, target lesion revascularization, definite stent thrombosis, and major adverse cardiac events were not significantly different between the groups at 1 year. As described earlier, PREPARE-CALC reported bail-out usage of RA after calcium modification using provisional cutting or scoring balloon [72].

OA is a newer percutaneous, endovascular system that incorporates the use of centrifugal force and differential sanding to modify calcified lesions [76]. In ORBIT II (Evaluate the Safety and Efficacy of OAS in Treating Severely Calcified Coronary Lesions), treatment of de novo severely calcified lesions resulted in a low rate of procedural and 1-year target vessel revascularization (5.9%), cardiac death (3.0%), and periprocedural MI (2% by Society for Cardiovascular Angiography and Interventions definition) [76].

In 1992, Mintz et al. [77] showed that after RA the surface of the intima was sharp, clear, and circumferential, which indicated a polished calcium surface by RA, that this appearance was different compared to the surface of nontreated calcium, and that calcium fracture was observed in 29% of atherectomy segments without severe soft plaque dissection. Recently, we reported IVUS findings after RA using 40 MHz IVUS in 38 severely calcified lesions [78]. By comparing pre- and post-RA IVUS, the angle of calcium reverberation increased significantly from 45 degrees [IQR, 31–67] to 96 degrees [50–148], $P = .003$, although calcium angle did not change. Calcium reverberation is a marker of a polished calcium surface after RA; there was a significant positive correlation between the amount of calcium and reverberation, suggesting RA modification occurred more effectively and frequently in more severe calcified lesions. Calcium fracture was observed in 58% (22/38) of lesions mostly after stenting following high-pressure ballooning (Fig. 5.9).

When the effect of ablation by RA versus OA was compared using OCT in severely calcified lesions (maximum calcium angle >270 degrees), calcified and

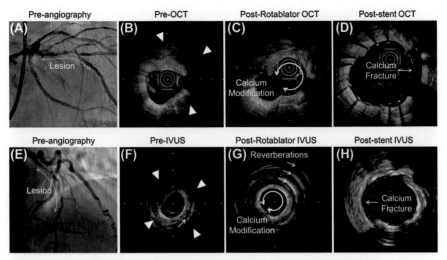

FIGURE 5.9 The Effect of Rotational Atherectomy Evaluated by OCT or IVUS.

The locations of (B–D) were matched and corresponded to the lesion in (A) (white arrow); and the locations of (F–H) were matched and corresponded to the lesion in (E) (white arrow). A large, thick calcium was observed (white triangles in B) in the pre-OCT where the round, polished, concave-shaped surface appeared (white double-headed arrow), indicating calcium modification by rotational atherectomy (RA). Poststent OCT (D) showed calcium fracture (yellow arrow) with good stent expansion. Pre-IVUS showed 360 degrees of calcium (white triangles in F) with a small arc of reverberations (white arrow). Post-RA IVUS (G) showed almost 360 degrees of reverberations (white arrows) indicating calcium modification by RA (white double-headed arrow). Poststent IVUS (H) showed calcium fracture (a separation of calcium resulting visualization of perivascular tissue at seven o'clock to 11 o'clock) with good stent expansion.

noncalcified plaque modification was always colocalized at the site of the guidewire [79]. Calcium modification had a similar appearance (round, concave, and polished lumen surface), and the ablated calcium area (comparing pre- to post-RA or pre- to post-OA was not different between RA and OA and was similarly small [OA 0.56 mm^2 (IQR, 0.38–0.76) versus RA 0.60 mm^2 (0.45–0.79), $P = .49$]); however, OA ablation was more diffusive than RA ablation at sites with intermediate lumen area (>4 mm^2) or intermediate size of calcium (calcium angle <180 degrees), but differential ablation seemed to be slightly better with RA than OA (Fig. 5.10).

After stenting with high-pressure ballooning following atherectomy, calcium fracture was observed within calcium with or without atherectomy modification [80,81]. In 58 severely calcified lesions treated by OA, calcium thickness at the fracture site was greater with versus without OA calcium modification (0.58 mm [IQR, 0.50–0.66] vs. 0.45 mm [0.38–0.52]) [82]. Severe calcified lesions treated by cutting balloon after RA ($n = 18$) were compared with conventional balloon after RA

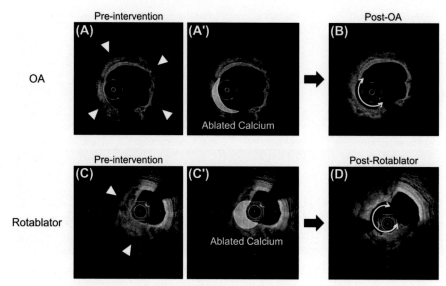

FIGURE 5.10 Similar calcium modification by OA or RA evaluated by OCT.

(A′) is the same image as (A) with annotation, and (C′) is the same image as (C) with annotation. Large, thick calcium (white triangles) is observed (A and C). Calcium modification (white double-headed arrow) is a round, concave, polished surface and was similar when using OA (B) or RA (D), which are always colocated with the OCT catheter. By comparing (A) and (B), the ablated calcium is estimated as the yellow area (A′), similarly in C′.

($n = 23$) [83]. The number of calcium fractures per lesion was greater in the cutting balloon group compared to conventional balloon (4 [IQR, 3−8] vs. 1 [0−2]), resulting in better stent expansion.

Calcium modification by RA or OA with or without adjunct cutting balloon facilitates calcium fracture even in regions of thick calcium. Greater calcium modification correlates with greater calcium fracture, in turn facilitating optimal stent expansion.

Lithotripsy

Lithotripsy, an established treatment for kidney stones, has evolved into a treatment for coronary calcium; multiple emitters mounted on a balloon catheter create diffusive circumferential pulsatile mechanical energy to disrupt calcified plaque [84]. Ali et al. [84] has reported OCT findings in 31 de novo severely calcified lesions treated by lithotripsy before stenting (Fig. 5.11).

After lithotripsy, calcium fracture was identified in 43% of lesions, a prevalence that increased to 55% after stenting and high-pressure balloon inflations, with >25% having multiple fractures in the same cross section. This could be promising even for treatment of stent underexpansion due to severe calcification behind the stent.

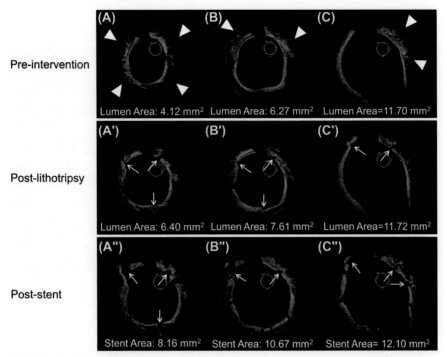

FIGURE 5.11 Effect of lithotripsy evaluated by OCT.

The location of (A), (A′), and (A″) is matched, as were (B) and (C). Large, thick calcium is seen (white triangles in A—C). Postlithotripsy OCT images (A′—C′) showed multiple incomplete calcium fractures including at the bifurcation (C′). Poststent OCT (A″—C″) images show complete calcium fracture accompanied by good stent expansion.

Reprinted from Ali ZA, Brinton TJ, Hill JM, Maehara A, Matsumura M, Karimi Galougahi K, Illindala U, Götberg M, Whitbourn R, Van Mieghem N, Meredith IT, Di Mario C, Fajadet J. Optical coherence tomography characterization of coronary lithoplasty for treatment of calcified lesions: first description. JACC Cardiovasc Imaging 2017;10:897—906, Copyright (2017), with permission from Elsevier Inc.

In-stent restenosis and its treatment
In-stent neoatherosclerosis and neointimal calcium

OCT-defined neoatherosclerosis has included lipidic neointima, similar to lipidic plaque in de novo atherosclerosis, neointimal rupture, and/or neointimal calcification [85—87] (Fig. 5.12).

Kang et al. [85] reported neoatherosclerosis in 50 ISR patients (mostly first-generation DES) evaluated by OCT and IVUS (median time from stent implantation = 2.7 years). Neointimal calcium was found in 1 of 24 (4%) patients within 20 months (median of this cohort) and 4 of 26 (15%) after 20 months. We evaluated 171 second-generation DES ISRs by OCT and found that neointimal calcium

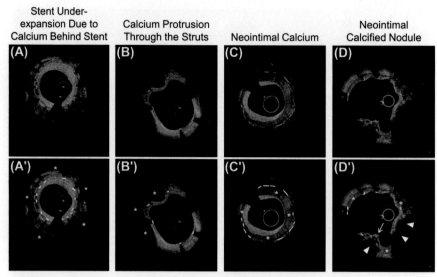

FIGURE 5.12 **Calcium in In-stent Restenosis Lesions.**

(A′−D′) are same images as (A−D) with annotation. Stent struts are shown as white dotted lines and calcium as white stars. In (A), there is large, thick calcium behind the stent causing stent underexpansion. In (B), there is protruding calcium through the struts. In (C), stent struts are located in a thick and large calcium deposit. In (D), there is an accumulation of small calcium fragments (white arrow) accompanying strong signal attenuation (white triangles) with adjacent calcium surrounding the stent struts.

was present in 2 of 57 (3.5%) within 1 year from stent implantation and 23 of 114 (20.2%) beyond 1 year (3.4 ± 1.9 years) [86]. When neoatherosclerosis was compared among bare metal stents, first-generation DES, and second-generation DES, the prevalence of neointimal calcium was similar [87], and the maximum angle of neointimal calcium was positively correlated with the maximum angle of calcium behind the stent, suggesting pan-coronary progression of calcium, as well as the duration from stent implantation. Although less sensitive, IVUS can diagnose neoatherosclerosis by detecting attenuated or calcified plaque within the stent. Attenuated plaque is defined as echo-signal attenuation without superficial bright echoes, thus indicating a fibroatheroma [18] (Fig. 5.13).

Treatment of ISR

Treatment of ISR is especially difficult in the setting of a severely calcified lesion that limited stent expansion at the time of the original PCI. We evaluated predictors of new stent expansion in the treatment of 143 ISR lesions (12% bare metal ISR, 31% first-generation DES ISR, 57% second-generation DES ISR) occurring 5.8 ± 4.8 years after the original stent implantation PCI [88]. Calcified neoatherosclerosis and native calcium behind stent was observed in 24% (34/143) and 23%

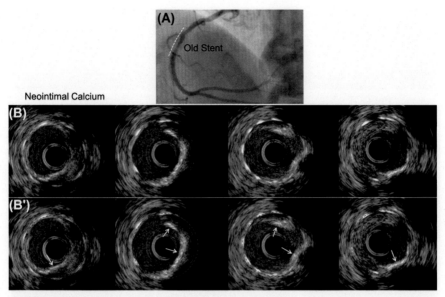

FIGURE 5.13 Nonatherosclerotic calcium evaluated by IVUS.

IVUS images (B and B′) correspond to the old stent segment in the angiogram (A). (B′) is the same as (B) but with annotation. The red dotted line shows old stent struts, and the white arrows indicate neoatherosclerotic calcium within the stent.

(32/143) of cases, respectively. Morphological parameters associated with new stent underexpansion included maximum calcium angle >180 degrees (odds ratio [OR] 6.7) and maximum calcium thickness >0.5 mm (OR 5.8) regardless of calcium location (within the stent or behind stent) in addition to old stent underexpansion (OR 6.3) and double layers of old stents (OR 8.8). It was notable that the effect of and criteria for severe calcium causing poor expansion when restenting an ISR lesion was remarkably similar to that in de novo lesions.

ELCA for ISR

Latib et al. [89] showed the efficacy of ELCA for the treatment of 28 ISR due to stent underexpansion using high energy during contrast injection but only within the underexpanded stent. IVUS showed an increase in MSA from 3.5 ± 1.1 mm^2 to 7.1 ± 1.9 mm^2 without major procedural complications with acceptable long-term outcomes (1 cardiac death and 2 target lesion revascularizations). We used OCT to evaluate the effect of ELCA to treat ISR due to peri-stent calcium (>90 degrees)-related stent underexpansion defined as old stent MSA divided by the average of the proximal and distal reference lumen area <0.8 and an MSA <5 mm^2 [90]. Twenty-three lesions treated with ELCA showed more peri-stent calcium fracture (61% vs. 12%) and greater minimum lumen area (4.8 mm^2 [3.3−5.6] vs. 3.5 mm^2 [2.8−4.1]) compared to 58 ISR lesions treated without ELCA (Fig. 5.14).

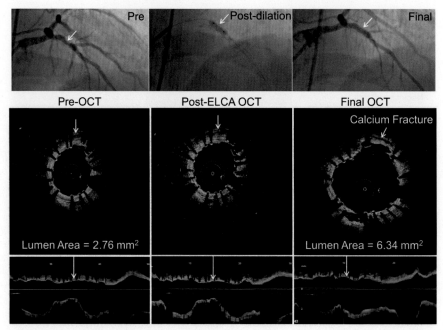

FIGURE 5.14 Excimer Laser Coronary Angioplasty (ELCA) for In-stent Restenosis due to Stent Underexpansion by Thick Calcium Behind Stent.

Pre-OCT, post-ELCA, and final OCT are matched. Pre-OCT shows an underexpanded old stent (minimum stent area = 2.76 mm^2) due to severe calcium behind the stent. Post-ELCA OCT shows no difference compared to the pre-OCT image. After high-pressure ballooning, final-OCT shows calcium fracture behind the stent resulting in good stent expansion (minimum stent area = 6.34 mm^2).

Reprinted from J Am Coll Cardiol Interv 8: Yin D et al. e137, Copyright (2015), with permission from Elsevier Inc.

In 171 second-generation DES ISR lesions [86], stent underexpansion (MSA <4 mm^2) was frequent (65% < 1 year and 53% > 1 year from stent implantation); if this was due to the stent underexpansion because of calcium behind the stent, it was very hard to achieve good expansion without modification of the calcium causing stent underexpansion.

RA for ISR

Hachinohe et al. [91] reported 200 ISR patients due to excessive hard neointimal tissue (neointimal calcium or dense fibrous tissue) in 57%, stent underexpansion in 33%, and the ISR lesions where no device could cross in 10%, and treated by IVUS- or OCT-guided RA to debulk old stent and/or hard tissue within the stent. Angiographic diffuse ISR was observed in 86% of lesions. Per operator discretion, 90 patients were treated by only balloon angioplasty resulting in a 5.4 ± 2.1 mm^2 MSA and a 49% rate of target vessel revascularization (TVR) at 1 year; 55 patients underwent new stent implantation resulting in a 7.2 ± 2.4 mm^2 MSA and a 46% rate of TVR at 1 year; and 55 patients were treated by drug-coated balloon, resulting in a 5.3 ± 2.5 mm^2 MSA and a 36% rate of TVR at 1 year. During the procedure, there were 2 slow-flow and one perforation with successful treatment and no RA burr entrapment.

References

[1] Mintz GS, Nissen SE, Anderson WD, Bailey SR, Erbel R, Fitzgerald PJ, Pinto FJ, Rosenfield K, Siegel RJ, Tuzcu EM, Yock PG. American college of cardiology clinical expert consensus document on standards for acquisition, measurement and reporting of intravascular ultrasound studies (IVUS). A report of the American college of cardiology task force on clinical expert consensus documents. J Am Coll Cardiol 2001;37: 1478−92.

[2] Kostamaa H, Donovan J, Kasaoka S, Tobis J, Fitzpatrick L. Calcified plaque cross-sectional area in human arteries: correlation between intravascular ultrasound and unde-calcified histology. Am Heart J 1999;137:482−8.

[3] Friedrich GJ, Moes NY, Mühlberger VA, Gabl C, Mikuz G, Hausmann D, Fitzgerald PJ, Yock PG. Detection of intralesional calcium by intracoronary ultrasound depends on the histologic pattern. Am Heart J 1994;128:435−41.

[4] Scott DS, Arora UK, Farb A, Virmani R, Weissman NJ. Pathologic validation of a new method to quantify coronary calcific deposits in vivo using intravascular ultrasound. Am J Cardiol 2000;85:37−40.

[5] Tearney GJ, Regar E, Akasaka T, Adriaenssens T, Barlis P, Bezerra HG, Bouma B, Bruining N, Cho JM, Chowdhary S, Costa MA, de Silva R, Dijkstra J, Di Mario C, Dudek D, Falk E, Feldman MD, Fitzgerald P, Garcia-Garcia HM, Gonzalo N, Granada JF, Guagliumi G, Holm NR, Honda Y, Ikeno F, Kawasaki M, Kochman J, Koltowski L, Kubo T, Kume T, Kyono H, Lam CC, Lamouche G, Lee DP, Leon MB, Maehara A, Manfrini O, Mintz GS, Mizuno K, Morel MA, Nadkarni S, Okura H, Otake H, Pietrasik A, Prati F, Räber L, Radu MD, Rieber J, Riga M, Rollins A, Rosenberg M, Sirbu V, Serruys PW, Shimada K, Shinke T, Shite J, Siegel E, Sonoda S, Suter M, Takarada S, Tanaka A, Terashima M, Thim T, Uemura S, Ughi GJ, van Beusekom HM, van der Steen AF, van Es GA, van Soest G, Virmani R, Waxman S, Weissman NJ, Weisz G. International working group for intravascular optical coherence tomography (IWG-IVOCT). Consensus standards for acquisition, measurement, and reporting of intravascular optical coherence tomography studies: a report from the international working group for intravascular optical coherence tomography standardization and validation. J Am Coll Cardiol 2012;59: 1058−72.

[6] Kume T, Okura H, Kawamoto T, Yamada R, Miyamoto Y, Hayashida A, Watanabe N, Neishi Y, Sadahira Y, Akasaka T, Yoshida K. Assessment of the coronary calcification by optical coherence tomography. EuroIntervention 2011;6:768−72.

[7] Mehanna E, Bezerra HG, Prabhu D, Brandt E, Chamié D, Yamamoto H, Attizzani GF, Tahara S, Van Ditzhuijzen N, Fujino Y, Kanaya T, Stefano G, Wang W, Gargesha M, Wilson D, Costa MA. Volumetric characterization of human coronary calcification by frequency-domain optical coherence tomography. Circ J 2013;77:2334−40.

[8] Saita T, Fujii K, Hao H, Imanaka T, Shibuya M, Fukunaga M, Miki K, Tamaru H, Horimatsu T, Nishimura M, Sumiyoshi A, Kawakami R, Naito Y, Kajimoto N, Hirota S, Masuyama T. Histopathological validation of optical frequency domain imaging to quantify various types of coronary calcifications. Eur Heart J Cardiovasc Imaging 2017;18:342−9.

[9] Mintz GS, Popma JJ, Pichard AD, Kent KM, Satler LF, Chuang YC, Ditrano CJ, Leon MB. Patterns of calcification in coronary artery disease. A statistical analysis of intravascular ultrasound and coronary angiography in 1155 lesions. Circulation 1995; 91:1959−65.

[10] Pompa JAA, Burke D. Qualitative and quantitative coronary angiography. In: Topol EJ, Teirstein PS, editors. Textbook of interventional cardiology. 6th ed. Philadelphia, PA: Saunders; 2011. p. 757−75.

[11] Wang X, Matsumura M, Mintz GS, Lee T, Zhang W, Cao Y, Fujino A, Lin Y, Usui E, Kanaji Y, Murai T, Yonetsu T, Kakuta T, Maehara A. In vivo calcium detection by comparing optical coherence tomography, intravascular ultrasound, and angiography. JACC Cardiovasc Imaging 2017;10:869−79.

[12] Sangiorgi G, Rumberger JA, Severson A, Edwards WD, Gregoire J, Fitzpatrick LA, Schwartz RS. Arterial calcification and not lumen stenosis is highly correlated with atherosclerotic plaque burden in humans: a histologic study of 723 coronary artery segments using nondecalcifying methodology. J Am Coll Cardiol 1998;31. 1261-1233.

[13] Mintz GS, Pichard AD, Popma JJ, Kent KM, Satler LF, Bucher TA, Leon MB. Determinants and correlates of target lesion calcium in coronary artery disease: a clinical, angiographic and intravascular ultrasound study. J Am Coll Cardiol 1997;29:268−74.

[14] Tinana A, Mintz GS, Weissman NJ. Volumetric intravascular ultrasound quantification of the amount of atherosclerosis and calcium in nonstenotic arterial segments. Am J Cardiol 2002;89:757−60.

[15] Cavalcante R, Bittencourt MS, Pinheiro TL, Falcao BA, Morais GR, Soares P, Mariani Jr J, Ribeiro E, Kalil-Filho R, Rochitte CE, Lemos PA. Validation of coronary computed tomography angiography scores for non-invasive assessment of atherosclerotic burden through a comparison with multivessel intravascular ultrasound. Atherosclerosis 2016;247:21−7.

[16] Ehara S, Kobayashi Y, Yoshiyama M, Shimada K, Shimada Y, Fukuda D, Nakamura Y, Yamashita H, Yamagishi H, Takeuchi K, Naruko T, Haze K, Becker AE, Yoshikawa J, Ueda M. Spotty calcification typifies the culprit plaque in patients with acute myocardial infarction: an intravascular ultrasound study. Circulation 2004;110:3424−9.

[17] Ong DS, Lee JS, Soeda T, Higuma T, Minami Y, Wang Z, Lee H, Yokoyama H, Yokota T, Okumura K, Jang IK. Coronary calcification and plaque vulnerability: an optical coherence tomographic study. Circ Cardiovasc Imaging 2016. https://doi.org/10.1161/CIRCIMAGING.115.003929.

[18] Pu J, Mintz GS, Biro S, Lee JB, Sum ST, Madden SP, Burke AP, Zhang P, He B, Goldstein JA, Stone GW, Muller JE, Virmani R, Maehara A. Insights into echo-attenuated plaques, echolucent plaques, and plaques with spotty calcification: novel findings from comparisons among intravascular ultrasound, near-infrared spectroscopy, and pathological histology in 2,294 human coronary artery segments. J Am Coll Cardiol 2014;63:2220−33.

[19] Dong L, Mintz GS, Witzenbichler B, Metzger DC, Rinaldi MJ, Duffy PL, Weisz G, Stuckey TD, Brodie BR, Yun KH, Xu K, Kirtane AJ, Stone GW, Maehara A. Comparison of plaque characteristics in narrowings with ST-elevation myocardial infarction

(STEMI), non-STEMI/unstable angina pectoris and stable coronary artery disease (from the ADAPT-DES IVUS Substudy). Am J Cardiol 2015;115:860−6.

[20] Shan P, Mintz GS, Witzenbichler B, Metzger DC, Rinaldi MJ, Duffy PL, Weisz G, Stuckey TD, Brodie BR, Généreux P, Crowley A, Kirtane AJ, Stone GW, Maehara A. Does calcium burden impact culprit lesion morphology and clinical results? An ADAPT-DES IVUS substudy. Int J Cardiol 2017;248:97−102.

[21] Krishnamoorthy P, Vengrenyuk Y, Ueda H, Yoshimura T, Pena J, Motoyama S, Baber U, Hasan C, Kesanakurthy S, Sweeny JM, Sharma SK, Narula J, Kovacic JC, Kini AS. Three-dimensional volumetric assessment of coronary artery calcification in patients with stable coronary artery disease by OCT. EuroIntervention 2017;13:312−9.

[22] Duissaillant GR, Mintz GS, Pichard AD, Kent KM, Satler LF, Popma JJ, Griffin J, Leon MB. Intravascular ultrasound identification of calcified intraluminal lesions misdiagnosed as thrombi by coronary angiography. Am Heart J 1996;132:687−9.

[23] Lee JB, Mintz GS, Lisauskas JB, Biro SG, Pu J, Sum ST, Madden SP, Burke AP, Goldstein J, Stone GW, Virmani R, Muller JE, Maehara A. Histopathologic validation of the intravascular ultrasound diagnosis of calcified coronary artery nodules. Am J Cardiol 2011;108:1547−51.

[24] Xu Y, Mintz GS, Tam A, McPherson JA, Iñiguez A, Fajadet J, Fahy M, Weisz G, De Bruyne B, Serruys PW, Stone GW, Maehara A. Prevalence, distribution, predictors, and outcomes of patients with calcified nodules in native coronary arteries: a 3-vessel intravascular ultrasound analysis from Providing Regional Observations to Study Predictors of Events in the Coronary Tree (PROSPECT). Circulation 2012;126:537−45.

[25] Lee T, Mintz GS, Matsumura M, Zhang W, Cao Y, Usui E, Kanaji Y, Murai T, Yonetsu T, Kakuta T, Maehara A. Prevalence, predictors, and clinical presentation of a calcified nodule as assessed by optical coherence tomography. JACC Cardiovasc Imaging 2017;10:883−91.

[26] Virmani R, Burke AP, Farb A, Kolodgie FD. Pathology of the vulnerable plaque. J Am Coll Cardiol 2006;47:C13−8.

[27] Gruberg L, Rai P, Mintz GS, Canos D, Pinnow E, Satler LF, Pichard AD, Kent KM, Waksman R, Lindsay J, Weissman NJ. Impact of renal function on coronary plaque morphology and morphometry in patients with chronic renal insufficiency as determined by intravascular ultrasound volumetric analysis. Am J Cardiol 2005;96:892−6.

[28] Chin CY, Mintz GS, Saito S, Witzenbichler B, Metzger DC, Rinaldi MJ, Mazzaferri Jr EL, Duffy PL, Weisz G, Stuckey TD, Brodie BR, Litherland C, Kirtane AJ, Stone GW, Maehara A. Relation between renal function and coronary plaque morphology (from the Assessment of Dual Antiplatelet Therapy with Drug-Eluting Stents virtual histology-intravascular ultrasound substudy). Am J Cardiol 2017;119: 217−24.

[29] Lanzer P, Boehm M, Sorribas V, Thiriet M, Janzen J, Zeller T, St Hilaire C, Shanahan C. Medial vascular calcification revisited: review and perspectives. Eur Heart J 2014;35: 1515−25.

[30] Chin CY, Matsumura M, Maehara A, Zhang W, Lee CT, Yamamoto MH, Song L, Parviz Y, Jhalani NB, Mohan S, Ratner LE, Cohen DJ, Ben-Yehuda O, Stone GW, Shlofmitz RA, Kakuta T, Mintz GS, Ali ZA. Coronary plaque characteristics in hemodialysis-dependent patients as assessed by optical coherence tomography. Am J Cardiol 2017;119:1313−9.

[31] Guthaner DF, Robert EW, Alderman EL, Wexler L. Long-term serial angiographic studies after coronary artery bypass surgery. Circulation 1979;60:250−9.

[32] Bourassa MG, Campeau L, Lespérance J, Grondin CM. Changes in grafts and coronary arteries after saphenous vein aortocoronary bypass surgery: results at repeat angiography. Circulation 1982;65:90−7.

[33] Shang Y, Mintz GS, Pu J, Guo J, Kobayashi N, Franklin-Bond T, Leon MB, Moses JW, Maehara A, Shimizu T, Yakushiji T. Bypass to the left coronary artery system may accelerate left main coronary artery negative remodeling and calcification. Clin Res Cardiol 2013;102:831−5.

[34] Wolny R, Mintz GS, Matsumura M, Ishida M, Fujino A, Lee T, Shlofmitz E, Goldberg A, Jeremias A, Haag E, Shlofmitz RA, Maehara A. Unique calcification patterns in the left main bifurcation in patients after coronary artery bypass grafting. J Am Coll Cardiol 2018;72(Suppl. 1). https://doi.org/10.1016/j.jacc.2018.08.2079.

[35] Castagna MT, Mintz GS, Ohlmann P, Kotani J, Maehara A, Gevorkian N, Cheneau E, Stabile E, Ajani AE, Suddath WO, Kent KM, Satler LF, Pichard AD, Weissman NJ. Incidence, location, magnitude, and clinical correlates of saphenous vein graft calcification: an intravascular ultrasound and angiographic study. Circulation 2005;111:1148−52.

[36] Roleder T, Pociask E, Wańha W, Dobrolińska M, Gąsior P, Smolka G, Walkowicz W, Jadczyk T, Bochenek T, Dudek D, Ochała A, Mizia-Stec K, Gąsior Z, Tendera M, Ali ZA, Wojakowski W. Optical coherence tomography of de novo lesions and in-stent restenosis in coronary saphenous vein grafts (OCTOPUS Study). Circ J 2016;80:1804−11.

[37] Sano K, Mintz GS, Carlier SG, Fujii K, Yasuda T, Kimura M, Costa Jr JR, Costa RA, Lui J, Weisz G, Moussa I, Dangas GD, Mehran R, Lansky AJ, Kreps EM, Collins M, Stone GW, Moses JW, Leon MB. Intravascular ultrasonic differences between aorto-ostial and shaft narrowing in saphenous veins used as aortocoronary bypass grafts. Am J Cardiol 2006;97:1463−6.

[38] Vavuranakis M, Toutouzas K, Stefanadis C, Chrisohou C, Markou D, Toutouzas P. Stent deployment in calcified lesions: can we overcome calcific restraint with high-pressure balloon inflations? Cathet Cardiovasc Interv 2001;52:164−72.

[39] Kobayashi Y, Okura H, Kume T, Yamada R, Kobayashi Y, Fukuhara K, Koyama T, Nezuo S, Neishi Y, Hayashida A, Kawamoto T, Yoshida K. Impact of target lesion coronary calcification on stent expansion. Circ J 2014;78:2209−14.

[40] Fujino A, Mintz GS, Matsumura M, Lee T, Kim SY, Hoshino M, Usui E, Yonetsu T, Haag ES, Shlofmitz RA, Kakuta T, Maehara A. A new optical coherence tomography-based calcium scoring system to predict stent underexpansion. EuroIntervention 2018;13:e2182−9.

[41] Fujino A, Mintz GS, Lee T, Hoshino M, Usui E, Kanaji Y, Murai T, Yonetsu T, Matsumura M, Ali ZA, Jeremias A, Moses JW, Shlofmitz RA, Kakuta T, Maehara A. Predictors of calcium fracture derived from balloon angioplasty and its effect on stent expansion assessed by optical coherence tomography. JACC Cardiovasc Interv 2018;11:1015−7.

[42] Fitzgerald PJ, Ports TA, Yock PG. Contribution of localized calcium deposits to dissection after angioplasty. An observational study using intravascular ultrasound. Circulation 1992;86:64−70.

[43] Kobayashi N, Mintz GS, Witzenbichler B, Metzger DC, Rinaldi MJ, Duffy PL, Weisz G, Stuckey TD, Brodie BR, Parvataneni R, Kirtane AJ, Stone GW, Maehara A. Prevalence, features, and prognostic importance of edge dissection after drug-eluting stent

implantation: an ADAPT-DES intravascular ultrasound substudy. Circ Cardiovasc Interv 2016. https://doi.org/10.1161/CIRCINTERVENTIONS.115.003553.

[44] Chamié D, Bezerra HG, Attizzani GF, Yamamoto H, Kanaya T, Stefano GT, Fujino Y, Mehanna E, Wang W, Abdul-Aziz A, Dias M, Simon DI, Costa MA. Incidence, predictors, morphological characteristics, and clinical outcomes of stent edge dissections detected by optical coherence tomography. JACC Cardiovasc Interv 2013;6:800—13.

[45] Grewal J, Ganz P, Selwyn A, Kinlay S. Usefulness of intravascular ultrasound in preventing stenting of hazy areas adjacent to coronary stents and its support of support spot-stenting. Am J Cardiol 2001;87:1246—9.

[46] Aoki J, Nakazawa G, Tanabe K, Hoye A, Yamamoto H, Nakayama T, Onuma Y, Higashikuni Y, Otsuki S, Yagishita A, Yachi S, Nakajima H, Hara K. Incidence and clinical impact of coronary stent fracture after sirolimus-eluting stent implantation. Cathet Cardiovasc Interv 2007;69:380—6.

[47] Kuramitsu S, Iwabuchi M, Haraguchi T, Domei T, Nagae A, Hyodo M, Yamaji K, Soga Y, Arita T, Shirai S, Kondo K, Ando K, Sakai K, Goya M, Takabatake Y, Sonoda S, Yokoi H, Toyota F, Nosaka H, Nobuyoshi M. Incidence and clinical impact of stent fracture after everolimus-eluting stent implantation. Circ Cardiovasc Interv 2012;5:663—71.

[48] Kuramitsu S, Hiromasa T, Enomoto S, Shinozaki T, Iwabuchi M, Mazaki T, Domei T, Yamaji K, Soga Y, Hyodo M, Shirai S, Ando K. Incidence and clinical impact of stent fracture after PROMUS element platinum chromium everolimus-eluting stent implantation. JACC Cardiovasc Interv 2015;8:1180—8.

[49] Doi H, Maehara A, Mintz GS, Tsujita K, Kubo T, Castellanos C, Lansky AJ, Witzenbichler B, Guagliumi G, Brodie B, Kellett Jr MA, Parise H, Mehran R, Leon MB, Moses JW, Stone GW. Intravascular ultrasound findings of stent fractures in patients with Sirolimus- and Paclitaxel-eluting stents. Am J Cardiol 2010;106: 952—7.

[50] Inaba S, Weisz G, Kobayashi N, Saito S, Dohi T, Dong L, Wang L, Moran JA, Rabbani LE, Parikh MA, Leon MB, Moses JW, Mintz GS, Maehara A. Prevalence and anatomical features of acute longitudinal stent deformation: an intravascular ultrasound study. Cathet Cardiovasc Interv 2014;84:388—96.

[51] Kubo T, Akasaka T, Shite J, Suzuki T, Uemura S, Yu B, Kozuma K, Kitabata H, Shinke T, Habara M, Saito Y, Hou J, Suzuki N, Zhang S. Optical coherence tomography compared to intravascular ultrasound in coronary lesion assessment study: OPUS-CLASS study. J Am Coll Cardiol Img 2013;6:1095—104.

[52] Ali ZA, Maehara A, Généreux P, Shlofmitz RA, Fabbiocchi F, Nazif TM, Guagliumi G, Meraj PM, Alfonso F, Samady H, Akasaka T, Carlson EB, Leesar MA, Matsumura M, Ozan MO, Mintz GS, Ben-Yehuda O, Stone GW. Optical coherence tomography compared with intravascular ultrasound and with angiography to guide coronary stent implantation (ILUMIEN III: OPTIMIZE PCI): a randomised controlled trial. Lancet 2016;388:2618—28.

[53] Guo N, Maehara A, Mintz GS, He Y, Xu K, Wu X, Lansky AJ, Witzenbichler B, Guagliumi G, Brodie B, Kellett Jr MA, Dressler O, Parise H, Mehran R, Stone GW. Incidence, mechanisms, predictors, and clinical impact of acute and late stent malapposition after primary intervention in patients with acute myocardial infarction: an intravascular ultrasound substudy of the Harmonizing Outcomes with Revascularization and Stents in Acute Myocardial Infarction (HORIZONS-AMI) trial. Circulation 2010;122: 1077—84.

[54] Steinberg DH, Mintz GS, Mandinov L, Yu A, Ellis SG, Grube E, Dawkins KD, Ormiston J, Turco MA, Stone GW, Weissman NJ. Long-term impact of routinely detected early and late incomplete stent apposition: an integrated intravascular ultrasound analysis of the TAXUS IV, V, and VI and TAXUS ATLAS Workhorse, long lesion, and direct stent studies. JACC Cardiovasc Interv 2010;3:486—94.

[55] Wang B, Mintz GS, Witzenbichler B, Souza CF, Metzger DC, Rinaldi MJ, Duffy PL, Weisz G, Stuckey TD, Brodie BR, Matsumura M, Yamamoto MH, Parvataneni R, Kirtane AJ, Stone GW, Maehara A. Predictors and long-term clinical impact of acute stent malapposition: an Assessment of Dual Antiplatelet Therapy with Drug-Eluting Stents (ADAPT-DES) Intravascular ultrasound substudy. J Am Heart Assoc 2016;5. https://doi.org/10.1161/JAHA.116.004438.

[56] Soeda T, Uemura S, Park SJ, Jang Y, Lee S, Cho JM, Kim SJ, Vergallo R, Minami Y, Ong DS, Gao L, Lee H, Zhang S, Yu B, Saito Y, Jang IK. Incidence and clinical significance of poststent optical coherence tomography findings: one-year follow-up study from a multicenter registry. Circulation 2015;132:1020—9.

[57] Prati F, Romagnoli E, Burzotta F, Limbruno U, Gatto L, La Manna A, Versaci F, Marco V, Di Vito L, Imola F, Paoletti G, Trani C, Tamburino C, Tavazzi L, Mintz GS. Clinical impact of OCT findings during PCI: the CLI-OPCI II study. J Am Coll Cardiol Img 2015;8:1297—305.

[58] Im E, Kim BK, Ko YG, Shin DH, Kim JS, Choi D, Jang Y, Hong MK. Incidence, predictors, and clinical outcomes of acute and late stent malapposition detected by optical coherence tomography after drug-eluting stent implantation. Circ Cardiovasc Interv 2014;7:88—96.

[59] Bourantas CV, Zhang YJ, Garg S, Iqbal J, Valgimigli M, Windecker S, Mohr FW, Silber S, Vries T, Onuma Y, Garcia-Garcia HM, Morel MA, Serruys PW. Prognostic implications of coronary calcification in patients with obstructive coronary artery disease treated by percutaneous coronary intervention: a patient-level pooled analysis of 7 contemporary stent trials. Heart 2014;100:1158—64.

[60] Huisman J, van der Heijden LC, Kok MM, Danse PW, Jessurun GA, Stoel MG, van Houwelingen KG, Löwik MM, Hautvast RW, IJzerman MJ, Doggen CJ, von Birgelen C. Impact of severe lesion calcification on clinical outcome of patients with stable angina, treated with newer generation permanent polymer-coated drug-eluting stents: a patient-level pooled analysis from TWENTE and Dutch PEERS (TWENTE II). Am Heart J 2016;175:121—9.

[61] Huisman J, van der Heijden LC, Kok MM, Louwerenburg JH, Danse PW, Jessurun GA, de Man FH, Löwik MM, Linssen GC, IJzerman MJ, Doggen CJ, von Birgelen C. Two-year outcome after treatment of severely calcified lesions with newer-generation drug-eluting stents in acute coronary syndromes: a patient-level pooled analysis from TWENTE and Dutch PEERS. J Cardiol 2017;69:660—5.

[62] Copeland-Halperin RS, Baber U, Aquino M, Rajamanickam A, Roy S, Hasan C, Barman N, Kovacic JC, Moreno P, Krishnan P, Sweeny JM, Mehran R, Dangas G, Kini AS, Sharma SK. Prevalence, correlates, and impact of coronary calcification on adverse events following PCI with newer-generation DES: findings from a large multiethnic registry. Cathet Cardiovasc Interv 2018;91:859—66.

[63] Nasu K, Oikawa Y, Habara M, Shirai S, Abe H, Kadotani M, Gotoh R, Hozawa H, Ota H, Suzuki T, Shibata Y, Tanabe M, Nakagawa Y, Serikawa T, Nagasaka S, Takeuchi Y, Fujimoto Y, Tamura H, Kobori Y, Yajima J, Aizawa T, Suzuki T, NO-RECOIL Registries Investigators. Efficacy of biolimus A9-eluting stent for treatment

of right coronary ostial lesion with intravascular ultrasound guidance: a multi-center registry. Cardiovasc Interv Ther 2018;33:321–7.

[64] Tsunoda T, Nakamura M, Wada M, Ito N, Kitagawa Y, Shiba M, Yajima S, Iijima R, Nakajima R, Yamamoto M, Takagi T, Yoshitama T, Anzai H, Nishida T, Yamaguchi T. Chronic stent recoil plays an important role in restenosis of the right coronary ostium. Coron Artery Dis 2004;15:39–44.

[65] Kim SW, Mintz GS, Ohlmann P, Hassani SE, Michalek A, Escolar E, Bui AB, Pichard AD, Satler LF, Kent KM, Suddath WO, Waksman R, Weissman NJ. Comparative intravascular ultrasound analysis of ostial disease in the left main versus the right coronary artery. J Invasive Cardiol 2007;19:377–80.

[66] Maehara A, Mintz GS, Castagna MT, Pichard AD, Satler LF, Waksman R, Laird Jr JR, Suddath WO, Kent KM, Weissman NJ. Intravascular ultrasound assessment of the stenoses location and morphology in the left main coronary artery in relation to anatomic left main length. Am J Cardiol 2001;88:1–4.

[67] Suzuki Y, Tsuchikane E, Katoh O, Muramatsu T, Muto M, Kishi K, Hamazaki Y, Oikawa Y, Kawasaki T, Okamura A. Outcomes of percutaneous coronary interventions for chronic total occlusion performed by highly experienced Japanese specialists: the first report from the Japanese CTO-PCI Expert Registry. JACC Cardiovasc Interv 2017;10:2144–54.

[68] Harding SA, Wu EB, Lo S, Lim ST, Ge L, Chen JY, Quan J, Lee SW, Kao HL, Tsuchikane E. A new algorithm for crossing chronic total occlusions from the Asia Pacific chronic total occlusion club. JACC Cardiovasc Interv 2017;10:2135–43.

[69] Fujii K, Ochiai M, Mintz GS, Kan Y, Awano K, Masutani M, Ashida K, Ohyanagi M, Ichikawa S, Ura S, Araki H, Stone GW, Moses JW, Leon MB, Carlier SG. Procedural implications of intravascular ultrasound morphologic features of chronic total coronary occlusions. Am J Cardiol 2006;97:1455–62.

[70] Tang Z, Bai J, Su SP, Wang Y, Liu MH, Bai QC, Tian JW, Xue Q, Gao L, An CX, Liu XJ. Cutting-balloon angioplasty before drug-eluting stent implantation for the treatment of severely calcified coronary lesions. J Geriatr Cardiol 2014;11:44–9.

[71] Sugawara Y, Ueda T, Soeda T, Watanabe M, Okura H, Saito Y. Plaque modification of severely calcified coronary lesions by scoring balloon angioplasty using Lacrosse nonslip element: insights from an optical coherence tomography evaluation. Cardiovasc Interv Ther 2018. https://doi.org/10.1007/s12928-018-0553-6.

[72] Abdel-Wahab M, Toelg R, Byrne RA, Geist V, El-Mawardy M, Allali A, Rheude T, Robinson DR, Abdelghani M, Sulimov DS, Kastrati A, Richardt G. High-speed rotational atherectomy versus modified balloons prior to drug-eluting stent implantation in severely calcified coronary lesions. Circ Cardiovasc Interv 2018;11:e007415.

[73] Mintz GS, Kovach JA, Javier SP, Pichard AD, Kent KM, Popma JJ, Salter LF, Leon MB. Mechanisms of lumen enlargement after excimer laser coronary angioplasty. An intravascular ultrasound study. Circulation 1995;92:3408–14.

[74] Ambrosini V, Sorropago G, Laurenzano E, Golino L, Casafina A, Schiano V, Gabrielli G, Ettori F, Chizzola G, Bernardi G, Spedicato L, Armigliato P, Spampanato C, Furegato M. Early outcome of high energy Laser (Excimer) facilitated coronary angioplasty ON hARD and complex calcified and balloOn-resistant coronary lesions: LEONARDO Study. Cardiovasc Revascularization Med 2015;16:141–6.

[75] Abdel-Wahab M, Richardt G, Joachim Büttner H, Toelg R, Geist V, Meinertz T, Schofer J, King L, Neumann FJ, Khattab AA. High-speed rotational atherectomy before paclitaxel-eluting stent implantation in complex calcified coronary lesions: the

randomized ROTAXUS (rotational atherectomy prior to Taxus stent treatment for complex native coronary artery disease) trial. JACC Cardiovasc Interv 2013;6:10—9.

[76] Généreux P, Lee AC, Kim CY, Lee M, Shlofmitz R, Moses JW, Stone GW, Chambers JW. Orbital atherectomy for treating de novo severely calcified coronary narrowing (1-year results from the pivotal ORBIT II trial). Am J Cardiol 2015;115:1685—90.

[77] Mintz GS, Potkin BN, Keren G, Satler LF, Pichard AD, Kent KM, Popma JJ, Leon MB. Intravascular ultrasound evaluation of the effect of rotational atherectomy in obstructive atherosclerotic coronary artery disease. Circulation 1992;86:1383—93.

[78] Kim SS, Yamamoto MH, Maehara A, Sidik N, Koyama K, Berry C, Oldroyd KG, Mintz GS, McEntegart M. Intravascular ultrasound assessment of the effects of rotational atherectomy in calcified coronary artery lesions. Int J Cardiovasc Imaging 2018;34:1365—71.

[79] Yamamoto MH, Maehara A, Karimi Galougahi K, Mintz GS, Parviz Y, Kim SS, Koyama K, Amemiya K, Kim SY, Ishida M, Losquadro M, Kirtane AJ, Haag E, Sosa FA, Stone GW, Moses JW, Ochiai M, Shlofmitz RA, Ali ZA. Mechanisms of orbital versus rotational atherectomy plaque modification in severely calcified lesions assessed by optical coherence tomography. JACC Cardiovasc Interv 2017;10:2584—6.

[80] Kubo T, Shimamura K, Ino Y, Yamaguchi T, Matsuo Y, Shiono Y, Taruya A, Nishiguchi T, Shimokado A, Teraguchi I, Orii M, Yamano T, Tanimoto T, Kitabata H, Hirata K, Tanaka A, Akasaka T. Superficial calcium fracture after PCI as assessed by OCT. J Am Coll Cardiol Img 2015;10:1228—9.

[81] Maejima N, Hibi K, Saka K, Akiyama E, Konishi M, Endo M, Iwahashi N, Tsukahara K, Kosuge M, Ebina T, Umemura S, Kimura K. Relationship between thickness of calcium on optical coherence tomography and crack formation after balloon dilatation in calcified plaque requiring rotational atherectomy. Circ J 2016;80:1413—9.

[82] Yamamoto MH, Maehara A, Kim SS, Koyama K, Kim SY, Ishida M, Fujino A, Haag ES, Alexandru D, Jeremias A, Sosa FA, Galougahi KK, Kirtane AJ, Moses JW, Ali ZA, Mintz GS, Shlofmitz RA. Effect of orbital atherectomy in calcified coronary artery lesions as assessed by optical coherence tomography. Cathet Cardiovasc Interv 2018. https://doi.org/10.1002/ccd.27902.

[83] Amemiya K, Yamamoto MH, Maehara A, Oyama Y, Igawa W, Ono M, Kido T, Ebara S, Okabe T, Yamashita K, Hoshimoto K, Saito S, Yakushiji T, Isomura N, Araki H, Mintz GS, Ochiai M. Effect of cutting balloon after rotational atherectomy in severely calcified coronary artery lesions assessed by optical coherence tomography. Cathet Cardiovasc Interv 2019. https://doi.org/10.1002/ccd.28278.

[84] Ali ZA, Brinton TJ, Hill JM, Maehara A, Matsumura M, Karimi Galougahi K, Illindala U, Götberg M, Whitbourn R, Van Mieghem N, Meredith IT, Di Mario C, Fajadet J. Optical coherence tomography characterization of coronary lithoplasty for treatment of calcified lesions: first description. JACC Cardiovasc Imaging 2017;10:897—906.

[85] Kang SJ, Mintz GS, Akasaka T, Park DW, Lee JY, Kim WJ, Lee SW, Kim YH, Whan Lee C, Park SW, Park SJ. Optical coherence tomographic analysis of in-stent neoatherosclerosis after drug-eluting stent implantation. Circulation 2011;123:2954—63.

[86] Song L, Mintz GS, Yin D, Yamamoto MH, Chin CY, Matsumura M, Kirtane AJ, Parikh MA, Moses JW, Ali ZA, Shlofmitz RA, Maehara A. Characteristics of early versus late in-stent restenosis in second- generation drug-eluting stents: an optical coherence tomography study. EuroIntervention 2017;13:294—302.

[87] Song L, Mintz GS, Yin D, Yamamoto MH, Chin CY, Matsumura M, Fall K, Kirtane AJ, Parikh MA, Moses JW, Ali ZA, Shlofmitz RA, Maehara A. Neoatherosclerosis assessed with optical coherence tomography in restenotic bare metal and first- and second-generation drug-eluting stents. Int J Cardiovasc Imaging 2017;33:1115−24.

[88] Yin D, Mintz GS, Song L, Lee T, Kirtane AJ, Parikh M, Moses JW, Fall K, Jeremias A, Sosa F, Haag E, Ali Z, Shlofmitz RA, Maehara A. In-stent restenosis lesion morphology related to new stent underexpansion as evaluated by optical coherence tomography. J Am Coll Cardiol 2017;70(Suppl. 1). https://doi.org/10.1016/j.jacc.2017.09.750.

[89] Latib A, Takagi K, Chizzola G, Tobis J, Ambrosini V, Niccoli G, Sardella G, DiSalvo ME, Armigliato P, Valgimigli M, Tarsia G, Gabrielli G, Lazar L, Maffeo D, Colombo A. Excimer Laser LEsion modification to expand non-dilatable stents: the ELLEMENT registry. Cardiovasc Revascularization Med 2014;15:8−12.

[90] Lee T, Shlofmitz RA, Song L, Tsiamtsiouris T, Pappas T, Madrid A, Jeremias A, Haag ES, Ali ZA, Moses JW, Matsumura M, Mintz GS, Maehara A. The effectiveness of excimer laser angioplasty to treat coronary in-stent restenosis with peri-stent calcium as assessed by optical coherence tomography. EuroIntervention 2018. https://doi.org/10.4244/EIJ-D-18-00139.

[91] Hachinohe D, Kashima Y, Hirata K, Kanno D, Kobayashi K, Kaneko U, Sugie T, Tadano Y, Watanabe T, Shitan H, Haraguchi T, Enomoto M, Sato K, Fujita T. Treatment for in-stent restenosis requiring rotational atherectomy. J Interv Cardiol 2018;31:747−54.

Coronary artery calcium in the general population, patients with chronic kidney disease and diabetes mellitus

Paolo Raggi, MD, PhD[1,2], Antonio Bellasi, MD, PhD[3], Nikolaos Alexopoulos, MD[4]

[1]*Mazankowski Alberta Heart Institute and University of Alberta, Edmonton, AB, Canada;*
[2]*Department of Medicine-Division of Cardiology, University of Alberta, Edmonton, AB, Canada;*
[3]*Research, Innovation and Brand Reputation, Ospedale di Bergamo, ASST-Papa Giovanni XXIII, Bergamo, Italy;* [4]*Division of Cardiology, Cardiovascular Imaging, Euroclinic Athens, Athens, Greece*

Coronary artery calcium as a marker of vascular disease

Atherosclerosis development is an almost a universal event beginning early in life as shown in autopsy studies of young recruits [1], and coronary artery calcium (CAC) has been known for centuries to be an innate component of the disease. Investigators demonstrated the high negative predictive value of CAC detected by fluoroscopy before the introduction and dissemination of cardiac computed tomography (CT), that made the detection and quantification of CAC more accurate [2]. The adverse prognosis associated with CAC was also reported before the introduction and dissemination of electron beam computed tomography (EBCT) or multidetector CTs (MDCTs) [3]. In the mid-1990s, EBCT was introduced as a breakthrough technology that due to its high imaging speed allowed the visualization of calcium in the relatively small and highly mobile coronary arteries [4]. In the early 2000s, another technological breakthrough brought to market MDCTs that, in spite of the lower imaging speed compared to EBCT, showed a higher spatial resolution [5]. Over time, EBCTs were completely replaced with MDCTs that are generally more affordable and available in almost every radiology imaging suite.

CAC detected on CT imaging is a sensitive marker of atherosclerosis. In histomorphological studies, it was demonstrated to hold a close association with the extent of atherosclerotic plaque, although it represents only approximately 20% of the total plaque burden [4,5]. However, the extent of CAC is not an accurate predictor of luminal stenosis, and therefore it should not be used to predict the presence of inducible myocardial ischemia. CAC detected on cardiac CT imaging (Fig. 6.1) is typically quantified with three different scores. The most frequently used is the Agatston score introduced by doctors Agatston and Janowitz in 1990 [2]. This

Coronary Calcium. **https://doi.org/10.1016/B978-0-12-816389-4.00006-2**

159

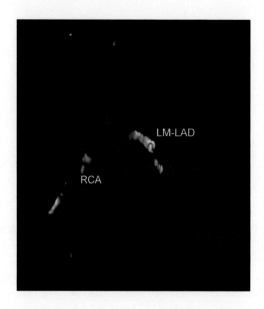

FIGURE 6.1

Volume-rendered computed tomography image of the heart showing calcium deposits in the left main trunk and left anterior descending (LAD) artery as well as the right coronary artery (RCA).

dimensionless number is calculated by multiplying the area of a calcified lesion by a coefficient, from 1 to 4, based on the peak density detected within the calcified area. The scalar nature of the measurement is such that minimal changes in peak density, that could be caused by a change in position or motion of the patient or a minor increase in calcium density, for example from 199 to 203 Hounsfield units, may cause a substantial change in the score measured multiple times in sequence. This reduces the reliability of the Agatston score for sequential scanning. The volume score was introduced by Callister et al. [6] to improve the reproducibility of the Agatston score, and it is calculated as the sum of all voxels within a calcified plaque with an attenuation greater than 130 Hounsfield units. This score is more reproducible than the Agatston score because it is not affected by the scalar coefficients used to calculate the Agatston score. The volume score has been utilized in several sequential CT studies and a few outcome studies. The final score is the mass score [7]; this renders a true measure of calcium content in the plaque. The proper calculation of this score requires the positioning of a calcium phantom underneath the patient's back. The mass score appears to be the most reproducible of the three quantitative scores [8], but it has been used minimally in clinical applications.

Besides the three classical scores, a modified Agatston score has been calculated and a visual qualitative assessment of CAC has been performed on nongated chest CT scans obtained for reasons other than cardiovascular risk assessment, such as lung cancer screening [9–11]. These scores showed a good correlation with the standard Agatston score obtained on gated chest CT scans. Additionally, the negative predictive value of no CAC on nongated CT scans to predict absence of CAC on a gated chest CT scan is over 90% [12]. Hence, the absence of CAC on nongated chest CT scans carries a high negative predictive value for cardiovascular events. In addition, CAC detected on nongated chest CT scans has been reported to be highly predictive of cardiovascular events [13]. Nonetheless, a very limited number of radiologists routinely report the presence of coronary artery and vascular calcification on nongated chest CT scans [14]. Another type of vascular calcification, breast arterial calcification seen on mammography exams, has also been reported to be associated with CAC [15] and is therefore considered a marker of cardiovascular risk [16−19].

For the purpose of assessing the risk of future events, CAC has been gauged in several more ways besides the methods described earlier. Among the methods that showed an improvement over the plain assessment of the total Agatston score were as follows: number and location of calcified lesions in the coronary artery tree [20], distribution [21], coverage of the coronary artery tree with calcific plaques [22], and low- versus high-density plaques [23]. These methods are discussed in more detail in the chapter dedicated to assessing risk of events with CAC in this book.

Coronary artery calcium and risk factors for atherosclerosis

In general, CAC scores increase with age and number of risk factors [24], or the number of points on a risk algorithm such as the Framingham risk score (FRS) [25]. Nonetheless, a large proportion of patients (40%−50%) considered at sufficient risk to warrant treatment with a statin have no CAC on a screening CT [26,27], and their event rate is extremely low [27,28]. This has potentially important implications for the selection of appropriate patients for treatment. The prevalence of a prognostically significant CAC score is too low to warrant screening in patients with a very low risk (<5% 10-year risk according to the FRS), but it increases significantly with increasing risk levels, rendering the screening for CAC more appropriate starting in patients at intermediate risk of events [25]. Of interest, the impact of risk factors is not equal among men and women and among subjects of different ethnic groups. A lower prevalence of vascular calcification in black individuals compared to white patients, and in some cases Asians, was reported in autopsy [29], fluoroscopy [30], and EBCT studies [31,32]. This difference was often noticeable despite more prevalent risk factors in black patients [33]. Initial observations also suggested a worse prognosis for black patients with CAC than for Whites [30]. However, the Dallas Heart Study investigators reported a similar prevalence of CAC in white and black patients although the patients selected for that study were recruited based on similar FRS levels in both races [34]. In a comparison of white North Americans, Brazilian, and Portuguese patients, Santos et al. [35] made a few interesting observations. First, the utilization of imaging data collected in white

Brazilian and Portuguese patients permitted the comparison of subjects with genetic similarities but different environmental exposures living on two different continents. Second, the prevalence and magnitude of CAC scores in North American patients was higher than in the Brazilian and Portuguese patients, in that order, even though North American patients reported fewer categorical risk factors than the patients from the other two nations. Aging and male sex appeared to have a greater effect on CAC accrual in the North American population than the other two populations. Among men, the CAC scores remained higher in North American patients compared to the other patients even after adjustment for risk factors. Among women, after risk factor adjustment, the CAC scores of North American and Brazilian patients were comparable but higher than those of Portuguese women. Importantly, the CAC scores in these three populations reflected the published cardiovascular disease (CVD) mortality rates in each of the three nations. Budoff et al. [36] reported the impact of risk factors on the prevalence and extent of CAC scores in 16,560 subjects of different ethnic background referred for CAC imaging by their primary care physicians. In men, after risk factor adjustment, the highest odds of any CAC were recorded in Whites, followed by Hispanics, Blacks, and Asians. Among women, black patients had the highest odds of any CAC, followed by Whites, Hispanics, and Asians. Therefore, Whites and Hispanics appeared to behave very similarly, and Asians were the group at the lowest risk overall in both sexes.

The investigators of the Multi-Ethnic Study of Atherosclerosis (MESA) provided seminal information on CAC collected in four ethnic groups living in the United States [37]. The most significant contribution of the MESA database is to have been collected among unselected, USA, free living individuals, 45−84-year-old patients of White, Hispanic, Black, and Chinese ethnicity. In these populations, Bild et al. [38] showed that the prevalence and magnitude of CAC were higher in Whites, followed by Chinese, Hispanics, and Blacks. Furthermore, the prevalence and extent of CAC were substantially higher in men than in women in all ethnicities. These statistics did not change after adjustment for several risk factors. The association of CAC with risk factors varied among ethnicities, although age, male sex, and hypertension were associated with CAC in all groups. The data summarized earlier show the complex association between traditional risk factors, age, gender, and race and development of subclinical atherosclerosis. A crucial and recurrent observation is that a sizable proportion of all populations have no detectable CAC, despite the presence of risk factors and even in the presence of high-risk disease states as discussed later in this chapter. Invariably, those patients have been reported to have a very low risk of events establishing the powerful significance of the absence of subclinical disease.

Age, sex, and race nomograms

On average, the coronary artery dimensions in women are smaller than in men [39−42]; therefore, the volume of atherosclerosis and CAC that can be accumulated in their coronary artery bed is smaller. There is also a substantial age difference between men and women in the development of atherosclerosis, and this is reflected in the reported 10—15-year delay in appearance of CAC in women compared to men [43]. These important differences are depicted in the nomograms of CAC scores published by several research groups (Table 6.1).

Nomograms of Agatston CAC scores have been used to describe the age and sex prevalence of subclinical atherosclerosis in several studies [43—45]. All showed a higher magnitude, and prevalence of CAC scores in men compared to women and a 10—15-year delay in appearance of CAC in women. In contrast with all prior publications that reported the CAC score distribution of a predominantly white population, the MESA investigators published the distribution of CAC scores in age and sex members of four different ethnicities [46]. The authors showed a clear age-by-ethnicity interaction. For 50—70-year-old women, Whites had the highest CAC scores, Hispanics had the lowest, and Blacks and Chinese were intermediate. In older age, however, Chinese women had the lowest scores of all four ethnicities. For men, again Whites had consistently the highest scores while Hispanics followed. Interestingly, in young age Blacks had the lowest scores but in older age groups Chinese men had the lowest scores. This alternating of scores among nonwhite ethnicities suggests that the progression of atherosclerosis may be very different between them, or that the impact of CAC (i.e., the associated morbidity and mortality) is different in patients of different etiologies at different ages. Data from the European equivalent of the MESA population, the Heinz Nixdorf Recall (HNR) study [47], showed a very similar distribution of CAC scores in white European and North American patients [48]. Of interest, both the MESA and the HNR investigators reported lower CAC scores in these population studies than all prior publications where data had been collected from self-referred patients, or patients referred for screening due to the presence of risk factors.

Raggi et al. [49] were the first to demonstrate the utility of CAC nomograms to differentiate risk among patients with a low absolute CAC score, but a high score relative to subjects of similar age and sex. They utilized data collected in 9728 asymptomatic subjects referred for EBCT and produced tables of percentiles for men and women grouped by quartiles of CAC scores. CAC was present in over 95% of the subjects who suffered a cardiovascular event during an average follow-up of 32 + 7 months, and the event rate was higher in patients with high absolute CAC scores. However, in a sizable number of patients the absolute CAC score was small, although the CAC score was >75th percentile in more than 70% of the patients who suffered an event. This was especially true for younger subjects where a high CAC score percentile suggests that they have accumulated too large an atherosclerotic plaque burden for their age, although the absolute CAC score may be small. The tables of percentiles used by Raggi et al. [49] were based on CAC volume scores collected in subjects predominantly of Caucasian ethnicity. In a subsequent publication, Nasir et al. [24] compared the numerical value of Agatston and volumetric CAC score percentiles in 12,936 patients referred by primary care physicians for cardiac CT imaging. Both scores increased with age and number of risk factors, but the

Table 6.1 Calcium score percentiles in 10,417 Caucasian subjects from middle Tennessee.

		Age categories								
		35–39	40–44	45–49	50–54	55–59	60–64	65–69	70–74	>74
Men										
Percentile	n	479	859	1066	1085	853	613	421	199	119
	10th	0	0	0	0	0	0	0	0	12
	25th	0	0	0	0	3	14	28	63	150
	50th	0	0	3	16	41	118	151	176	457
	75th	2	11	44	101	187	434	569	595	925
	90th	21	64	176	320	502	804	1178	1470	1582
	95th	55	155	392	587	884	1477	1452	2012	1739
Women										
Percentile	n	288	589	822	903	693	515	398	310	205
	10th	0	0	0	0	0	0	0	0	0
	25th	0	0	0	0	0	0	0	1	15
	50th	0	0	0	0	0	4	24	35	135
	75th	0	0	0	10	33	87	123	159	406
	90th	4	9	23	66	140	310	362	462	888
	95th	24	28	77	130	271	583	529	728	1422

Total number of patients: men 5694; women 4723.
Modified from Raggi P, Callister TQ, Cooil B, et al. Identification of patients at increased risk of first unheralded acute myocardial infarction by electron-beam computed tomography. Circulation 2000; 101:850-855.

volume scores were significantly smaller than the Agatston scores at the upper quartile ranges. This observation carries an important weight from the risk estimation point of view. In fact, since the upper quartile of CAC score is associated with a high risk of coronary events [49], it is important to realize what score is being reported in an individual patient; a score >75th percentile with one method may not correspond to the same value with another method.

Shaw et al., [50] Sirineni et al. [51], and McClelland et al. [52] suggested that CAC scores could be utilized to estimate the vascular age of asymptomatic subjects submitted to CT imaging. Shaw et al. [50] used the data collected in 10,377 asymptomatic, predominantly Caucasian, subjects followed for an average of 5 years for the occurrence of all-cause mortality. The vascular age was calculated with linear regression on the individual CAC score. They concluded that a CAC score >400 in a middle-aged subject adds 15—28 years of age to that patient, while a score <10 lowers the vascular age of a hypothetical 70-year-old patient by 8—10 years. Hence, a 52-year-old Caucasian man with a CAC score of 450 is 67—80-year-old as far as his vascular system is concerned. Sirineni et al. [51] used data from the MESA trial to calculate a vascular age based on the median CAC score for each sex and race. If a 58-year-old man presents with a CAC score of 150 that corresponds to the median CAC score of a 65—74-year-old man of the same race, his biological age should be considered 7—16 years higher than his chronological age. Finally, McClelland et al. [52] obtained the arterial age of MESA patients by calculating it as a linear function of the log-transformed CAC score. They then compared arterial age to observed age to predict incident coronary heart disease events in the MESA cohort, and concluded that arterial age was superior to chronological age for risk assessment. Similarly, Shaw et al. [50] concluded that arterial age significantly improved the area under the curve for prediction of all-cause death compared to observed age. Besides improving risk assessment, the estimation of vascular age facilitates the explanation of the risk inherent with a certain CAC score to an individual patient. Although it may be hard for a patient to understand the significance of a 12% risk of hard events at 10 year, it is probably easier to grasp the concept that "your actual age is 10 years older than your chronological age" and therefore you have accumulated additional risk of morbidity and mortality. An online calculator derived from the MESA database is available at: https://www.mesa-nhlbi.org/Calcium/ArterialAge.aspx (last accessed on September 25, 2018).

Cardiovascular calcifications in chronic kidney disease
Pathogenesis of vascular calcification in CKD

The risk of cardiovascular (CV) events in chronic kidney disease (CKD) patients is exceptionally high [53]. Large epidemiological studies documented a stepwise increase in the risk of CV events as glomerular filtration rate (GFR) declines [53]. About half of the patients receiving maintenance dialysis die due to CV diseases, despite the use of drugs proven to provide CV protection in the general population [54–56]. The pathophysiology of this high risk has not been fully elucidated [53]. With renal function decline, the influence of nontraditional risk factors increases and cardiac arrhythmias and congestive heart failure become more prevalent compared to ischemic heart disease [53]. Survival after a CV event is usually poorer in patients with CKD than patients with normal renal function [53]. Although validated in the general population, tools commonly used to assess mid- and long-term CV risk often fail to accurately predict risk in CKD [57]. Hence, one of the hardest challenges in CKD is how to stratify risk and individualize care in this fragile population.

Accelerated CV senescence and increased prevalence of cardiovascular calcifications have been reported in subjects with impaired renal function [58] (Fig. 6.2). These are associated with progressively increasing arterial stiffness and left ventricular hypertrophy as the estimated glomerular filtration rate (eGFR) declines [59]. Ex vivo studies reported the coexistence of subintimal and medial calcification in the large arteries of patients with CKD, although the latter may not occur in the coronary arteries. Although subintimal calcification develops in connection

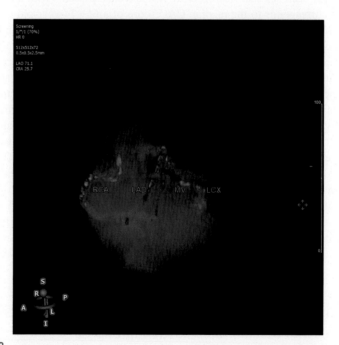

FIGURE 6.2

Volume-rendered computed tomography image of the heart showing extensively calcified coronary artery tree and mitral valve in a patient receiving hemodialysis for 4 years. *LAD*, left anterior descending artery; *LCX*, left circumflex artery; *MV*, mitral valve; *RCA*, right coronary artery.

with atherosclerosis and traditional CV risk factors, medial calcification has been associated with derangement of bone and mineral metabolism typical of advanced renal failure [60]. Beyond traditional risk factors, numerous nontraditional risk factors have been associated with CV aging and vascular calcification in CKD [58], namely inflammation, oxidative stress, metabolic derangements, and accumulation of uremic toxins [58,61].

Abnormalities of mineral and bone metabolism have received ample attention as they may drive development of cardiovascular calcification. In physiologic conditions, blood vessels are protected from calcium and phosphate precipitation by a number of inhibitors such as phyrophosphate, matrix-GLA protein (MGP), or fetuin-A [60]. These prevent transformation of amorphous calcium-phosphate complexes into insoluble crystals of hydroxyapatite and their precipitation in soft tissues [61]. In vitro and in vivo data also suggest the role of micronutrients in vascular calcification propagation [62]. Animal and in vitro experiments showed that high serum levels of calcium and phosphate induce an osteocondrogenic phenotypic switch of vascular smooth muscle cells (VSMCs) that become capable of secreting bone matrix in the context of the arterial wall, promoting vascular calcification deposition and progression [60]. Furthermore, precipitation of nanocrystals of hydroxyapatite due to chronically elevated serum concentration of calcium and phosphate attracts and stimulates the production of proinflammatory cytokines by resident macrophages. In an attempt to eliminate calcium phosphate, these cells may also undergo apoptosis further promoting inflammation and calcification [61].

Experimental and clinical data suggest that renal failure is characterized by an imbalance of pro- and antiinflammatory cytokines. Proinflammatory cytokines such as interleukin 6 (IL-6) or tumor necrosis factor alpha (TNFα) reduce the hepatic expression of fetuin-A. In turn, fetuin-A is required for the formation of the highly soluble calciproteins (a complex of fetuin-A and plasma calcium-phosphate crystals), as well as kidney expression of α-klotho, an antiinflammatory agent that modulates renal excretion of phosphate [61]. Although the roles of α-klotho and its coreceptor fibroblast growth factor 23 (FGF23) need to be fully elucidated, dysregulation of the FGF23-Klotho axis has been implicated in vascular calcification [61]. Klotho and FGF23 control phosphate excretion through the kidneys and may have a direct vascular protective role by modulating different signaling pathways such as FGF-receptor 1 and mTOR [61]. However, researchers have reported conflicting clinical data on the role of these factors and future efforts are needed to establish the contribution of the FGF23-Klotho axis in the development of vascular calcifications in CKD patients.

Oxidative stress and advanced glycation end-products (AGEs) generation, for which oxidative stress is responsible in part, have further been implicated in the pathogenesis of vascular calcifications [61]. AGEs have been shown to induce VSMC osteogenic differentiation through p38/mitogen-activated protein kinase and Wnt/β catenin signaling. AGEs promote receptor activator of nuclear factor kappa-B ligand (RANKL) activation in osteoblasts and calcium/phosphate removal from the bone. Finally, AGEs together with some uremic toxins synergistically trigger inflammation by inducing the synthesis of proinflammatory cytokines (IL-1, IL-6, TNFα) linked to endothelial dysfunction and vascular calcification [61].

Uremic toxins such as indoxyl sulfate (IS) accumulate when renal function declines and may directly affect the vasculature [61]. Experimental evidence suggests that IS increases the expression of the sodium-phosphate cotransporter Pit-1 in VSMC, favoring the osteogenic differentiation of these cells induced by calcium and phosphorus. It has also been reported that IS suppresses liver synthesis of fetuin-A further predisposing CKD patients to vascular calcification development [61]. Finally, epidemiological observations suggest that deficiency of vitamin K (responsible for MGP carboxylation and activation), and pyrophosphate (a major endogenous inhibitor of calcium-phosphate crystals formation) are prevalent in CKD [63] weakening the physiological defenses against calcium-phosphorus crystal precipitation in soft tissues.

Epidemiology, outcome, and therapeutic interventions for cardiovascular calcification in CKD

Several publications reported a graded association between prevalence of vascular calcification and severity of renal impairment. In a large series of 572 nondialysis-dependent CKD patients, vascular calcifications detected by simple in office imaging tools (abdomen, hips, and hands planar X-rays) were present in 79% of the cases and deemed prominent in about half (47%) of the subjects studied [64]. Although age was a predictor of vascular calcification, there was a progressive increase in prevalence of vascular calcification as CKD worsened. Similar conclusions were drawn by the MESA working group; when compared with individuals with normal renal function ($N = 5269$), subjects with CKD ($N = 1284$) had a higher prevalence and severity of CAC [65]. Consistently, the prevalence of CAC in patients starting renal replacement therapy (RRT) is about 60% and continues to increase during dialysis (the prevalence in maintenance dialysis patients is about 80%). This is likely the result of complete loss of renal function along with the oxidative/inflammatory stress induced by RRT [66]. Data on vascular calcification after kidney transplantation are contradictory, and it is still unclear whether renal function restoration prevents vascular calcification progression or induces its regression [67].

Other markers of cardiovascular risk have been associated with CAC in CKD patients. These include an association of CAC with myocardial repolarization abnormalities [68], increased pulse wave velocity, thoracic aorta and abdominal aorta calcification, cardiac valves calcification, left ventricular hypertrophy, and epicardial adipose tissue (EAT) [66,69]. In a series of 411 patients with CKD in different stages, presence and extent of CAC and EAT predicted the development of abnormal myocardial perfusion on nuclear stress testing [70].

In the MESA study, CAC predicted CV events or death in both normal renal function as well as CKD subjects. In particular, prediction of unfavorable events was independent of age, sex, race, and comorbid conditions. Among markers of vascular disease, CAC was a better predictor of events than ankle-brachial index and intima-media thickness, supporting the usefulness of this marker in this high-risk population [65]. Data from 1541 nondialysis-dependent patients with CKD enrolled in the Chronic Renal Insufficiency Cohort study confirmed these findings by showing that CAC independently predicted myocardial infarction, congestive heart failure, and all-cause mortality, independent of baseline CV risk evaluated by traditional risk score algorithms. Additionally, inclusion of CAC score in the statistical model led to a significant, albeit small, increase in the accuracy of cardiovascular events prediction [71]. Similarly, CAC in patients receiving hemodialysis, peritoneal dialysis, or after kidney transplantation was shown to be an independent predictor of all-cause mortality irrespective of demographic, clinical and laboratory characteristics [66,67,72]. In a cohort of 141 consecutive patients receiving maintenance hemodialysis, a simple cardiovascular calcification index (CCI) that included patient's age, dialysis vintage, calcification of the cardiac valves, and abdominal aorta was linearly associated with risk of all-cause mortality, such that the unadjusted hazard risk (HR) increased by 12% for each point increase in CCI ($P < .001$) [73]. Adjustment for confounders did not substantially change the strength of the association [73].

Whether calcium deposition is a repair mechanism or promotes vascular damage has long been debated. Preliminary data in hemodialysis patients suggest that not only presence and extent of vascular calcification but also mineral content in the plaque predicts survival. In a series of 140 consecutive hemodialysis patients, the mean CAC density was directly and independently associated with all-cause mortality [74]. Of note, both plaque density and baseline CAC were independent predictors of survival and plaque density mitigated the risk associated with CAC burden at study inception (significant interaction effect) [74]. These findings are in conflict with data published in the general population. In fact, Criqui et al. [23] reported an inverse association of plaque density and survival in the general population. Inverse epidemiology could explain these findings. Only selected subjects survive for a long time with CKD and most patients die in the course of CKD mainly due to CV events. Hence, patients receiving dialysis may not be comparable to age- and sex-matched individuals with normal renal function.

Similar to vascular calcification, the prevalence of valvular calcification in patients with CKD is higher than in the general population. Aortic and mitral valve calcification have been associated with dysfunctional leaflet motility, increase pressure gradient across the mitral valve and enlargement of the left atrium [75,76], generating a plausible rationale to explain the observed link between valve calcification and poor survival beyond the reported correlation with coronary artery or aortic calcification [77,78].

In view of its prognostic significance, efforts have been ongoing to find therapies to delay or reverse cardiovascular calcification in patients with CKD. As in the general population, statins seem to accelerate rather than reduce calcification progression [79]. However, in vitro experiments suggest that this may be restricted to lipophilic statins that appear to inhibit vitamin K synthesis [79]. Because vitamin K is important for activation of MGP, a potent inhibitor of cardiovascular calcification, ongoing randomized clinical trials aimed at testing the effect of vitamin K supplementation on CAC are awaited with interest. Similarly, trials designed to compare the effects of new direct oral anticoagulants with warfarin in patients with atrial fibrillation will shed light on the potential of vitamin K metabolism modulation to prevent vascular calcification deposition and progression [80]. Calcium supplements have been associated with CAC progression in the general population [81] as well as in nondialysis-dependent patients with CKD [82], and CKD patients receiving renal replacement therapy [83]. Calcium salts are effective phosphate binders commonly used in advanced CKD to correct hyperphosphatemia. However, several studies showed that patients with advanced CKD treated with binders containing calcium are often exposed to an excess load resulting in a positive calcium balance [84]. A chronic calcium overload may trigger crystal formation and deposition in soft tissues as demonstrated in a recent placebo-controlled trial that showed a worrisome increase in CAC in patients with moderate CKD treated with calcium-based phosphate binders [85]. A meta-analysis of several randomized trials showed convincing evidence that calcium-based binders have a detrimental effect on CAC progression and mortality in patients suffering from end-stage renal disease (ESRD) undergoing hemodialysis, as opposed to calcium neutral phosphate binders [83]. Although based on preliminary observations, the concomitant prescription of calcium supplements with drugs that modulate calcium metabolism, such as calcimimetics or vitamin D, may further promote vascular calcification progression and affect survival in ESRD patients [86–88]. Consequently, current guidelines on mineral metabolism management in patients with CKD suggest restricting the dose of calcium-based phosphate binders in all stages of renal impairment [55]. Besides noncalcium-containing phosphate binders, already shown to slow the progression of vascular calcification and reduce mortality, newer compounds are currently under investigation and hold promise for the future. SNF472 [89], a compound that shares chemical properties with pyrophosphate, has just moved to phase II clinical development and the results of an ongoing randomized clinical trial are expected by the end of 2019. Other compounds are of potential interest to tackle vascular calcification. Sotatercept, an antianemia compound that inhibits the Activin A receptor, has shown an inverse effect on bone and vasculature increasing mineralization in the former and decreasing it in the latter. Bortezomib appears to increase Wnt/B-catenin signaling and exert some protective effect against vascular calcification progression. Everolimus by inhibiting mTOR could increase the synthesis of the antiaging factor Klotho. Finally, Wnt inhibitor antagonists (sclerostin, DKK1-secreted frizzled-related proteins) are under preclinical development [90].

In summary, cardiovascular calcifications are prevalent and portend a poor prognosis in patients with CKD. Extensive research efforts have been devoted to understanding the pathogenesis of vascular calcification to inhibit its deposition and progression in this high-risk population. Contemporary evidence suggests that nontraditional and uremia-specific factors contribute to the high propensity toward developing cardiovascular calcification demonstrated by patients with declining renal function. Current guidelines call for careful management of mineral metabolism, and caution against the indiscriminate use of calcium-based binders and/or calcium supplements to curb serum levels of phosphorus, to limit the development and progression of cardiovascular calcification in patients with CKD.

Coronary artery calcium in patients with diabetes mellitus

Diabetes mellitus (DM) is a major risk factor for CVD, and its prevalence is steadily increasing. Coronary heart disease is involved in most deaths in patients with DM, and it is at least two times more common in patients with DM than in the general population [91,92]. Furthermore, the prognosis of coronary heart disease in patients with DM is worse than in the general population. Thus, DM has traditionally been regarded as an equivalent of coronary heart disease. However, in light of recent evidence that not all patients with DM carry the same risk for atherosclerotic disease, current prevention guidelines emphasize the heterogeneity in cardiovascular risk among these patients [93]. Identification of subclinical coronary atherosclerosis may help to better risk stratify patients with DM.

The prevalence and extent of CAC are higher in patients with DM than in the general population [94,95]; and the same has been reported in patients with impaired glucose metabolism and the metabolic syndrome [96,97]. CAC is prevalent even in young ($\leq$28 years) patients with type-1 DM, especially in smokers or with concomitant dyslipidemia [98]. In a large cross-sectional study in 25,564 Korean adults without DM, higher HbA1c levels (i.e., 6.0%−6.4%) were associated with a greater odds-ratio for the presence of CAC than low HbA1c levels (i.e., <5.5%), especially in women [99]. Furthermore, in the CARDIA Study, in adults without overt DM higher HbA1c levels were associated with higher CAC scores at baseline, although the association disappeared after adjustment for other established cardiovascular risk factors [100]. Higher HbA1c levels were also associated with a faster CAC score progression, defined as an increase >100 Agatston units between examinations, even after adjustment for other established cardiovascular risk factors [100]. In an analysis of the MESA Study with more than 10 years of follow-up, the mean CAC score progression was 23.9 Agatston units/year in the general population, whereas the mean CAC score progression was 31.3 Agatston units/year in patients with DM ($P < .001$) [101]. Among 3453 patients with DM followed for 5 years in the Heinz Nixdorf Recall Study, poorly controlled DM was associated with greater CAC progression between the baseline and the follow-up EBT scan [102]. Finally, the effect of DM treatment on CAC incidence was studied in the Diabetes Prevention Program Outcome Study where CAC imaging was performed in 2029 patients with prediabetes after 14 years on average from the time of randomization to placebo or metformin. Men, but not women, receiving long-term metformin showed a significantly lower prevalence and severity of CAC score [103].

The propensity of patients with DM to develop more extensive CAC deposits is not fully understood. It appears however that patients with DM tend to have a larger atherosclerotic plaque burden [104] than patients in the general population.

Furthermore, advanced glycation end-products (AGEs) appear to induce the expression of genes regulating the secretion of bone morphogenic proteins such as osteopontin in vascular smooth muscle cells [105]. Osteopontin may also induce the expression of platelet-derived growth factor [105], hence promoting a proatherogenic and prothrombotic environment that ultimately results in calcification of the plaque. An additional source of calcification of the vascular wall in DM is the media layer of the vessel wall [106]. Although not associated with typical risk factors for atherosclerosis, this form of calcification is also associated with a substantially increased risk of cardiovascular events [106].

The value of CAC score as a prognosticator of morbidity and mortality has repeatedly been demonstrated in patients with DM. In the first report by Raggi et al. [95] any CAC score was associated with a higher risk of all-cause mortality in 903 asymptomatic patients with type-2 DM than in 9474 patients from the general population, during a follow-up of 5 years (Fig. 6.3). Of note, the survival of patients with zero CAC score was approximately 99% after 5 years of follow-up irrespective of the presence of DM; this finding denotes that the high negative predictive value of CAC score observed in the general population is applicable to patients with DM as well. All subsequent cohort studies reproduced and further expanded these results. In a study of 510 asymptomatic patients with DM followed for 2.2 years, the investigators recorded 20 cardiovascular events, none of which occurred in patients with

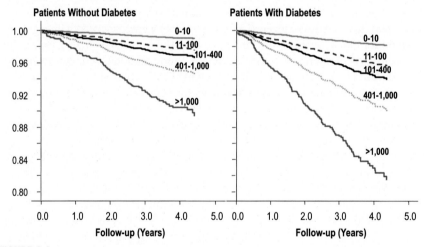

FIGURE 6.3

Five-year survival in 903 patients with and 9474 patients without diabetes mellitus after coronary artery calcium screening. The survival of patients without coronary artery calcium is similar for all patients independent of the presence of metabolic abnormalities. However, the mortality is higher for patients with diabetes mellitus at all levels of calcification ($P < .0001$).

Modified from Raggi P, Shaw LJ, Berman DS, et al. Prognostic value of coronary artery calcium screening in subjects with and without diabetes. J Am Coll Cardiol 2004; 43:1663-1669.

a baseline CAC score of 0—10 [107]. In the PREDICT study, 589 asymptomatic patients with type-2 DM were followed for a median of 4 years after CAC screening. Doubling of the CAC score was associated with a 32% increase in risk of first coronary heart disease events and stroke (29% after adjustment for confounders) [108]. CAC provided incremental prognostic value over the United Kingdom Prospective Diabetes Study risk score [108]. A similar trend was reported in other cohorts where the predictive value of CAC was compared with the Framingham risk score [109,110]. In a more recent analysis of the MESA study involving 6751 participants 45—84 years without known CVD, followed for 11.1 years after CAC screening, CAC was independently associated with incident coronary heart disease events in patients with DM (HR, 1.30; 95% CI, 1.19—1.43) or the metabolic syndrome (HR, 1.30; 95% CI, 1.20—1.41) [111]. The addition of CAC improved classification of patients beyond traditional risk score models, such as the FRS (net reclassification index (NRI) 0.23 for the DM group and 0.22 for the metabolic syndrome group) [111]. As shown in prior series, in patients with DM and a zero CAC score at baseline the coronary heart disease event rate was as low as 3.7 per 1000 person-years. Another important finding of this study was that the low event rate in the absence of CAC was also observed in patients with long-standing (>10 years) DM [111]. In a 5.6 ± 2.6 year follow-up for all-cause mortality of 2384 asymptomatic patients with DM, Silverman et al. [112] showed that CAC could be used to select asymptomatic patients who may benefit from low-dose aspirin for primary prevention.

The presence and extend of CAC is associated with a higher probability of abnormal results on stress myocardial perfusion imaging (MPI) in asymptomatic patients with DM [107]. This probability is particularly high in those with a CAC score ≥400, similar to the general population [113], but it may also be clinically relevant for lower CAC scores. Wong et al. [114] compared the results of CAC imaging and stress MPI in 1043 patients without known CAD in 313 patients with the metabolic syndrome or diabetes mellitus and 730 patients without metabolic abnormalities. The probability of an abnormal stress MPI was similarly low for patients with and without DM when the CAC score was <100. Among subjects with a CAC score ≥100, patients with DM or the metabolic syndrome had a significantly higher risk of ischemic defects on stress MPI compared to patients without DM: 13% versus 3.6% for a CAC score of 100%—399%, and 23.4% versus 13.6% for a score ≥400, respectively [114]. In the study by Anand et al. [107], patients with DM were referred for MPI only if their CAC score was >100; there was a significant interaction between CAC scores and abnormal MPI for the prediction of adverse cardiovascular events [107]. These findings suggest that if CAC screening were to be implemented for asymptomatic patients with type-2 DM, it would be reasonable to restrict stress MPI to patients with CAC scores ≥100 in the hope of detecting silent ischemia. In a more conservative approach, a recent position statement issued by the imaging council of the American College of Cardiology recommended the performance of CAC screening to assess risk of CVD in asymptomatic patients with DM above age 40, and the performance of a stress test to rule out obstructive coronary artery disease if the CAC score is ≥400 [115].

Conclusions

CAC is an excellent marker of atherosclerosis and can help assess the prevalence of this universal disease in patients of different race, sex, and age. As described elsewhere in this textbook, it is the best predictor of adverse events among all biomarkers tested so far in clinical medicine. Its ability to predict events extends to patients at very high risk such as patients with diabetes mellitus, the metabolic syndrome, and advanced chronic kidney disease. Alternatively, its absence is systematically associated with a very low probability of myocardial ischemia, cardiovascular events, and mortality in all patient subsets.

References

[1] Enos WF, Holmes RH, Beyer J. Coronary disease among United States soldiers killed in action in Korea; preliminary report. J Am Med Assoc 1953;152:1090–3.

[2] Agatston AS, Janowitz WR, Hildner FJ, et al. Quantification of coronary artery calcium using ultrafast computed tomography. J Am Coll Cardiol 1990;15:827–32.

[3] Detrano RC, Wong ND, Doherty TM, et al. Prognostic significance of coronary calcific deposits in asymptomatic high-risk subjects. Am J Med 1997;102:344–9.

[4] Mautner GC, Mautner SL, Froehlich J, et al. Coronary artery calcification: assessment with electron beam CT and histomorphometric correlation. Radiology 1994;192: 619–23.

[5] Rumberger JA, Sheedy 3rd PF, Breen JF, et al. Coronary calcium, as determined by electron beam computed tomography, and coronary disease on arteriogram. Effect of patient's sex on diagnosis. Circulation 1995;91:1363–7.

[6] Callister TQ, Cooil B, Raya SP, et al. Coronary artery disease: improved reproducibility of calcium scoring with an electron-beam CT volumetric method. Radiology 1998;208:807–14.

[7] Hoffmann U, Siebert U, Bull-Stewart A, et al. Evidence for lower variability of coronary artery calcium mineral mass measurements by multi-detector computed tomography in a community-based cohort–consequences for progression studies. Eur J Radiol 2006;57:396–402.

[8] McCollough CH, Ulzheimer S, Halliburton SS, et al. Coronary artery calcium: a multi-institutional, multimanufacturer international standard for quantification at cardiac CT. Radiology 2007;243:527–38.

[9] Wu MT, Yang P, Huang YL, et al. Coronary arterial calcification on low-dose ungated MDCT for lung cancer screening: concordance study with dedicated cardiac CT. AJR Am J Roentgenol 2008;190:923–8.

[10] Budoff MJ, Nasir K, Kinney GL, et al. Coronary artery and thoracic calcium on non-contrast thoracic CT scans: comparison of ungated and gated examinations in patients from the COPD Gene cohort. J Cardiovasc Comput Tomogr 2011;5:113–8.

[11] Kim SM, Chung MJ, Lee KS, et al. Coronary calcium screening using low-dose lung cancer screening: effectiveness of MDCT with retrospective reconstruction. AJR Am J Roentgenol 2008;190:917–22.

[12] Azour L, Kadoch MA, Ward TJ, et al. Estimation of cardiovascular risk on routine chest CT: ordinal coronary artery calcium scoring as an accurate predictor of Agatston score ranges. J Cardiovasc Comput Tomogr 2017;11:8−15.

[13] Hughes-Austin JM, Dominguez 3rd A, Allison MA, et al. Relationship of coronary calcium on standard chest CT scans with mortality. JACC Cardiovasc Imaging 2016;9: 152−9.

[14] Williams Sr. KA, Kim JT, Holohan KM. Frequency of unrecognized, unreported, or underreported coronary artery and cardiovascular calcification on noncardiac chest CT. J Cardiovasc Comput Tomogr 2013;7:167−72.

[15] Yoon YE, Kim KM, Han JS, et al. Prediction of subclinical coronary artery disease with breast arterial calcification and low bone mass in asymptomatic women: registry for the women health cohort for breast, bone, and coronary artery disease study. JACC Cardiovasc Imaging 2018 Aug 15. pii: S1936-878X(18)30551-5, https://doi.org/10. 1016/j.jcmg.2018.07.004.

[16] Topal U, Kaderli A, Topal NB, et al. Relationship between the arterial calcification detected in mammography and coronary artery disease. Eur J Radiol 2007;63:391−5.

[17] Iribarren C, Go AS, Tolstykh I, et al. Breast vascular calcification and risk of coronary heart disease, stroke, and heart failure. J Women's Health 2004;13:381−9. discussion 390-382.

[18] Schnatz PF, Marakovits KA, O'Sullivan DM. The association of breast arterial calcification and coronary heart disease. Obstet Gynecol 2011;117:233−41.

[19] Fiuza Ferreira EM, Szejnfeld J, Faintuch S. Correlation between intramammary arterial calcifications and CAD. Acad Radiol 2007;14:144−50.

[20] Williams M, Shaw LJ, Raggi P, et al. Prognostic value of number and site of calcified coronary lesions compared with the total score. JACC Cardiovasc Imaging 2008;1: 61−9.

[21] Tota-Maharaj R, Joshi PH, Budoff MJ, et al. Usefulness of regional distribution of coronary artery calcium to improve the prediction of all-cause mortality. Am J Cardiol 2015;115:1229−34.

[22] Brown ER, Kronmal RA, Bluemke DA, et al. Coronary calcium coverage score: determination, correlates, and predictive accuracy in the Multi-Ethnic Study of Atherosclerosis. Radiology 2008;247:669−75.

[23] Criqui MH, Denenberg JO, Ix JH, et al. Calcium density of coronary artery plaque and risk of incident cardiovascular events. J Am Med Assoc 2014;311:271−8.

[24] Nasir K, Raggi P, Rumberger JA, et al. Coronary artery calcium volume scores on electron beam tomography in 12,936 asymptomatic adults. Am J Cardiol 2004;93:1146−9.

[25] Okwuosa TM, Greenland P, Ning H, et al. Distribution of coronary artery calcium scores by Framingham 10-year risk strata in the MESA (Multi-Ethnic Study of Atherosclerosis) potential implications for coronary risk assessment. J Am Coll Cardiol 2011; 57:1838−45.

[26] Nasir K, Bittencourt MS, Blaha MJ, et al. Implications of coronary artery calcium testing among statin candidates according to American college of cardiology/American heart association cholesterol management guidelines: MESA (Multi-Ethnic study of atherosclerosis). J Am Coll Cardiol 2015;66:1657−68.

[27] Mortensen MB, Falk E, Li D, et al. Statin trials, cardiovascular events, and coronary artery calcification: implications for a trial-based approach to statin therapy in MESA. JACC Cardiovasc Imaging 2018;11:221−30.

[28] Mahabadi AA, Mohlenkamp S, Lehmann N, et al. CAC score improves coronary and CV risk assessment above statin indication by ESC and AHA/ACC primary prevention guidelines. JACC Cardiovasc Imaging 2017;10:143−53.

[29] Eggen DA, Strong JP, McGill Jr HC. Coronary calcification. Relationship to clinically significant coronary lesions and race, sex, and topographic distribution. Circulation 1965;32:948−55.

[30] Tang W, Detrano RC, Brezden OS, et al. Racial differences in coronary calcium prevalence among high-risk adults. Am J Cardiol 1995;75:1088−91.

[31] Newman AB, Naydeck BL, Whittle J, et al. Racial differences in coronary artery calcification in older adults. Arterioscler Thromb Vasc Biol 2002;22:424−30.

[32] Budoff MJ, Yang TP, Shavelle RM, et al. Ethnic differences in coronary atherosclerosis. J Am Coll Cardiol 2002;39:408−12.

[33] Lee TC, O'Malley PG, Feuerstein I, et al. The prevalence and severity of coronary artery calcification on coronary artery computed tomography in black and white subjects. J Am Coll Cardiol 2003;41:39−44.

[34] Jain T, Peshock R, McGuire DK, et al. African Americans and Caucasians have a similar prevalence of coronary calcium in the Dallas Heart Study. J Am Coll Cardiol 2004;44:1011−7.

[35] Santos RD, Nasir K, Rumberger JA, et al. Difference in atherosclerosis burden in different nations and continents assessed by coronary artery calcium. Atherosclerosis 2006;187:378−84.

[36] Budoff MJ, Nasir K, Mao S, et al. Ethnic differences of the presence and severity of coronary atherosclerosis. Atherosclerosis 2006;187:343−50.

[37] Bild DE, Bluemke DA, Burke GL, et al. Multi-ethnic study of atherosclerosis: objectives and design. Am J Epidemiol 2002;156:871−81.

[38] Bild DE, Detrano R, Peterson D, et al. Ethnic differences in coronary calcification: the multi-ethnic study of atherosclerosis (MESA). Circulation 2005;111:1313−20.

[39] Dodge Jr JT, Brown BG, Bolson EL, et al. Lumen diameter of normal human coronary arteries. Influence of age, sex, anatomic variation, and left ventricular hypertrophy or dilation. Circulation 1992;86:232−46.

[40] Roberts CS, Roberts WC. Cross-sectional area of the proximal portions of the three major epicardial coronary arteries in 98 necropsy patients with different coronary events. Relationship to heart weight, age and sex. Circulation 1980;62:953−9.

[41] Kucher N, Lipp E, Schwerzmann M, et al. Gender differences in coronary artery size per 100 g of left ventricular mass in a population without cardiac disease. Swiss Med Wkly 2001;131:610−5.

[42] Hiteshi AK, Li D, Gao Y, et al. Gender differences in coronary artery diameter are not related to body habitus or left ventricular mass. Clin Cardiol 2014;37:605−9.

[43] Janowitz WR, Agatston AS, Kaplan G, et al. Differences in prevalence and extent of coronary artery calcium detected by ultrafast computed tomography in asymptomatic men and women. Am J Cardiol 1993;72:247−54.

[44] Hoff JA, Chomka EV, Krainik AJ, et al. Age and gender distributions of coronary artery calcium detected by electron beam tomography in 35,246 adults. Am J Cardiol 2001;87:1335−9.

[45] Mitchell TL, Pippin JJ, Devers SM, et al. Age- and sex-based nomograms from coronary artery calcium scores as determined by electron beam computed tomography. Am J Cardiol 2001;87:453−6. A456.

[46] McClelland RL, Chung H, Detrano R, et al. Distribution of coronary artery calcium by race, gender, and age: results from the Multi-Ethnic Study of Atherosclerosis (MESA). Circulation 2006;113:30−7.

[47] Schmermund A, Mohlenkamp S, Stang A, et al. Assessment of clinically silent atherosclerotic disease and established and novel risk factors for predicting myocardial infarction and cardiac death in healthy middle-aged subjects: rationale and design of the Heinz Nixdorf RECALL Study. Risk Factors, Evaluation of Coronary Calcium and Lifestyle. Am Heart J 2002;144:212−8.

[48] Schmermund A, Mohlenkamp S, Berenbein S, et al. Population-based assessment of subclinical coronary atherosclerosis using electron-beam computed tomography. Atherosclerosis 2006;185:177−82.

[49] Raggi P, Callister TQ, Cooil B, et al. Identification of patients at increased risk of first unheralded acute myocardial infarction by electron-beam computed tomography. Circulation 2000;101:850−5.

[50] Shaw LJ, Raggi P, Berman DS, et al. Coronary artery calcium as a measure of biologic age. Atherosclerosis 2006;188:112−9.

[51] Sirineni GK, Raggi P, Shaw LJ, et al. Calculation of coronary age using calcium scores in multiple ethnicities. Int J Cardiovasc Imaging 2008;24:107−11.

[52] McClelland RL, Nasir K, Budoff M, et al. Arterial age as a function of coronary artery calcium (from the Multi-Ethnic Study of Atherosclerosis [MESA]). Am J Cardiol 2009;103:59−63.

[53] Thompson S, James M, Wiebe N, et al. Cause of death in patients with reduced kidney function. J Am Soc Nephrol 2015;26:2504−11.

[54] Luscher T. Frontiers in cardiovascular prevention. Eur Heart J 2018;39:329−32.

[55] Ketteler M, Block GA, Evenepoel P, et al. Executive summary of the 2017 KDIGO chronic kidney disease-mineral and bone disorder (CKD-MBD) guideline update: what's changed and why it matters. Kidney Int 2017;92:26−36. Kidney Int 2017; 92:1558.

[56] Foley RN, Parfrey PS, Sarnak MJ. Epidemiology of cardiovascular disease in chronic renal disease. J Am Soc Nephrol 1998;9:S16−23.

[57] Ravera M, Bussalino E, Paoletti E, et al. Haemorragic and thromboembolic risk in CKD patients with non valvular atrial fibrillation: do we need a novel risk score calculator? Int J Cardiol 2019;274:179−85.

[58] Kooman JP, Kotanko P, Schols AM, et al. Chronic kidney disease and premature ageing. Nat Rev Nephrol 2014;10:732−42.

[59] Briet M, Boutouyrie P, Laurent S, et al. Arterial stiffness and pulse pressure in CKD and ESRD. Kidney Int 2012;82:388−400.

[60] Raggi P, Giachelli C, Bellasi A. Interaction of vascular and bone disease in patients with normal renal function and patients undergoing dialysis. Nat Clin Pract Cardiovasc Med 2007;4:26−33.

[61] Henaut L, Chillon JM, Kamel S, et al. Updates on the mechanisms and the care of cardiovascular calcification in chronic kidney disease. Semin Nephrol 2018;38:233−50.

[62] Bellasi A, Di Lullo L, Raggi P. Cardiovascular calcification: the emerging role of micronutrients. Atherosclerosis 2018;273:119−21.

[63] Villa-Bellosta R, Egido J. Phosphate, pyrophosphate, and vascular calcification: a question of balance. Eur Heart J 2017;38:1801−4.

[64] Gorriz JL, Molina P, Cerveron MJ, et al. Vascular calcification in patients with nondialysis CKD over 3 years. Clin J Am Soc Nephrol 2015;10:654−66.

[65] Matsushita K, Sang Y, Ballew SH, et al. Subclinical atherosclerosis measures for cardiovascular prediction in CKD. J Am Soc Nephrol 2015;26:439—47.

[66] Bellasi A, Raggi P. Vascular calcification in chronic kidney disease: usefulness of a marker of vascular damage. J Nephrol 2011;24(Suppl. 18):S11—5.

[67] D'Marco L, Bellasi A, Mazzaferro S, et al. Vascular calcification, bone and mineral metabolism after kidney transplantation. World J Transplant 2015;5:222—30.

[68] Di Iorio B, Nargi O, Cucciniello E, et al. Coronary artery calcification progression is associated with arterial stiffness and cardiac repolarization deterioration in hemodialysis patients. Kidney Blood Press Res 2011;34:180—7.

[69] D'Marco LG, Bellasi A, Kim S, et al. Epicardial adipose tissue predicts mortality in incident hemodialysis patients: a substudy of the Renagel in New Dialysis trial. Nephrol Dial Transplant 2013;28:2586—95.

[70] Karohl C, D'Marco L, Bellasi A, et al. Hybrid myocardial imaging for risk stratification prior to kidney transplantation: added value of coronary calcium and epicardial adipose tissue. J Nucl Cardiol 2013;20:1013—20.

[71] Chen J, Budoff MJ, Reilly MP, et al. Coronary artery calcification and risk of cardiovascular disease and death among patients with chronic kidney disease. JAMA Cardiol 2017;2:635—43.

[72] Xie Q, Ge X, Shang D, et al. Coronary artery calcification score as a predictor of all-cause mortality and cardiovascular outcome in peritoneal dialysis patients. Perit Dial Int 2016;36:163—70.

[73] Bellasi A, Block GA, Ferramosca E, et al. Integration of clinical and imaging data to predict death in hemodialysis patients. Hemodial Int 2013;17:12—8.

[74] Bellasi A, Ferramosca E, Ratti C, et al. The density of calcified plaques and the volume of calcium predict mortality in hemodialysis patients. Atherosclerosis 2016;250:166—71.

[75] Pressman GS, Movva R, Topilsky Y, et al. Mitral annular dynamics in mitral annular calcification: a three-dimensional imaging study. J Am Soc Echocardiogr 2015;28:786—94.

[76] Movva R, Murthy K, Romero-Corral A, et al. Calcification of the mitral valve and annulus: systematic evaluation of effects on valve anatomy and function. J Am Soc Echocardiogr 2013;26:1135—42.

[77] Bellasi A, Ferramosca E, Muntner P, et al. Correlation of simple imaging tests and coronary artery calcium measured by computed tomography in hemodialysis patients. Kidney Int 2006;70:1623—8.

[78] Raggi P, Bellasi A, Gamboa C, et al. All-cause mortality in hemodialysis patients with heart valve calcification. Clin J Am Soc Nephrol 2011;6:1990—5.

[79] Chen Z, Qureshi AR, Parini P, et al. Does statins promote vascular calcification in chronic kidney disease? Eur J Clin Investig 2017;47:137—48.

[80] Caluwe R, Pyfferoen L, De Boeck K, et al. The effects of vitamin K supplementation and vitamin K antagonists on progression of vascular calcification: ongoing randomized controlled trials. Clin Kidney J 2016;9:273—9.

[81] Anderson JJ, Kruszka B, Delaney JA, et al. Calcium intake from diet and supplements and the risk of coronary artery calcification and its progression among older adults: 10-year follow-up of the multi-ethnic study of atherosclerosis (MESA). J Am Heart Assoc 2016;5.

[82] Di Iorio B, Bellasi A, Russo D. Mortality in kidney disease patients treated with phosphate binders: a randomized study. Clin J Am Soc Nephrol 2012;7:487—93.

[83] Jamal SA, Vandermeer B, Raggi P, et al. Effect of calcium-based versus non-calcium-based phosphate binders on mortality in patients with chronic kidney disease: an updated systematic review and meta-analysis. Lancet 2013;382:1268—77.

[84] Hill KM, Martin BR, Wastney ME, et al. Oral calcium carbonate affects calcium but not phosphorus balance in stage 3-4 chronic kidney disease. Kidney Int 2013;83: 959—66.

[85] Block GA, Wheeler DC, Persky MS, et al. Effects of phosphate binders in moderate CKD. J Am Soc Nephrol 2012;23:1407—15.

[86] Bellasi A, Cozzolino M, Russo D, et al. Cinacalcet but not vitamin D use modulates the survival benefit associated with sevelamer in the INDEPENDENT study. Clin Nephrol 2016;86:113—24.

[87] Raggi P, Chertow GM, Torres PU, et al. The ADVANCE study: a randomized study to evaluate the effects of cinacalcet plus low-dose vitamin D on vascular calcification in patients on hemodialysis. Nephrol Dial Transplant 2011;26:1327—39.

[88] Urena-Torres PA, Floege J, Hawley CM, et al. Protocol adherence and the progression of cardiovascular calcification in the ADVANCE study. Nephrol Dial Transplant 2013; 28:146—52.

[89] Perello J, Gomez M, Ferrer MD, et al. SNF472, a novel inhibitor of vascular calcification, could be administered during hemodialysis to attain potentially therapeutic phytate levels. J Nephrol 2018;31:287—96.

[90] Wu M, Rementer C, Giachelli CM. Vascular calcification: an update on mechanisms and challenges in treatment. Calcif Tissue Int 2013;93:365—73.

[91] Stamler J, Vaccaro O, Neaton JD, et al. Diabetes, other risk factors, and 12-yr cardiovascular mortality for men screened in the Multiple Risk Factor Intervention Trial. Diabetes Care 1993;16:434—44.

[92] Emerging Risk Factors C, Sarwar N, Gao P, et al. Diabetes mellitus, fasting blood glucose concentration, and risk of vascular disease: a collaborative meta-analysis of 102 prospective studies. Lancet 2010;375:2215—22.

[93] Piepoli MF, Hoes AW, Agewall S, et al. 2016 European Guidelines on cardiovascular disease prevention in clinical practice. Rev Esp Cardiol 2016;69:939.

[94] Iwasaki K, Matsumoto T, Aono H, et al. Prevalence of subclinical atherosclerosis in asymptomatic diabetic patients by 64-slice computed tomography. Coron Artery Dis 2008;19:195—201.

[95] Raggi P, Shaw LJ, Berman DS, et al. Prognostic value of coronary artery calcium screening in subjects with and without diabetes. J Am Coll Cardiol 2004;43:1663—9.

[96] Wong ND, Sciammarella MG, Polk D, et al. The metabolic syndrome, diabetes, and subclinical atherosclerosis assessed by coronary calcium. J Am Coll Cardiol 2003; 41:1547—53.

[97] Meigs JB, Larson MG, D'Agostino RB, et al. Coronary artery calcification in type 2 diabetes and insulin resistance: the framingham offspring study. Diabetes Care 2002;25:1313—9.

[98] Starkman HS, Cable G, Hala V, et al. Delineation of prevalence and risk factors for early coronary artery disease by electron beam computed tomography in young adults with type 1 diabetes. Diabetes Care 2003;26:433—6.

[99] Chang Y, Yun KE, Jung HS, et al. A1C and coronary artery calcification in nondiabetic men and women. Arterioscler Thromb Vasc Biol 2013;33:2026—31.

[100] Carson AP, Steffes MW, Carr JJ, et al. Hemoglobin a1c and the progression of coronary artery calcification among adults without diabetes. Diabetes Care 2015;38:66–71.

[101] Gassett AJ, Sheppard L, McClelland RL, et al. Risk factors for long-term coronary artery calcium progression in the multi-ethnic study of atherosclerosis. J Am Heart Assoc 2015;4:e001726.

[102] Kowall B, Lehmann N, Mahabadi AA, et al. Progression of coronary artery calcification is stronger in poorly than in well controlled diabetes: results from the Heinz Nixdorf Recall Study. J Diabet Complicat 2017;31:234–40.

[103] Goldberg RB, Aroda VR, Bluemke DA, et al. Effect of long-term metformin and lifestyle in the diabetes prevention Program and its outcome study on coronary artery calcium. Circulation 2017;136:52–64.

[104] Goraya TY, Leibson CL, Palumbo PJ, et al. Coronary atherosclerosis in diabetes mellitus: a population-based autopsy study. J Am Coll Cardiol 2002;40:946–53.

[105] Mori S, Takemoto M, Yokote K, et al. Hyperglycemia-induced alteration of vascular smooth muscle phenotype. J Diabet Complicat 2002;16:65–8.

[106] Lehto S, Niskanen L, Suhonen M, et al. Medial artery calcification. A neglected harbinger of cardiovascular complications in non-insulin-dependent diabetes mellitus. Arterioscler Thromb Vasc Biol 1996;16:978–83.

[107] Anand DV, Lim E, Hopkins D, et al. Risk stratification in uncomplicated type 2 diabetes: prospective evaluation of the combined use of coronary artery calcium imaging and selective myocardial perfusion scintigraphy. Eur Heart J 2006;27:713–21.

[108] Elkeles RS, Godsland IF, Feher MD, et al. Coronary calcium measurement improves prediction of cardiovascular events in asymptomatic patients with type 2 diabetes: the PREDICT study. Eur Heart J 2008;29:2244–51.

[109] Anand DV, Lim E, Lahiri A, et al. The role of non-invasive imaging in the risk stratification of asymptomatic diabetic subjects. Eur Heart J 2006;27:905–12.

[110] Yeboah J, Erbel R, Delaney JC, et al. Development of a new diabetes risk prediction tool for incident coronary heart disease events: the multi-ethnic study of atherosclerosis and the Heinz Nixdorf Recall study. Atherosclerosis 2014;236:411–7.

[111] Malik S, Zhao Y, Budoff M, et al. Coronary artery calcium score for long-term risk classification in individuals with type 2 diabetes and metabolic syndrome from the multi-ethnic study of atherosclerosis. JAMA Cardiol 2017;2:1332–40.

[112] Silverman MG, Blaha MJ, Budoff MJ, et al. Potential implications of coronary artery calcium testing for guiding aspirin use among asymptomatic individuals with diabetes. Diabetes Care 2012;35:624–6.

[113] He ZX, Hedrick TD, Pratt CM, et al. Severity of coronary artery calcification by electron beam computed tomography predicts silent myocardial ischemia. Circulation 2000;101:244–51.

[114] Wong ND, Rozanski A, Gransar H, et al. Metabolic syndrome and diabetes are associated with an increased likelihood of inducible myocardial ischemia among patients with subclinical atherosclerosis. Diabetes Care 2005;28:1445–50.

[115] Budoff MJ, Raggi P, Beller GA, et al. Noninvasive cardiovascular risk assessment of the asymptomatic diabetic patient: the imaging council of the American College of Cardiology. JACC Cardiovasc Imaging 2016;9:176–92.

Vascular calcification in response to pharmacological interventions

Stephen J. Nicholls, MBBS, PhD [1,2,3], **Emma Akers, BS** [4], **Belinda Di Bartolo, PhD** [5]

[1]*Monash Cardiovascular Research Centre, Monash University, Melbourne, Australia;* [2]*Professor of Cardiology, South Australian Health & Medical Research Institute, Australia;* [3]*SAHMRI Heart Foundation Heart Disease Theme Leader, South Australian Health & Medical Research Institute, Australia;* [4]*South Australian Health and Medical Research Institute, University of Adelaide, Adelaide, SA, Australia;* [5]*Kolling Research Institute, University of Sydney, NSW, Australia*

Introduction

Atherosclerotic cardiovascular disease remains the leading cause of morbidity and mortality in the world. Despite use of intensive control of established cardiovascular risk factors, there remains a substantial residual risk of cardiovascular events [1]. This highlights the need to identify and develop new therapeutic strategies that will more effectively target the underlying atherosclerotic disease process and reduce residual cardiovascular risk in the patient with clinically manifest coronary artery disease.

Calcium and atherosclerosis

The last quarter century has witnessed considerable pathological insights that have elucidated the factors involved in the generation and progression of atherosclerotic plaque and its subsequent transition from the quiescent to symptomatic state. One component of the atherosclerosis continuum that has proven to be increasingly complex involves the deposition of calcium within atherosclerotic plaque. Although calcification of the artery wall can involve different vascular components in a range of clinical settings, its presence within the atherosclerotic plaque has been traditionally regarded as a late stage event, reflecting a lesion that contains less vulnerable material and therefore less likely to underlie an ischemic event. From a therapeutic perspective, such lesions present their greatest challenge to the interventional cardiologist, by virtue of its presence to obscure interpretation of angiographic stenoses and to increase the difficulty of successful percutaneous coronary interventions.

More recently, investigations have revealed that calcification within the atherosclerotic plaque typically results from the same cellular factors that drive bone calcification. With increasing work, it has been discovered that these processes are present not only at the end stage of the disease, but can be found in the early, inflammatory stages of atherosclerosis. This microcalcific pattern is considered to be more reflective of vulnerable, rather than stable, disease and associates with a greater likelihood of plaque progression and rupture. This suggests that calcification may have implications for the variable impacts of therapeutic interventions at different stages of the disease process [2].

Imaging as a tool to monitor changes in plaque vascular calcification

The increase in pathology insights into the atherosclerotic disease process has been paralleled by technological advances in arterial wall imaging. This has permitted visualization of atherosclerosis, beyond simply demonstration of luminal stenoses on angiography. The ability to directly image the full thickness of the artery wall in a range of vascular territories has enabled measurement of plaque burden and delineation of individual plaque components. With serial imaging of anatomically matched arterial segments, clinical studies have been able to characterize the factors associated with disease progression and changes in plaque phenotype. This has been further extended to clinical trials, which have evaluated the impact of medical therapies on atherosclerosis. These studies have revealed that intensive statin therapy [3–5], antihypertensive therapy [6], high-density lipoprotein infusions [7], and pioglitazone in patients with diabetes [8] each have favorable effects on changes in plaque burden. The clinical importance of these findings is supported by observations that the burden and progression of atherosclerosis directly associate with cardiovascular event rates [9]. As imaging modalities have increasingly focused on plaque components, there have been emerging reports of favorable effects of systemic therapies on fibrous cap thickness [10] and size of both the necrotic core [11] and lipid pool [12] within plaques.

A number of imaging modalities have been developed to image plaque calcification across the disease continuum. The calcification depicted on coronary angiography typically is extensive by the time it can be visualized. Intravascular imaging modalities including ultrasound, optical coherence tomography, and virtual histology can each demonstrate both minor (spotty) and more extensive calcification, largely due to the imaging artifacts that are generated by impedance of both sound and light to the presence of calcium. The development of computed tomography coronary artery imaging can distinguish calcified and noncalcified plaque components, with simple measurement of coronary calcification scores consistently reported to independently predict the risk of cardiovascular events in a range of clinical settings [13–17]. On the basis of these findings, coronary artery calcium scoring has been increasingly integrated into treatment guidelines as an approach to triage individuals to use of more intensive preventive therapies [18].

Radioisotope imaging permits the ability to directly target specific factors within atherosclerotic plaque. Although early studies have largely employed nonspecific glucose-based approaches [19], increasing effort has been concentrated on individual plaque components. Sodium fluoride imaging has been demonstrated to visualize microcalcifications within atherosclerotic plaque [20], identifying vulnerable plaques and patients more likely to experience acute ischemic events [21]. It is likely that this imaging modality will be subsequently applied to evaluation of medical therapies in future clinical trials.

Pharmacological interventions
Statins

Given the importance of coronary artery calcium scoring in risk prediction, there was considerable interest in determining the impact of statin therapy on changes in calcium scores over time. This question was examined in three clinical trials. A study of 66 patients with coronary calcification and baseline low-density lipoprotein cholesterol (LDL-C) greater than 130 mg/dL revealed that 12 months of treatment with cerivastatin resulted in a lower increase in calcification compared with patients who received no lipid-lowering therapy [22]. In contrast, two subsequent studies failed to demonstrate a benefit of intensive lipid lowering on changes in coronary artery calcification. The Beyond Endorsed Lipid Lowering with EBT Scanning (BELLES) trial compared the effects of treatment with atorvastatin 80 mg versus pravastatin 40 mg daily for 12 months in 615 postmenopausal women with hyperlipidemia [23]. A greater degree of LDL-C lowering with atorvastatin (46% vs. 24%) did not result in any difference in the increase in calcium score over the course of the 12-month treatment period. A subsequent study of asymptomatic individuals with elevated coronary calcium scores similarly failed to demonstrate any differential effect on changes in calcium with 12 months of treatment of atorvastatin 80 mg compared with 10 mg daily and its associated greater lipid lowering [24]. Given the observed relationship between coronary calcium scores and both disease burden and cardiovascular risk, the findings of these studies were initially viewed as disappointing, although their true meaning was subsequently uncertain with reports that more intensive statin therapy favorably modifies plaque burden and cardiovascular event rates.

Intravascular imaging has been employed in a number of clinical trials to demonstrate that statin therapy can either slow plaque progression or promote disease regression, typically when LDL-C levels are lowered below 70 mg/dL, with evidence of a direct relationship between achieved LDL-C levels and the rate of disease progression [25–27]. Although the primary focus of these studies has been to determine the impact of statins on changes in plaque burden, they have revealed a number of important insights regarding plaque calcification. The first observation determined that plaque containing more extensive calcification at baseline was less likely to undergo both progression and regression on serial imaging in these studies [28]. This supported the concept that more advanced calcification identified plaques with less material required to promote plaque progression and at the same time likely to undergo therapeutic modification.

The second observation from studies of intensive lipid lowering with high-intensity statin therapy revealed that plaque regression was accompanied by an increase in plaque calcification. This provided a second mechanistic effect that may underscore that more intensive lipid lowering with statin therapy reduces cardiovascular event rates. Such a finding is important in terms of its potential for the role of statins in plaque stabilization, suggesting that it was unlikely that early studies involving coronary artery calcium score measurements were less likely to demonstrate reductions in scores with statin therapy. The findings also have important implications for the use of serial calcium score measurements in statin-treated patients in clinical practice. Although an elevated calcium score, as an isolated measure, associates with increased cardiovascular risk [13–17] and an increase in calcium score on serial imaging has been reported to also associate with a greater likelihood of clinical events [29], it is uncertain how to interpret therapeutic induced increases in calcium scores. Given that it is plausible that such changes do reflect stabilization, serial changes with and without treatment are likely to reflect diverse states.

With the reports that intravascular ultrasound can also demonstrate more limited, spotty calcification, the impact of statin therapy in these patients may be different. The presence of spotty calcification associates with accelerated disease progression on serial ultrasound imaging [30] and a more vulnerable plaque phenotype on optical coherence tomography [31]. Of particular interest, the increased rate of disease progression with spotty calcification is reversed with the use of statin therapy, suggesting the presence of modifiable plaque [32]. This provides a clear contrast to the early reports that advanced calcification identifies less modifiable disease. With the proliferation of noninvasive imaging in risk prediction and triage of intensive therapies, it is possible that such factors will be incorporated into clinical algorithms for preventive cardiology.

PCSK9 inhibitors

Proprotein convertase subtilisin kexin type 9 (PCSK9) inhibitors have emerged as a novel strategy to lower LDL-C to very low levels [33], promote plaque regression [34], and reduce cardiovascular event rates [35] in statin-treated patients. Post hoc analysis of the serial intravascular ultrasound study of the PCSK9 inhibitor, evolocumab, demonstrated that plaque regression associated with an increase in plaque calcification over 24-month follow-up [36]. An increase in plaque calcium was also observed in the statin monotherapy group, continuing to support the concept that long-term therapy with statins will ultimately promote plaque calcification. An inverse relationship was observed between achieved LDL-C levels and changes in plaque calcification. The finding of this study extends the concept of plaque calcification with lipid-lowering therapy beyond statins and suggests that this impact within the artery wall is not likely to be a pleiotropic effect of statins. Rather, it is likely to reflect a reduction in lipid within the atherosclerotic plaque reducing the mechanistic substrate that is required for deposition of calcium over time. Whether the increase in plaque calcium observed with these studies of intensive statins and PCSK9 inhibitors directly contributes to the clinical benefit of these agents in large clinical trials remains to be determined.

Modification of other risk factors

The impact of medical therapies targeting other cardiovascular risk factors on atherosclerotic plaque calcium has not been well studied. Although blood pressure lowering, HDL infusions, and pioglitazone in diabetes have each been reported to favorably modulate progression of coronary atherosclerosis, their impact on plaque calcium has not been well studied. Pioglitazone was demonstrated to reduce early microcalcification in a hypercholesterolemia, nondiabetic rabbit model of atherosclerosis [37]. However, in a small observational study of patients with diabetes, pioglitazone treatment was not associated with a reduction in plaque calcification [38]. Whether plaque regression induced by nonlipid-lowering therapies associates with an increase in plaque calcification has yet to be demonstrated.

Other cardiovascular therapies, beyond the atherosclerosis field, have been reported to influence plaque calcification. A pooled analysis of more than 4000 patients undergoing serial intravascular ultrasound imaging reported that concomitant use of warfarin independently associated with increases in plaque calcification, regardless of changes in plaque burden [39]. This is consistent with observations that warfarin inhibits synthesis and activity of matrix Gla protein, a vitamin K-dependent inhibitor of arterial calcification [40]. The clinical significance of this finding is unknown. In contrast, pooled analyses of patients treated with either calcium supplements or vitamin D did not demonstrate an increase in plaque calcification, compared with placebo [41]. In a similar fashion, novel therapies that have the potential to directly target factors implicated in plaque calcification, such as carbonic hydrase inhibitors [42], may modulate the natural history of atherosclerosis and its translation to clinically manifest disease, although none of these approaches have yet to be investigated in advanced preclinical or early human studies.

Summary

With increasing use of arterial imaging in humans and greater elucidation of molecular mediators of atherosclerosis, increasing interest has focused on the complexity of calcium in patients with coronary artery disease. It is now likely that calcification plays an important role in the disease process, spanning from microcalcification present in inflammatory plaque through to more advanced macrocalcification that is typically associated with more stable disease. The relative pattern and content of plaque calcium is likely to influence the degree of response to intensive risk factor modification. The benefits of multiple lipid-lowering agents have supported the importance of reducing lipid content within the artery wall as an important factor involved in increasing the degree of plaque calcification. Although it is likely that such effects are likely to result in more clinically stable disease, this has yet to be fully elucidated in large-scale clinical studies.

References

[1] Libby P. The forgotten majority: unfinished business in cardiovascular risk reduction. J Am Coll Cardiol 2005;46:1225—8.

[2] Mori H, Torii S, Kutyna M, Sakamoto A, Finn AV, Virmani R. Coronary artery calcification and its progression: what does it really mean? JACC Cardiovasc Imaging 2018; 11:127—42.

[3] Nicholls SJ, Ballantyne CM, Barter PJ, Chapman MJ, Erbel RM, Libby P, Raichlen JS, Uno K, Borgman M, Wolski K, Nissen SE. Effect of two intensive statin regimens on progression of coronary disease. N Engl J Med 2011;365:2078—87.

[4] Nissen SE, Nicholls SJ, Sipahi I, Libby P, Raichlen JS, Ballantyne CM, Davignon J, Erbel R, Fruchart JC, Tardif JC, Schoenhagen P, Crowe T, Cain V, Wolski K, Goormastic M, Tuzcu EM. Effect of very high-intensity statin therapy on regression of coronary atherosclerosis: the ASTEROID trial. JAMA 2006;295:1556—65.

[5] Nissen SE, Tuzcu EM, Schoenhagen P, Brown BG, Ganz P, Vogel RA, Crowe T, Howard G, Cooper CJ, Brodie B, Grines CL, DeMaria AN. Effect of intensive compared with moderate lipid-lowering therapy on progression of coronary atherosclerosis: a randomized controlled trial. J Am Med Assoc 2004;291:1071—80.

[6] Nissen SE, Tuzcu EM, Libby P, Thompson PD, Ghali M, Garza D, Berman L, Shi H, Buebendorf E, Topol EJ. Effect of antihypertensive agents on cardiovascular events in patients with coronary disease and normal blood pressure: the CAMELOT study: a randomized controlled trial. J Am Med Assoc 2004;292:2217—25.

[7] Nissen SE, Tsunoda T, Tuzcu EM, Schoenhagen P, Cooper CJ, Yasin M, Eaton GM, Lauer MA, Sheldon WS, Grines CL, Halpern S, Crowe T, Blankenship JC, Kerensky R. Effect of recombinant ApoA-I Milano on coronary atherosclerosis in patients with acute coronary syndromes: a randomized controlled trial. J Am Med Assoc 2003;290:2292—300.

[8] Nissen SE, Nicholls SJ, Wolski K, Nesto R, Kupfer S, Perez A, Jure H, De Larochelliere R, Staniloae CS, Mavromatis K, Saw J, Hu B, Lincoff AM, Tuzcu EM. Comparison of pioglitazone vs glimepiride on progression of coronary atherosclerosis in patients with type 2 diabetes: the PERISCOPE randomized controlled trial. J Am Med Assoc 2008;299:1561—73.

[9] Nicholls SJ, Hsu A, Wolski K, Hu B, Bayturan O, Lavoie A, Uno K, Tuzcu EM, Nissen SE. Intravascular ultrasound-derived measures of coronary atherosclerotic plaque burden and clinical outcome. J Am Coll Cardiol 2010;55:2399—407.

[10] Komukai K, Kubo T, Kitabata H, Matsuo Y, Ozaki Y, Takarada S, Okumoto Y, Shiono Y, Orii M, Shimamura K, Ueno S, Yamano T, Tanimoto T, Ino Y, Yamaguchi T, Kumiko H, Tanaka A, Imanishi T, Akagi H, Akasaka T. Effect of

atorvastatin therapy on fibrous cap thickness in coronary atherosclerotic plaque as assessed by optical coherence tomography: the EASY-FIT study. J Am Coll Cardiol 2014;64:2207−17.

[11] Serruys PW, Garcia-Garcia HM, Buszman P, Erne P, Verheye S, Aschermann M, Duckers H, Bleie O, Dudek D, Botker HE, von Birgelen C, D'Amico D, Hutchinson T, Zambanini A, Mastik F, van Es GA, van der Steen AF, Vince DG, Ganz P, Hamm CW, Wijns W, Zalewski A. Effects of the direct lipoprotein-associated phospholipase A(2) inhibitor darapladib on human coronary atherosclerotic plaque. Circulation 2008;118:1172−82.

[12] Kini AS, Baber U, Kovacic JC, Limaye A, Ali ZA, Sweeny J, Maehara A, Mehran R, Dangas G, Mintz GS, Fuster V, Narula J, Sharma SK, Moreno PR. Changes in plaque lipid content after short-term intensive versus standard statin therapy: the YELLOW trial (reduction in yellow plaque by aggressive lipid-lowering therapy). J Am Coll Cardiol 2013;62:21−9.

[13] Budoff MJ, Shaw LJ, Liu ST, Weinstein SR, Mosler TP, Tseng PH, Flores FR, Callister TQ, Raggi P, Berman DS. Long-term prognosis associated with coronary calcification: observations from a registry of 25,253 patients. J Am Coll Cardiol 2007;49:1860−70.

[14] Budoff MJ, McClelland RL, Nasir K, Greenland P, Kronmal RA, Kondos GT, Shea S, Lima JA, Blumenthal RS. Cardiovascular events with absent or minimal coronary calcification: the Multi-Ethnic Study of Atherosclerosis (MESA). Am Heart J 2009;158:554−61.

[15] Greenland P, Alpert JS, Beller GA, Benjamin EJ, Budoff MJ, Fayad ZA, Foster E, Hlatky MA, Hodgson JM, Kushner FG, Lauer MS, Shaw LJ, Smith Jr SC, Taylor AJ, Weintraub WS, Wenger NK, Jacobs AK, Smith Jr SC, Anderson JL, Albert N, Buller CE, Creager MA, Ettinger SM, Guyton RA, Halperin JL, Hochman JS, Kushner FG, Nishimura R, Ohman EM, Page RL, Stevenson WG, Tarkington LG, Yancy CW, American College of Cardiology F, American Heart A. 2010 ACCF/AHA guideline for assessment of cardiovascular risk in asymptomatic adults: a report of the American college of cardiology foundation/American heart association task force on practice guidelines. J Am Coll Cardiol 2010;56:e50−103.

[16] Raggi P, Callister TQ, Cooil B, He ZX, Lippolis NJ, Russo DJ, Zelinger A, Mahmarian JJ. Identification of patients at increased risk of first unheralded acute myocardial infarction by electron-beam computed tomography. Circulation 2000;101:850−5.

[17] Budoff MJ, Nasir K, McClelland RL, Detrano R, Wong N, Blumenthal RS, Kondos G, Kronmal RA. Coronary calcium predicts events better with absolute calcium scores than age-sex-race/ethnicity percentiles: MESA (Multi-Ethnic Study of Atherosclerosis). J Am Coll Cardiol 2009;53:345−52.

[18] Piepoli MF, Hoes AW, Agewall S, Albus C, Brotons C, Catapano AL, Cooney MT, Corra U, Cosyns B, Deaton C, Graham I, Hall MS, Hobbs FD, Lochen ML, Lollgen H, Marques-Vidal P, Perk J, Prescott E, Redon J, Richter DJ, Sattar N, Smulders Y, Tiberi M, van der Worp HB, van Dis I, Verschuren WM, Authors/Task Force M. 2016 European guidelines on cardiovascular disease prevention in clinical practice: the sixth joint task force of the European society of cardiology and other societies on cardiovascular disease prevention in clinical practice (constituted by representatives of 10 societies and by invited experts) Developed with the special contribution of the European association for cardiovascular prevention & rehabilitation (EACPR). Eur Heart J 2016;37:2315−81.

[19] Rudd JHF, Warburton EA, Fryer TD, Jones HA, Clark JC, Antoun N, Johnstrom P, Davenport AP, Kirkpatrick PJ, Arch BN, Pickard JD, Weissberg PL. Imaging

atherosclerotic plaque inflammation with [18F]-Fluorodeoxyglucose positron emission tomography. Circulation 2002;105:2708—11.

[20] Irkle A, Vesey AT, Lewis DY, Skepper JN, Bird JL, Dweck MR, Joshi FR, Gallagher FA, Warburton EA, Bennett MR, Brindle KM, Newby DE, Rudd JH, Davenport AP. Identifying active vascular microcalcification by (18)F-sodium fluoride positron emission tomography. Nat Commun 2015;6:7495.

[21] Joshi NV, Vesey AT, Williams MC, Shah AS, Calvert PA, Craighead FH, Yeoh SE, Wallace W, Salter D, Fletcher AM, van Beek EJ, Flapan AD, Uren NG, Behan MW, Cruden NL, Mills NL, Fox KA, Rudd JH, Dweck MR, Newby DE. 18F-fluoride positron emission tomography for identification of ruptured and high-risk coronary atherosclerotic plaques: a prospective clinical trial. Lancet 2014;383:705—13.

[22] Achenbach S, Ropers D, Pohle K, Leber A, Thilo C, Knez A, Menendez T, Maeffert R, Kusus M, Regenfus M, Bickel A, Haberl R, Steinbeck G, Moshage W, Daniel WG. Influence of lipid-lowering therapy on the progression of coronary artery calcification: a prospective evaluation. Circulation 2002;106:1077—82.

[23] Raggi P, Davidson M, Callister TQ, Welty FK, Bachmann GA, Hecht H, Rumberger JA. Aggressive versus moderate lipid-lowering therapy in hypercholesterolemic postmenopausal women: beyond Endorsed Lipid Lowering with EBT Scanning (BELLES). Circulation 2005;112:563—71.

[24] Schmermund A, Achenbach S, Budde T, Buziashvili Y, Forster A, Friedrich G, Henein M, Kerkhoff G, Knollmann F, Kukharchuk V, Lahiri A, Leischik R, Moshage W, Schartl M, Siffert W, Steinhagen-Thiessen E, Sinitsyn V, Vogt A, Wiedeking B, Erbel R. Effect of intensive versus standard lipid-lowering treatment with atorvastatin on the progression of calcified coronary atherosclerosis over 12 months: a multicenter, randomized, double-blind trial. Circulation 2006;113:427—37.

[25] Puri R, Nicholls SJ, Shao M, Kataoka Y, Uno K, Kapadia SR, Tuzcu EM, Nissen SE. Impact of statins on serial coronary calcification during atheroma progression and regression. J Am Coll Cardiol 2015;65:1273—82.

[26] Puri R, Libby P, Nissen SE, Wolski K, Ballantyne CM, Barter PJ, Chapman MJ, Erbel R, Raichlen JS, Uno K, Kataoka Y, Tuzcu EM, Nicholls SJ. Long-term effects of maximally intensive statin therapy on changes in coronary atheroma composition: insights from SATURN. Eur Heart J Cardiovasc Imaging 2014;15:380—8.

[27] Raber L, Taniwaki M, Zaugg S, Kelbaek H, Roffi M, Holmvang L, Noble S, Pedrazzini G, Moschovitis A, Luscher TF, Matter CM, Serruys PW, Juni P, Garcia-Garcia HM, Windecker S, Investigators IT. Effect of high-intensity statin therapy on atherosclerosis in non-infarct-related coronary arteries (IBIS-4): a serial intravascular ultrasonography study. Eur Heart J 2015;36:490—500.

[28] Nicholls SJ, Tuzcu EM, Wolski K, Sipahi I, Schoenhagen P, Crowe T, Kapadia SR, Hazen SL, Nissen SE. Coronary artery calcification and changes in atheroma burden in response to established medical therapies. J Am Coll Cardiol 2007;49:263—70.

[29] Budoff MJ, Young R, Lopez VA, Kronmal RA, Nasir K, Blumenthal RS, Detrano RC, Bild DE, Guerci AD, Liu K, Shea S, Szklo M, Post W, Lima J, Bertoni A, Wong ND. Progression of coronary calcium and incident coronary heart disease events: MESA (Multi-Ethnic Study of Atherosclerosis). J Am Coll Cardiol 2013;61:1231—9.

[30] Kataoka Y, Wolski K, Uno K, Puri R, Tuzcu EM, Nissen SE, Nicholls SJ. Spotty calcification as a marker of accelerated progression of coronary atherosclerosis: insights from serial intravascular ultrasound. J Am Coll Cardiol 2012;59:1592—7.

[31] Kataoka Y, Puri R, Hammadah M, Duggal B, Uno K, Kapadia SR, Tuzcu EM, Nissen SE, Nicholls SJ. Spotty calcification and plaque vulnerability in vivo: frequency-domain optical coherence tomography analysis. Cardiovasc Diagn Ther 2014;4:460−9.

[32] Kataoka Y, Wolski K, Balog C, Uno K, Puri R, Tuzcu EM, Nissen SE, Nicholls SJ. Progression of coronary atherosclerosis in stable patients with ultrasonic features of high-risk plaques. Eur Heart J Cardiovasc Imaging 2014;15:1035−41.

[33] Sabatine MS, Giugliano RP, Wiviott SD, Raal FJ, Blom DJ, Robinson J, Ballantyne CM, Somaratne R, Legg J, Wasserman SM, Scott R, Koren MJ, Stein EA, Open-Label Study of Long-Term Evaluation against LDLCI. Efficacy and safety of evolocumab in reducing lipids and cardiovascular events. N Engl J Med 2015;372:1500−9.

[34] Nicholls SJ, Puri R, Anderson T, Ballantyne CM, Cho L, Kastelein JJ, Koenig W, Somaratne R, Kassahun H, Yang J, Wasserman SM, Scott R, Ungi I, Podolec J, Ophuis AO, Cornel JH, Borgman M, Brennan DM, Nissen SE. Effect of evolocumab on progression of coronary disease in statin-treated patients: the GLAGOV randomized clinical trial. J Am Med Assoc 2016;316:2373−84.

[35] Sabatine MS, Giugliano RP, Keech AC, Honarpour N, Wiviott SD, Murphy SA, Kuder JF, Wang H, Liu T, Wasserman SM, Sever PS, Pedersen TR, Committee FS, Investigators. Evolocumab and clinical outcomes in patients with cardiovascular disease. N Engl J Med 2017;376:1713−22.

[36] Nicholls SJ, Puri R, Anderson T, Ballantyne CM, Cho L, Kastelein JJP, Koenig W, Somaratne R, Kassahun H, Yang J, Wasserman SM, Honda S, Shishikura D, Scherer DJ, Borgman M, Brennan DM, Wolski K, Nissen SE. Effect of evolocumab on coronary plaque composition. J Am Coll Cardiol 2018;72:2012−21.

[37] Xu J, Nie M, Li J, Xu Z, Zhang M, Yan Y, Feng T, Zhao X, Zhao Q. Effect of pioglitazone on inflammation and calcification in atherosclerotic rabbits: an (18)F-FDG-PET/CT in vivo imaging study. Herz 2018;43:733−40.

[38] Clementi F, Di Luozzo M, Mango R, Luciani G, Trivisonno A, Pizzuto F, Martuscelli E, Mehta JL, Romeo F. Regression and shift in composition of coronary atherosclerotic plaques by pioglitazone: insight from an intravascular ultrasound analysis. J Cardiovasc Med 2009;10:231−7.

[39] Andrews J, Psaltis PJ, Bayturan O, Shao M, Stegman B, Elshazly M, Kapadia SR, Tuzcu EM, Nissen SE, Nicholls SJ, Puri R. Warfarin use is associated with progressive coronary arterial calcification: insights from serial intravascular ultrasound. JACC Cardiovasc Imaging 2018;11(9):1315−23. https://doi.org/10.1016/j.jcmg.2017.04.010. Epub 2017 Jul 19.

[40] Lomashvili KA, Wang X, Wallin R, O'Neill WC. Matrix Gla protein metabolism in vascular smooth muscle and role in uremic vascular calcification. J Biol Chem 2011; 286:28715−22.

[41] Manson JE, Allison MA, Carr JJ, Langer RD, Cochrane BB, Hendrix SL, Hsia J, Hunt JR, Lewis CE, Margolis KL, Robinson JG, Rodabough RJ, Thomas AM, Women's Health I and Women's Health Initiative-Coronary Artery Calcium Study I. Calcium/vitamin D supplementation and coronary artery calcification in the Women's Health Initiative. Menopause 2010;17:683−91.

[42] Gram J, Bollerslev J, Nielsen HK, Larsen HF, Mosekilde L. The effect of carbonic anhydrase inhibition on calcium and bone homeostasis in healthy postmenopausal women. J Intern Med 1990;228:367−71.

Coronary artery calcium: biology and clinical relevance

8

Petra Zubin Maslov, MD, Jagat Narula, MD, PhD, MACC, Harvey Hecht, MD

Icahn School of Medicine at Mount Sinai, New York, NY, United States

Biology

Calcium mineral deposits within the arterial wall are one of the most prominent features of the atherosclerotic process. Initially, vascular calcification was thought to be a passive degenerative process of calcium accumulation in the arteries associated with decreased elastance and compliance of the affected arteries [1]. Nowadays, pathogenesis of well recognized as a complex series of events including inflammation [2], oxidative stress and metabolic changes such as hyperlipidemia [3] and hyperglycemia [4]. Formation of calcified plaque is influenced by several different factors inside the arterial wall that promote calcium mineralization [5]. Master transcription factors such as Msx2, Runx2, Osterix, and Sox9 [6], apolipoproteins, and oxidized phospholipids are only some of the key players in coronary artery calcification (CAC). Inflammatory cytokines and oxidized lipids found in atheromatous portion of the plaque induce matrix calcification and osteogenic differentiation in subpopulations of vascular cells resulting in atherosclerotic calcifications [1,7].

Similar inflammatory and metabolic signals driving skeletal mineralization are present in the arterial wall [1]. Calcified vascular cells within the arterial wall are able to undergo osteoblastic differentiation and mineralization resembling bone development and metabolism [1]. Calcified vascular cells (CVC) are found to have chondrogenic potential that was evidenced by expression of types II and IX collagen as well as leiomyogenic potential evidenced by expression of smooth muscle-alpha actin, calponin, caldesmon, and myosin heavy chain [6]. Osteogenesis is rarely seen in CAC and is mostly observed in heavily calcified segments of the arterial wall.

There are three major histological categories of arterial calcification: osteomorphic, chondromorphic, and amorphic. Anatomically arterial calcification can be classified as intimal (patchy) atherosclerotic calcification and medial calcification (diffuse and independent of atherosclerosis) [1]. CAC is mainly localized in the intima of the coronary arterial wall and starts as microcalcification within the area of smooth muscle cell (SMC) apoptosis [8] and macrophage-derived matrix vesicles. SMC apoptosis results in phospholipid-rich medium that is infiltrated with macrophages that eventually also undergo apoptosis, leading to formation of the necrotic core. Microcalcifications fuse together into larger areas and form speckles and fragments of calcifications [9].

Increase in plaque size and luminal narrowing are often related to the extent of the area of calcification. However, the extent of calcification is found to be different in different plaque types. Histologically, diffuse calcification in the form of so-called sheet calcification is often seen in healed ruptured plaques and fibrocalcific plaques, while thin-cap fibroatheroma have less diffuse calcifications [9].

The presence of calcium is a marker of subclinical atherosclerosis and by itself does not predict plaque future (stability vs. instability) as different patterns of calcifications have been associated with different types of plaque and subsequently different clinical outcomes [10]. When CT angiography was performed in patients with acute coronary syndromes (ACS) and compared with patients with stable angina before undergoing percutaneous coronary intervention, spotty calcifications were more commonly observed in ACS patients [11]. Generally, unstable plaque arises from spotty calcifications, while extensive calcifications are associated with more stable plaques. A post hoc patient-level analysis of eight prospective randomized trials using coronary intravascular ultrasound (IVUS) revealed a serial change in coronary atheroma volume and calcium across matched coronary segments in patients with coronary artery disease who were taking high-dose or low-dose statin therapy when compared to no-statin therapy [12]. High-dose statin therapy was associated with decrease atheroma volume. Unexpectedly, increase in calcium index was reported in groups taking high-dose and low-dose statin therapy. These controversial findings suggest that CAC potentially could have a role in stabilization of the plaque progression. This statement is supported by another study where CAC density was inversely associated with coronary heart disease (CHD) [13] suggesting greater calcium density in plaque is associated with decreased CVD risk.

Use in screening

CAC scanning obtained by noncontrast computed tomography is a noninvasive diagnostic modality used for quantification of atherosclerosis. It only requires 3—5 s of breath holding and identifies calcium deposits through the entire epicardial coronary system. CAC scan is an anatomic imagining modality that quantifies coronary calcium as lesions above 130 Hounsfield units with an area of ≥ 3 adjacent pixels (at least 1 mm^2) (Fig. 8.1). Standardized quantification of CAC is based on the Agatson score that is the product of the calcified plaque area and maximal calcium lesion density [14]. The risk categories are defined as 0 indicating the absence of calcified plaque, 1—10 as minimal plaque, 11—100 mild plaque, 101—400 as moderate plaque and >400 as severe plaque [15]. CAC scanning has been widely recognized as highly useful imaging modality for cardiovascular (CV) risk stratification in asymptomatic populations without known CHD, particularly in the intermediate-risk group. Unfortunately, application of CAC scanning in everyday clinical practice has not been widely implemented, even though its role in primary prevention of CHD has

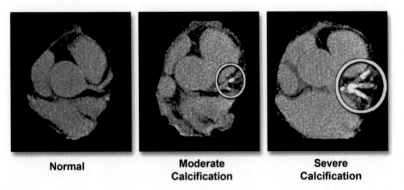

Normal	**Moderate**	**Severe**
	Calcification	**Calcification**

FIGURE 8.1

Examples of coronary artery calcium scans. (Left) Normal scan without calcified plaque. (Middle) Moderate calcified plaque in the left anterior descending and left circumflex coronary arteries. (Right) Severe calcified plaque involving the left main, left anterior descending, and left circumflex coronary arteries.

been well justified in thousands of publications. Concerns regarding radiation exposure to large patient populations and insurance coverage issues due to cost are some of the potential obstacles for lack of clinical use in primary prevention. Newer CT protocols have reduced radiation exposure without compromising image quality [16]. CAC scanning is associated with radiation exposure of less than 1 mSv, which is comparable to mammography (0.8 mSv) [17]. Mean effective dose of radiation was evaluated in 3442 participants from the Multi-Ethnic Study of Atherosclerosis (MESA) [18,19] undergoing CAC scoring with four different modern CT scanners (Siemens64, Siemens Somatom Definition, GE64, and Toshiba 320). Among all participants, the mean effective dose was 1.05 mSv, a median dose of 0.95 mSv. The mean effective dose across all six centers in the MESA cohort ranged from 0.74 to 1.26 mSv [16].

Other noninvasive imaging modalities such as cardiac stress test (stress electrocardiography and echocardiography, nuclear stress test) detect the changes in the heart function that are consequences of a hemodynamically significant coronary stenosis and so are not ideal tests for the detection of subclinical atherosclerosis [20]. CAC burden positively correlates with histologic plaque areas in the coronary arteries [21,22]. By measuring calcium burden within the atherosclerotic plaque, noncontrast CT scan estimates the overall extent of the total atherosclerotic plaque within coronary arteries. Knowing that atherosclerotic plaque is the main precursor of most CHD events, CAC is considered a measure of subclinical disease and not a risk factor for disease [23]. Although standard CV risk factors only predict the risk of developing coronary artery disease, CAC scanning detects the subclinical disease process and as such is a much better predictor of future clinical coronary artery disease. Moreover, the progression of CAC over a time has shown to be a strong predictor of mortality [24].

The Framingham risk score (FRS) is based on the assessment of nonmodifiable risk factors such as age and gender and presence or absence of modifiable risk factors for CHD such as smoking, hypertension, total cholesterol, and diabetes to statistically predict who is likely to develop myocardial infarction in 10 years' time [25]. However, these traditional risk factors often overlap in individuals with and without CHD and adding an imaging modality can improve risk prediction and stratification through direct quantification of atherosclerotic burden [25].

The prognostic value of CAC in asymptomatic individuals has been addressed in numerous reports involving 84,182 individuals [26−30]. The results have shown that CAC scores are proportionally related to clinical outcomes (cardiac and all-cause mortality, myocardial infarction, stroke, unstable angina, transient ischemic attacks). CAC score of 0 was associated with annual major adverse cardiovascular event (MACE) of 0.1%, CAC 1−99 (mild-risk CAC) with MACE of 0.4%, CAC 100−399 (moderate-risk CAC) with MACE of 0.7% and CAC >400 (high-risk CAC) with risk of 1.6%. Traditional risk factors (smoking, diabetes, and hypertension) have been shown to further increase the MACE risk [28,31−33].

It is now well established that individuals with a CAC score of 0 have a very low risk of future CV events over the next 10−15 years. In a study by Blaha and colleagues, negative risk markers: CAC = 0, carotid intima-media thickness (CIMT) < 25th percentile, absence of carotid plaque, brachial flow-mediated dilation >5% change, ankle brachial index (ABI) >0.9 and <1.3, high sensitivity C reactive protein (hsCRP) <2 mg/L, homocysteine <10 mmol/L, NTpro-BP <100 pg/mL, no microalbuminuria, no family history of CHD, and no metabolic syndrome were compared for CVD events over the 10-year follow-up in 6814 participants from the MESA [34]. Among negative risk markers, CAC = 0 was the strongest negative predictor of CVD. To evaluate the accuracy of change in risk classification, net reclassification improvement (NRI) was calculated and CAC = 0 resulted in the greatest, most accurate downward risk reclassification [34]. The JUPITER population was used to compare a biochemical marker of CV disease, hsCRP, to CAC in the prediction of CHD; CAC was found to be superior. CAC prevalence and increasing CAC burden were strong clinical predictors of CV events after full multivariable adjustment [23]. When comparing individuals with CAC = 0 and those with CAC scores 1−10 in a study of 44 052 patients referred for CAC scanning, individuals with CAC 1−10 had higher CV risk than those with CAC = 0 with hazard ratio for all-cause mortality of 1.99 (95% confidence interval [CI]: 1.44−2.75) after adjustment for traditional risk factors. This was the first study to identify low CAC score as a CHD risk predictor compared to CAC = 0 [35]. In the MESA study that included 6809 individuals, those with CAC 1−10 had three times higher relative risk ratio for CHD events compared to individuals with CAC 0 [36].

In a 10-year follow-up, MESA showed benefits of including CAC score into CV risk assessment together with traditional risk factors. McClelland and colleagues proposed a MESA CHD risk score, the first available algorithm incorporating CAC with traditional risk factors. The MESA CHD risk score can be used in clinical practice to communicate a patient's risk factors and to guide risk-based treatment decisions. Moreover, the MESA risk score can be used to calculate and provide a posttest 10 years CHD risk after CAC scanning [37].

The net reclassification index represents the percentage of correctly reclassified CV events into higher or lower risk groups when CAC is added to traditional risk factors. Three primary prevention outcome studies, MESA [38], Heinz Nixdorf [39], and Rotterdam [40], addressed reclassifications of FRS by CAC in asymptomatic population. These studies have shown that 52%−65.6% of the patients in the intermediate-risk group, 34%−35.8% in the high-risk group, and 11.6%−15% in the low-risk group were correctly reclassified by using the CAC score [17]. Thus, CAC should be used to upgrade risk in the low and intermediate FRS groups. These data emphasize the importance of incorporating CAC into CV risk predictions in primary practice. Whether, a high-risk FRS patient should be downgraded based on low or 0 CAC to a lower risk group is still controversial, although there is no data showing that treating patients with 0 CAC improves the outcomes [17,22]. It has been shown that 0 CAC score is associated with 0.3% 5 year all-cause mortality in individuals with 0 traditional risk factors (hypertension, diabetes, smoking), 0.7% in those with one or two traditional risk factors and 1% for those with three or more risk factors over 5.6 years of follow-up. These findings identify the importance of looking at traditional risk factors as a treatable target in patients with elevated CAC, rather than using just traditional risk factors to guide the prevention and treatment of CVD [22]. In another prospective cohort study, the Rotterdam study [41,42] involving 7983 older than 55 years participants, a strong association was found between CAC and myocardial infarction. In the population-based Heinz Nixdorf Recall (HNR [Risk factors, Evaluation of Coronary Calcium, and Lifestyle]) study, CAC was a similar predictor of events as it was in MESA and Rotterdam study [43].

The Pooled Cohort Equation (PCE) was established by Risk Assessment Work Group of ACC/AHA Guidelines and is incorporated in the 2013 ACC/AHA Cholesterol Guidelines and 2016 United States Preventive Services Task Force (USPSTF) guidelines. PCE uses traditional risk factors to estimate the 10-year risk of atherosclerotic cardiovascular disease (ASCVD) in healthy population. Calculated 10-year ASCVD risk >7.5% is considered elevated and identifies candidates for statin treatment. However, PCE has been shown to overestimate the risk of MACE especially in women, younger men, and diverse racial and ethnic groups [44]. Incorporating CAC scores into CV risk assessment can more precisely detect statin therapy candidates as well as those who do not require treatment [45−47].

In a study of 4758 subjects, 41% of subjects who had ASCVD 10-year risk >7.5% and thus were statin eligible had CAC 0% and 57% of subjects with ASCVD 10-year risk 5%−7.5% who were "consider" statin therapy group had CAC 0. In HNR study that enrolled 3745 subjects, CAC score differentiated risk for statin-eligible individuals by AHA/ACC guidelines (2.7 vs. 9.1 coronary events per 1000 person-years for CAC 0 vs. > 100) [48]. Similar results have been reported for the Framingham Heart Study [46] and BioImage study [47]. All these studies suggest that CAC score can help to identify those individuals deemed as statin eligible by PCE who have CAC 0 and can be down-reclassified to a lower risk category with no need for statins.

Opposite is also true; CAC score has been shown to reclassify PCE low-risk individuals to a higher risk group requiring statin treatment. In a MESA analysis, 6.8% of individuals who were classified to ASCVD risk <7.5%, were reclassified to higher risk statin-eligible group based on CAC $\geq$300 or CAC $\geq$75th percentile for age, sex, and ethnicity [30].

Interventional management—CAC implementation in the guidelines

In 2006, CAC was recognized for the first time as an important tool for risk assessment based on subclinical atherosclerosis. Screening for Heart Attack Prevention and Education (SHAPE) guidelines issued the approach to CHD risk analysis based on CIMT and CAC rather than traditional risk factors [49]. SHAPE was based on the concept that the quantity of atherosclerosis measured as CIMT and CAC can direct the decision about appropriate treatment. Moderate-to-high intensity statin therapy was recommended if CAC $\geq$100, along with lifestyle changes [50].

Knowing that CAC scoring is the strongest clinical predictor of future CV events, the 2013 ACC/AHA risk assessment guidelines [51] and the 2013 Cholesterol guideline emphasis on shared decision-making [52] have implemented the option of CAC screening in certain populations. CAC scoring is used to guide recommendations regarding initiation and intensity of statin therapy [53]. More specifically, CAC scoring should be considered in shared decision-making discussions with the individuals in whom the quantitative risk assessment using traditional risk factors is non-conclusive and treatment recommendation is *"consider treatment."*

The most appropriate use of CAC scanning is in asymptomatic individuals without clinical CV disease with 10-year atherosclerotic cardiovascular disease score (ASCVD) 5%—15% who are considered or recommended for statin therapy. CAC = 0 in this group means lower CV risk and can reclassify this group downward to a lower category in which statin therapy can be deferred. Exceptions to this rule are patients with familial hypercholesterolemia or diabetes, in whom statin therapy is still recommended. In clinical practice, among the patients with a 10-year ASCVD risk of 5%—20%, those with CAC 0 do not need a statin treatment, those with CAC 1—99 or CAC <75th percentile for age/gender/race should be on moderate-intensity statin treatment, those with CAC 100—299 or $\geq$75th percentile for age/gender/race should be on high intensity statin treatment and aspirin 81 mg, while those with CAC >300 should be on high statin plus aspirin 81 mg. In a 10-year ASCVD 15%—20% group, downgrading the CV risk can also be considered with CAC = 0 or low scores [53]. In patients with 10-year ASCVD risk >20% CAC scoring is not recommended as it would not change the management (reclassification not recommended). Another scenario when CAC scoring is not recommended is in patients with known CHD or who are already on aggressive medical therapy to prevent CHD. Patients who refuse the treatment despite the CAC results should not undergo CAC testing [50].

CAC progression

CAC progression is defined as a change >15%/year from the baseline CAC score [54] or ≥ 2.5 mm^3 of the square root of the initial volume score [55,56] (Fig. 8.2). Progression of CAC has been associated with higher risk of CHD events, independent of traditional risk factors and baseline CAC scores [57] (add 97). The relationship between CAC progression and CHD event risk is linear, suggesting that the greater the CAC progression, the greater the CHD event risk. Any progression of CAC in those with CAC = 0 at baseline as well as progression of at least 100 units in those with CAC >0 at baseline were associated with increased CHD event risk. Numerous observational reports have shown attenuation of CAC progression by ~20% [58,59]. However, four randomized clinical trials (RCTs) failed to confirm any effects of statins on CAC progression [56,60−63]. These trials have been summarized in a 2008 statement from the European Society of Cardiac Radiology and North American Society for Cardiovascular Imaging showed that statin therapy did not statistically significantly decrease CAC progression [56]. A relatively small sample size and need for longer follow-up are potential explanation for the negative results. Post hoc analysis of eight RCTs that examined the effect of statin on CAC using coronary IVUS showed

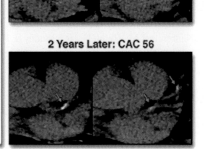

43 Year Old Asymptomatic Male, Father MI 41

	Baseline	2 Years
Lipids:		
TC	244	163
LDL	149	92
HDL	39	52
TG	280	94
Plaque:		
Calcium Score	12	56
Calcium Percentile	75	89
Treatment:		
Statin	None	20 mg
Niacin	None	2000 mg

Baseline: CAC 12

2 Years Later: CAC 56

FIGURE 8.2

Progression of CAC and risk of first MI in 495 asymptomatic patients receiving cholesterol-lowering therapy. CAC score increased from 12 to 56 over the course of 2 years of intensive lipid-lowering therapy, despite normalization of lipid values. *CAC,* coronary artery calcium; *MI,* myocardial infarction; *TC,* total cholesterol; *TG,* triglyceride.

increased CAC that was explained as plaque stabilization [12]. Additional trials using virtual histology IVUS have shown a decrease in volume of necrotic core and increase in fibrofatty plaque with statin treatment [64].

Taken together, repeating CAC scanning offers the ability to identify failure of lipid-lowering agents and to modify treatment toward more aggressive options. In a study that included patients with CAC = 0 who underwent CAC testing every year for 5 years, conversion from CAC 0 to CAC >0 occurred in 25% patients in ~4 years [65]. In 2017, the Society of Cardiovascular Computed Tomography recommended considering repeat CAC scanning at an interval of 5 years for patients with CAC = 0 and 3−5-year interval for patients with CAC >0 in patients for whom the development or progression of CAC would support intensification of the management [66]. CAC scanning and presentation of the CAC score and CAC images to the patients can improve medication adherence as shown in numerous reports that were summarized in a systematic review [48].

References

[1] Demer LL, Tintut Y. Vascular calcification: pathobiology of a multifaceted disease. Circulation 2008;117(22):2938−48.

[2] Li H, Hong S, Qian J, Zheng Y, Yang J, Yi Q. Cross talk between the bone and immune systems: osteoclasts function as antigen-presenting cells and activate CD4+ and CD8+ T cells. Blood 2010;116(2):210−7.

[3] Tintut Y, Morony S, Demer LL. Hyperlipidemia promotes osteoclastic potential of bone marrow cells ex vivo. Arterioscler Thromb Vasc Biol 2004;24(2):e6−10.

[4] Chen NX, Duan D, O'Neill KD, Moe SM. High glucose increases the expression of Cbfa1 and BMP-2 and enhances the calcification of vascular smooth muscle cells. Nephrol Dial Transplant 2006;21(12):3435−42.

[5] Kalampogias A, Siasos G, Oikonomou E, Tsalamandris S, Mourouzis K, Tsigkou V, et al. Basic mechanisms in atherosclerosis: the role of calcium. Med Chem 2016; 12(2):103−13.

[6] Tintut Y, Alfonso Z, Saini T, Radcliff K, Watson K, Boström K, et al. Multilineage potential of cells from the artery wall. Circulation 2003;108(20):2505−10.

[7] Nahrendorf M, Zhang H, Hembrador S, Panizzi P, Sosnovik DE, Aikawa E, et al. Nanoparticle PET-CT imaging of macrophages in inflammatory atherosclerosis. Circulation 2008;117(3):379−87.

[8] Vengrenyuk Y, Carlier S, Xanthos S, Cardoso L, Ganatos P, Virmani R, et al. A hypothesis for vulnerable plaque rupture due to stress-induced debonding around cellular microcalcifications in thin fibrous caps. Proc Natl Acad Sci USA 2006; 103(40):14678−83.

[9] Mori H, Torii S, Kutyna M, Sakamoto A, Finn AV, Virmani R. Coronary artery calcification and its progression: what does it really mean? JACC Cardiovasc Imaging 2018; 11(1):127−42.

[10] Shaw LJ, Narula J, Chandrashekhar Y. The never-ending story on coronary calcium: is it predictive, punitive, or protective? J Am Coll Cardiol 2015;65(13):1283−5.

[11] Motoyama S, Kondo T, Sarai M, Sugiura A, Harigaya H, Sato T, et al. Multislice computed tomographic characteristics of coronary lesions in acute coronary syndromes. J Am Coll Cardiol 2007;50(4):319−26.

[12] Puri R, Nicholls SJ, Shao M, Kataoka Y, Uno K, Kapadia SR, et al. Impact of statins on serial coronary calcification during atheroma progression and regression. J Am Coll Cardiol 2015;65(13):1273−82.

[13] Criqui MH, Denenberg JO, Ix JH, McClelland RL, Wassel CL, Rifkin DE, et al. Calcium density of coronary artery plaque and risk of incident cardiovascular events. J Am Med Assoc 2014;311(3):271−8.

[14] Qazi AH, Zallaghi F, Torres-Acosta N, Thompson RC, O'Keefe JH. Computed tomography for coronary artery calcification scoring: mammogram for the heart. Prog Cardiovasc Dis 2016;58(5):529−36.

[15] Callister TQ, Cooil B, Raya SP, Lippolis NJ, Russo DJ, Raggi P. Coronary artery disease: improved reproducibility of calcium scoring with an electron-beam CT volumetric method. Radiology 1998;208(3):807−14.

[16] Messenger B, Li D, Nasir K, Carr JJ, Blankstein R, Budoff MJ. Coronary calcium scans and radiation exposure in the multi-ethnic study of atherosclerosis. Int J Cardiovasc Imaging 2016;32(3):525−9.

[17] Hecht HS. Coronary artery calcium scanning: past, present, and future. JACC Cardiovasc Imaging 2015;8(5):579−96.

[18] Detrano R, Guerci AD, Carr JJ, Bild DE, Burke G, Folsom AR, et al. Coronary calcium as a predictor of coronary events in four racial or ethnic groups. N Engl J Med 2008; 358(13):1336−45.

[19] Hecht H, Blaha MJ, Berman DS, Nasir K, Budoff M, Leipsic J, et al. Clinical indications for coronary artery calcium scoring in asymptomatic patients: expert consensus statement from the Society of Cardiovascular Computed Tomography. J Cardiovasc Comput Tomogr 2017;11(2):157−68.

[20] Rozanski A, Muhlestein JB, Berman DS. Primary prevention of CVD: the role of imaging trials. JACC Cardiovasc Imaging 2017;10(3):304−17.

[21] Sangiorgi G, Rumberger JA, Severson A, Edwards WD, Gregoire J, Fitzpatrick LA, et al. Arterial calcification and not lumen stenosis is highly correlated with atherosclerotic plaque burden in humans: a histologic study of 723 coronary artery segments using nondecalcifying methodology. J Am Coll Cardiol 1998;31(1):126−33.

[22] Hecht HS. Coronary artery calcium scanning: the key to the primary prevention of coronary artery disease. Endocrinol Metab Clin N Am 2014;43(4):893−911.

[23] Blaha MJ, Budoff MJ, DeFilippis AP, Blankstein R, Rivera JJ, Agatston A, et al. Associations between C-reactive protein, coronary artery calcium, and cardiovascular events: implications for the JUPITER population from MESA, a population-based cohort study. Lancet 2011;378(9792):684−92.

[24] Budoff MJ, Hokanson JE, Nasir K, Shaw LJ, Kinney GL, Chow D, et al. Progression of coronary artery calcium predicts all-cause mortality. JACC Cardiovasc Imaging 2010; 3(12):1229−36.

[25] Blaha MJ, Silverman MG, Budoff MJ. Is there a role for coronary artery calcium scoring for management of asymptomatic patients at risk for coronary artery disease?: clinical risk scores are not sufficient to define primary prevention treatment strategies among asymptomatic patients. Circ Cardiovasc Imaging 2014;7(2):398−408 [discussion].

[26] Shaw LJ, Giambrone AE, Blaha MJ, Knapper JT, Berman DS, Bellam N, et al. Long-term prognosis after coronary artery calcification testing in asymptomatic patients: a cohort study. Ann Intern Med 2015;163(1):14–21.

[27] Yeboah J, McClelland RL, Polonsky TS, Burke GL, Sibley CT, O'Leary D, et al. Comparison of novel risk markers for improvement in cardiovascular risk assessment in intermediate-risk individuals. J Am Med Assoc 2012;308(8):788–95.

[28] Tota-Maharaj R, Blaha MJ, McEvoy JW, Blumenthal RS, Muse ED, Budoff MJ, et al. Coronary artery calcium for the prediction of mortality in young adults <45 years old and elderly adults >75 years old. Eur Heart J 2012;33(23):2955–62.

[29] Nasir K, Rubin J, Blaha MJ, Shaw LJ, Blankstein R, Rivera JJ, et al. Interplay of coronary artery calcification and traditional risk factors for the prediction of all-cause mortality in asymptomatic individuals. Circ Cardiovasc Imaging 2012;5(4):467–73.

[30] Yeboah J, Young R, McClelland RL, Delaney JC, Polonsky TS, Dawood FZ, et al. Utility of nontraditional risk markers in atherosclerotic cardiovascular disease risk assessment. J Am Coll Cardiol 2016;67(2):139–47.

[31] Raggi P, Shaw LJ, Berman DS, Callister TQ. Prognostic value of coronary artery calcium screening in subjects with and without diabetes. J Am Coll Cardiol 2004;43(9): 1663–9.

[32] McEvoy JW, Blaha MJ, Rivera JJ, Budoff MJ, Khan AN, Shaw LJ, et al. Mortality rates in smokers and nonsmokers in the presence or absence of coronary artery calcification. JACC Cardiovasc Imaging 2012;5(10):1037–45.

[33] Schulman-Marcus J, Valenti V, Hartaigh B, Gransar H, Truong Q, Giambrone A, et al. Prognostic utility of coronary artery calcium scoring in active smokers: a 15-year follow-up study. Int J Cardiol 2014;177(2):581–3.

[34] Blaha MJ, Cainzos-Achirica M, Greenland P, McEvoy JW, Blankstein R, Budoff MJ, et al. Role of coronary artery calcium score of zero and other negative risk markers for cardiovascular disease: the multi-ethnic study of atherosclerosis (MESA). Circulation 2016;133(9):849–58.

[35] Blaha M, Budoff MJ, Shaw LJ, Khosa F, Rumberger JA, Berman D, et al. Absence of coronary artery calcification and all-cause mortality. JACC Cardiovasc Imaging 2009; 2(6):692–700.

[36] Budoff MJ, McClelland RL, Nasir K, Greenland P, Kronmal RA, Kondos GT, et al. Cardiovascular events with absent or minimal coronary calcification: the Multi-Ethnic Study of Atherosclerosis (MESA). Am Heart J 2009;158(4):554–61.

[37] McClelland RL, Jorgensen NW, Budoff M, Blaha MJ, Post WS, Kronmal RA, et al. 10-Year coronary heart disease risk prediction using coronary artery calcium and traditional risk factors: derivation in the MESA (Multi-Ethnic study of atherosclerosis) with validation in the HNR (Heinz Nixdorf Recall) study and the DHS (Dallas heart study). J Am Coll Cardiol 2015;66(15):1643–53.

[38] Polonsky TS, McClelland RL, Jorgensen NW, Bild DE, Burke GL, Guerci AD, et al. Coronary artery calcium score and risk classification for coronary heart disease prediction. J Am Med Assoc 2010;303(16):1610–6.

[39] Erbel R, Möhlenkamp S, Moebus S, Schmermund A, Lehmann N, Stang A, et al. Coronary risk stratification, discrimination, and reclassification improvement based on quantification of subclinical coronary atherosclerosis: The Heinz Nixdorf Recall study. J Am Coll Cardiol 2010;56(17):1397–406.

[40] Elias-Smale SE, Proença RV, Koller MT, Kavousi M, van Rooij FJ, Hunink MG, et al. Coronary calcium score improves classification of coronary heart disease risk in the elderly: the Rotterdam study. J Am Coll Cardiol 2010;56(17):1407−14.

[41] Vliegenthart R, Oudkerk M, Hofman A, Oei HH, van Dijck W, van Rooij FJ, et al. Coronary calcification improves cardiovascular risk prediction in the elderly. Circulation 2005;112(4):572−7.

[42] Greenland P, Blaha MJ, Budoff MJ, Erbel R, Watson KE. Coronary calcium score and cardiovascular risk. J Am Coll Cardiol 2018;72(4):434−47.

[43] Lehmann N, Erbel R, Mahabadi AA, Rauwolf M, Möhlenkamp S, Moebus S, et al. Value of progression of coronary artery calcification for risk prediction of coronary and cardiovascular events: Result of the HNR study (Heinz Nixdorf Recall). Circulation 2018;137(7):665−79.

[44] Cook NR, Ridker PM. Further insight into the cardiovascular risk calculator: the roles of statins, revascularizations, and underascertainment in the Women's Health Study. JAMA Intern Med 2014;174(12):1964−71.

[45] Nasir K, Bittencourt MS, Blaha MJ, Blankstein R, Agatson AS, Rivera JJ, et al. Implications of coronary artery calcium testing among statin candidates according to American college of cardiology/American heart association cholesterol management guidelines: MESA (Multi-Ethnic study of atherosclerosis). J Am Coll Cardiol 2015; 66(15):1657−68.

[46] Mahabadi AA, Möhlenkamp S, Lehmann N, Kälsch H, Dykun I, Pundt N, et al. CAC score improves coronary and CV risk assessment above statin indication by ESC and AHA/ACC primary prevention guidelines. JACC Cardiovasc Imaging 2017;10(2): 143−53.

[47] Mortensen MB, Fuster V, Muntendam P, Mehran R, Baber U, Sartori S, et al. A simple disease-guided approach to personalize ACC/AHA-recommended statin allocation in elderly people: the BioImage study. J Am Coll Cardiol 2016;68(9):881−91.

[48] Mamudu HM, Paul TK, Veeranki SP, Budoff M. The effects of coronary artery calcium screening on behavioral modification, risk perception, and medication adherence among asymptomatic adults: a systematic review. Atherosclerosis 2014;236(2):338−50.

[49] Naghavi M, Falk E, Hecht HS, Shah PK, Force ST. The first SHAPE (screening for heart attack prevention and education) guideline. Crit Pathw Cardiol 2006;5(4):187−90.

[50] Blankstein R, Gupta A, Rana JS, Nasir K. The implication of coronary artery calcium testing for cardiovascular disease prevention and diabetes. Endocrinol Metab (Seoul) 2017;32(1):47−57.

[51] Stone NJ, Robinson JG, Lichtenstein AH, Bairey Merz CN, Blum CB, Eckel RH, et al. 2013 ACC/AHA guideline on the treatment of blood cholesterol to reduce atherosclerotic cardiovascular risk in adults: a report of the American College of Cardiology/ American Heart Association Task Force on Practice Guidelines. J Am Coll Cardiol 2014;63(25 Pt B):2889−934.

[52] Martin SS, Sperling LS, Blaha MJ, Wilson PWF, Gluckman TJ, Blumenthal RS, et al. Clinician-patient risk discussion for atherosclerotic cardiovascular disease prevention: importance to implementation of the 2013 ACC/AHA Guidelines. J Am Coll Cardiol 2015;65(13):1361−8.

[53] Hecht HS, Cronin P, Blaha MJ, Budoff MJ, Kazerooni EA, Narula J, et al. 2016 SCCT/ STR guidelines for coronary artery calcium scoring of noncontrast noncardiac chest CT scans: a report of the Society of Cardiovascular Computed Tomography and Society of Thoracic Radiology. J Thorac Imaging 2017;32(5):W54−66.

[54] Raggi P, Cooil B, Ratti C, Callister TQ, Budoff M. Progression of coronary artery calcium and occurrence of myocardial infarction in patients with and without diabetes mellitus. Hypertension 2005;46(1):238−43.

[55] Hokanson JE, MacKenzie T, Kinney G, Snell-Bergeon JK, Dabelea D, Ehrlich J, et al. Evaluating changes in coronary artery calcium: an analytic method that accounts for interscan variability. AJR Am J Roentgenol 2004;182(5):1327−32.

[56] Oudkerk M, Stillman AE, Halliburton SS, Kalender WA, Möhlenkamp S, McCollough CH, et al. Coronary artery calcium screening: current status and recommendations from the European Society of cardiac radiology and North American Society for cardiovascular imaging. Eur Radiol 2008;18(12):2785−807.

[57] Budoff MJ, Young R, Lopez VA, Kronmal RA, Nasir K, Blumenthal RS, et al. Progression of coronary calcium and incident coronary heart disease events: MESA (Multi-Ethnic Study of Atherosclerosis). J Am Coll Cardiol 2013;61(12):1231−9.

[58] Budoff MJ, Lane KL, Bakhsheshi H, Mao S, Grassmann BO, Friedman BC, et al. Rates of progression of coronary calcium by electron beam tomography. Am J Cardiol 2000; 86(1):8−11.

[59] Callister TQ, Raggi P, Cooil B, Lippolis NJ, Russo DJ. Effect of HMG-CoA reductase inhibitors on coronary artery disease as assessed by electron-beam computed tomography. N Engl J Med 1998;339(27):1972−8.

[60] Arad Y, Spadaro LA, Goodman K, Newstein D, Guerci AD. Prediction of coronary events with electron beam computed tomography. J Am Coll Cardiol 2000;36(4): 1253−60.

[61] Raggi P, Davidson M, Callister TQ, Welty FK, Bachmann GA, Hecht H, et al. Aggressive versus moderate lipid-lowering therapy in hypercholesterolemic postmenopausal women: beyond Endorsed Lipid Lowering with EBT Scanning (BELLES). Circulation 2005;112(4):563−71.

[62] Schmermund A, Achenbach S, Budde T, Buziashvili Y, Förster A, Friedrich G, et al. Effect of intensive versus standard lipid-lowering treatment with atorvastatin on the progression of calcified coronary atherosclerosis over 12 months: a multicenter, randomized, double-blind trial. Circulation 2006;113(3):427−37.

[63] Achenbach S, Ropers D, Pohle K, Leber A, Thilo C, Knez A, et al. Influence of lipid-lowering therapy on the progression of coronary artery calcification: a prospective evaluation. Circulation 2002;106(9):1077−82.

[64] Hong MK, Park DW, Lee CW, Lee SW, Kim YH, Kang DH, et al. Effects of statin treatments on coronary plaques assessed by volumetric virtual histology intravascular ultrasound analysis. JACC Cardiovasc Interv 2009;2(7):679−88.

[65] Al Rifai M, Cainzos-Achirica M, Blaha MJ. Establishing the warranty of a coronary artery calcium score of zero. Atherosclerosis 2015;238(1):1−3.

[66] Matthew B, Stephan A, Jagat N, Harvey H. Atlas of cardiovascular computed tomography. 2 ed. London: Springer-Verlag; 2018.

Imaging vascular calcification: where are we headed

Maaz B.J. Syed, MBChB, MSc [1,2], **Mhairi Doris**[1], **Marc Dweck, MD, PhD** [1,3],
Rachael Forsythe, MBChB, MRCS [1,4],
David E. Newby, BA, BSc (Hons), PhD, BM, DM, FRCP, FESC, FRSE, FMedSci [1,5]

[1]*British Heart Foundation Department of Cardiovascular Sciences, Queens Medical Research Institute, University of Edinburgh, Edinburgh, United Kingdom;* [2]*Clinical Research Fellow, Department of Cardiovascular Sciences, University of Edinburgh, Edinburgh, United Kingdom;* [3]*BHF Reader in Cardiology, Consultant Cardiologist, Centre for Cardiovascular Science, University of Edinburgh, Edinburgh, United Kingdom;* [4]*Speciality Registrar in Vascular Surgery, Edinburgh Vascular Unit, Royal Infirmary of Edinburgh, Edinburgh, United Kingdom;* [5]*Professor, British Heart Foundation Centre for Cardiovascular Diseases, University of Edinburgh, Edinburgh, United Kingdom*

Introduction

Arterial calcification is a chronic process that affects all components of the vessel wall and is intimately related to inflammation, atherosclerosis, and cellular necrosis. Calcified plaque exerts a spectrum of effects on the arterial wall including lesions that alter the biomechanical properties of vessels, cause vessel stenosis, and impede tissue perfusion. On the other hand, an abundance of calcium in atherosclerotic caps may protect it from rupture. The relationship between calcium and cardiovascular disease is, therefore, complex.

Risk stratification predominantly relies on clinical factors such as age, smoking habit, hypertension, and hypercholesterolemia. Modern imaging can identify and characterize calcified lesions. Owing to these properties, imaging has become indispensable to cardiovascular risk prediction. Advances in molecular imaging mean that we can detect disease activity at a far earlier stage than was conventionally possible. This is particularly relevant to the imaging of vascular calcification, where the assessment of anatomical calcium burden can be combined with complementary information regarding calcification activity. Together, structural and biological investigations inform us about the contribution of calcifying processes to cardiovascular disease.

Novel imaging techniques improve disease process monitoring and our ability to study therapeutic effects. Combining imaging modalities will ultimately provide the tools to improve risk stratification and effectively direct intervention. This chapter will discuss how such approaches apply to the imaging of calcification in cardiovascular disease.

Coronary Calcium. https://doi.org/10.1016/B978-0-12-816389-4.00009-8

Microcalcification versus macrocalcification

Vascular calcification starts at a microscopic level and at a scale that is beyond the resolution of anatomical imaging modalities. Pathology such as fibrous cap rupture causes increased metabolic activity in the vessel wall, which is reflected in the deposition of tiny calcium-containing crystals [1]. This early-stage microcalcification signifies intense biological activity and exerts its own effects on vascular tissue.

Eventually, calcium coalesces into larger mature fragments that can be readily detected using computed tomography [2]. Inflammatory activity within the arterial tree is cyclical—it undergoes periods of elevated and reduced metabolic activity. Hence, established calcified plaque, in itself, is not a marker of active arterial disease. It may equally have formed many years ago and instead may reflect a dormant disease process.

Calcium deposition is dynamic and present throughout the arterial tree. The mechanism of vessel wall calcification differs based on predisposing risk factors. It can take many forms and can affect the vessel intima or media. Rather than being a passive by-product of degradation, calcium deposition is active and controlled [3]. Atherosclerotic injury predominantly affects the intima and is associated with intimal calcification [4]. Conversely, calcification in chronic kidney disease and diabetes mellitus adopts a crescentic transmural morphology involving the tunica media. Although the triggers for intimal and medial calcification may be different, many of the subsequent pathways appear to be shared [4].

Protective effects of macroscopic arterial calcification

In atherosclerotic disease, the lipid-rich necrotic core is isolated from luminal blood flow by a fibrous cap. Disruption of the fibrous cap results in exposure of the necrotic core to luminal blood flow. Consequently, intense prothrombotic activity abruptly occludes vessels and can cause catastrophic visceral infarction and organ impairment. The fibrous plaque is normally composed of smooth muscle cells organized within an extracellular matrix. The structural integrity of the fibrous cap is critical to prevent plaque disruption—a finding reflected in the observation of thin-capped fibroatheroma being more prone to rupture [5].

Inflammation within the fibrous cap is inherently a reparative process. Multiple pathways ultimately cause the formation of calcified plaque within the fibrous cap. This calcification is mature and structurally stable. It reinforces the strength within the fibrous cap and walls off the thrombogenic components of the necrotic core from the blood pool [6]. In this regard, macrocalcification can be considered protective [7,8].

Microcalcification as a marker of disease activity

Necrosis and inflammation are the key pathological processes that ultimately cause atheroma formation and vascular calcification. Detecting microcalcification allows early identification of cell death associated with increased inflammatory activity that is likely to cause morphological changes over time.

Microcalcification colocalizes with tissue necrosis and can be considered a marker of necrosis-derived inflammatory activity [9,10]. Necrosis causes the spillage of intracellular calcium and phosphate. This triggers the formation of inflammatory mediators that drive microcalcification. The lipid-rich core is one such example of dense necrotic material. Macrophage apoptosis induced by high lipid concentrations creates a toxic environment of high calcium and phosphate ions. These induce endothelial and vascular smooth muscle cell necrosis. A positive feedback loop is created where ongoing cellular necrosis maintains high ion concentrations that increase cell death and, with it, microcalcification [10]. This mechanism ties high atherosclerotic lipid content with cell death and the consequent propagation of aggressive microcalcification.

Inflammation within the vessel wall represents an active phase of disease that affects the fibrous cap. Atherosclerotic plaques that are metabolically active in this way exhibit marked microcalcification due to the expression of bone matrix regulatory proteins by calcifying smooth muscle cells [11]. Fibrous caps with rich macrophage infiltration and reduced smooth muscle cells have a disorganized cellular architecture and increased degradation of the extracellular matrix [5,12]. The loss of cellular architecture weakens the fibrous cap and it starts to thin [13,14]. This happens at an extremely small scale [5,15] and is associated with dense microcalcification [16].

Microcalcification as a mediator of plaque rupture

Vascular smooth muscle cells within the fibrous cap produce components of the extracellular matrix. Macrophage infiltration and cellular necrosis causes the release of calcium. This triggers a functional change in vascular smooth muscle cells, which undergo vesicle-mediated calcification themselves [17]. A reduction in vascular smooth muscle cell numbers further destabilizes plaque through a reduction in extracellular matrix formation. Here, microcalcification propagates further calcification of the cellular components within atheroma. This positive feedback loop may explain the mechanism by which microcalcification reinforces structural vulnerability within atheromatous plaque.

Advances in ex vivo tissue analysis allow the detection of minute (sub 10-μm) structures of atherosclerotic plaque in exceptionally high detail. This allows the mapping of tiny calcium-containing vesicles in the fibrous cap. Finite element analysis using advanced computational algorithms can estimate the structural effects of these vesicles on the fibrous cap itself. Calcified macrophages and smooth muscles are thought to cause areas of marked focal stress when the fibrous cap is thin [18]. This relationship is most pronounced in areas where larger calcium-containing vesicles are concentrated in a small area [2]. Microcalcification vesicles that are fully embedded within the fibrous cap present higher wall stress concentrations and propagation of stress through the fibrous cap [19,20]. The presence of microcalcification itself may directly compromise the physical stability of the fibrous cap.

Increased microcalcification in vulnerable plaque is well established. Microcalcification mediates this process through encouraging further inflammation and altering the cellular composition of the fibrous cap. The structural integrity of the fibrous cap is further compromised by the physical disturbances caused by tiny calcium crystals within it. A combination of the biological and structural properties of microcalcification mediates plaque rupture.

Calcification and high-risk plaque features

Atherogenesis leads to arterial calcification. Atherosclerosis is a multifocal immune-mediated condition of medium and large-sized arteries. An intimal injury is caused by exposure to hypertension, hypercholesterolemia, and smoking. These create a systemic environment that encourages endothelial dysfunction [21], oxidation of lipoproteins [22,23], and the production of free oxygen radicals [24]. Oxidized lipoproteins coalesce into fatty streaks. The resultant endothelial injury promotes macrophage infiltration within the vessel wall [3]. The lipid-rich necrotic core produces a microenvironment of localized hypoxia [25]. The production of hypoxia-related factors promotes angiogenesis. This stimulates macrophages and smooth muscle cells to express $\alpha_v\beta_3$ integrin glycoproteins on their surfaces [26]. Ultimately, the high lipid content of macrophages results in cellular apoptosis. This debris, along with necrotic endothelial and smooth muscle cells, is the primary constituents of the atherosclerotic lipid-rich core [27,28]. Macrophage also invades the fibrous cap, whose presence is associated with focal cap thinning and colocalizes with microcalcification causing structural instability [3].

Inflammatory activity in the plaque precedes destabilization. The sequence of events leading to rupture involves necrotic core—driven microcalcification and abnormal calcific differentiation of smooth muscle cells in the fibrous plaque. Exposure of the lipid core to luminal blood flow triggers a prothrombotic response. The resulting thrombus may cause sudden occlusion of the vessel or embolize distal to the rupture site. Over time, macrophage-mediated reparative processes incorporate adherent thrombus within the plaque [29]. Further calcification occurs because of this [30]. An unstable plaque may rupture multiple times before exerting any clinical effects. Episodic subclinical plaque rupture results in sequential stenosis of the vessel at the rupture site and deposition of calcium at each interval.

Glycolytic activity is a proxy for generalized metabolic activity within tissue and is typically increased in response to inflammation. Structures with a high demand for glucose metabolism have higher baseline glycolytic activity within their cells. Areas with high metabolic activity within the arterial wall are likely to have increased glycolysis. However, this glucose metabolism is not specific to arterial disease. Indeed, structures with high baseline metabolic demands, such as the heart, have an immense capacity for glycolysis to function adequately [31]. Glycolysis is a useful marker of overall cellular activity but is compromised by its nonspecific nature.

Imaging vascular calcification
Principles of anatomical imaging

Calcification most commonly causes nonobstructive lesions but with advanced atheroma can impede blood flow. Anatomical imaging relies broadly on detecting stenotic disease in two ways: imaging the arterial lumen, or visualizing the vascular calcium directly.

Catheter delivered angiography was long considered the gold standard to image the arterial lumen. However, this paradigm has shifted in modern practice. It is possible to get highly detailed images of the coronary and peripheral vessels using noninvasive angiographic techniques. Chief among these is computed tomography angiography (CTA) [32,33]. Contrast media during CTA can delineate the boundaries of the vessel lumen with exceptional detail as well as image the plaque within the arterial wall. As computed tomography (CT) image acquisition uses X-rays, it has the added benefit of directly visualizing areas of established calcification and quantifies the overall burden of calcium in vascular beds [34].

Magnetic resonance angiography (MRA) can also be used to detect the boundaries of the arterial lumen. A gadolinium-based contrast medium is used for these purposes. However, recent concerns with gadolinium accumulation in the brain are leading to changes in the use of this imaging modality [35]. Novel agents such as manganese or iron oxide particle-based contrast media are being evaluated for their role in detecting vessel and tissue quality [36]. Nonetheless, MRA can detect vessel narrowing in the carotid and lower limb arteries. Indeed, MRA is already the first-line recommended imaging modality to evaluate specific conditions [37,38]. Improvements in the spatial resolution and reduced acquisition times of MRA are increasing its application in detecting coronary artery disease [39].

Duplex ultrasound has the added advantage of combining structural and functional data to quantify the degree of vessel stenosis. Ultrasound is noninvasive, portable, widely available, and inexpensive. It is particularly useful to study accessible vessels such as the aorta, carotid arteries, and arteries of the upper and lower limbs. Despite its widespread use, ultrasound does have some drawbacks. A chief limitation is that ultrasound cannot image calcified plaque directly because of acoustic shadowing. This means that the degree of stenosis is inferred by changes in velocity rather than direct assessment of the lumen.

Detecting microcalcification using ^{18}F-sodium fluoride positron emission tomography

Detecting microcalcification has added value because it provides a new perspective on the biological activity within tissue. Areas of microcalcification do not always coincide with established macrocalcification [40]. Microcalcification occurs at too small a scale to be observed by conventional anatomical imaging techniques and thus requires additional imaging techniques.

Positron emission tomography (PET) is a powerful imaging technique that can map radioactivity in the body following the administration of positron-emitting radiotracers. This technique can be used to detect specific disease or pathological processes by targeting a pathway using a selective radiotracer. Radiotracers are compounds that generally contain two components. The first is a ligand that binds to the target structure or molecule of interest. The second component holds a biologically compatible radio-isotope with a predictable rate of decay. The PET scanner can detect this radioactivity and produce a three-dimensional map of radiotracer binding in the body.

Most conventional imaging techniques provide high-resolution morphological data. In contrast, PET is designed to detect individual disease processes and gives relatively poor anatomical information with a resolution of only 3–5 mm. Data on biological activity obtained from PET are typically combined with morphological data from CT or magnetic resonance imaging (MRI). Hybrid PET/CT and PET/MR imaging offers unprecedented insight into the assessment of vascular structures. This extends beyond merely characterizing the distribution of vascular "macro"-pathology. PET informs on the activity of disease processes that are associated with vascular structural changes.

^{18}F-sodium fluoride (^{18}F—NaF) is a well-tolerated radiolabeled biological tracer that is soluble and has a predictable rate of decay. Dissociated ^{18}F—NaF ions bind to exposed hydroxyapatite crystals that are deposited as the primary components of microcalcification [41]. New calcium mineralization occurs within vascular tissue as a reparative response to injury from harmful external stimuli. Hence, by binding to areas of microcalcification, ^{18}F—NaF can detect injured areas that are undergoing an immune-mediated reparative process (Fig. 9.1). As ^{18}F—NaF binds to areas of microcalcification as opposed to general tissue metabolism, it is not taken up by the myocardium [42]. This property makes ^{18}F—NaF a particularly useful tracer for assessing structures close to the heart, such as the coronary vessels.

An ^{18}F—NaF PET scan can acquire a three-dimensional map of radiotracer binding. Combining PET data with images from CT or MRI scans provides highly detailed anatomical context to sites of ^{18}F—NaF uptake. ^{18}F—NaF is a promising radiotracer in the field of cardiovascular medicine and has proven useful in detecting disease in multiple vascular beds [40,43–46].

Positron emission tomography—computed tomography

Hybrid PET/CT is the most widely used modality to study the heart and major arteries. CT has excellent spatial resolution and can detect differences in tissue down to half a millimeter. Timed contrast agents can delineate the arterial lumen from its wall and adjacent atheroma. This is particularly useful to detect arterial stenosis or occlusion in a noninvasive way. CT has the benefit of visualizing the entire vessel from origin to target structure. CT is also widely accessible, and images are obtained with short acquisition times. This makes CT an ideal modality to detect diseased arteries and plan interventions [47,48].

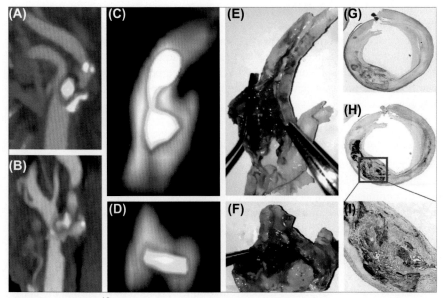

FIGURE 9.1 Carotid ¹⁸F—NaF uptake and carotid plaque rupture.

In vivo (A and B) and ex vivo (C and D) positron emission and computed tomograms showing colocalization of ¹⁸F-sodium fluoride (¹⁸F—NaF) uptake (yellow-orange) to the site of plaque rupture with adherent thrombus on excised carotid endarterectomy tissue (E and F). Histology of the ¹⁸F—NaF-positive region shows a large necrotic core (Movat's pentachrome, magnification 4×, G), within which increased staining for tissue nonspecific alkaline phosphatase can be seen as a marker of calcification activity on immunohistochemistry (magnification 4×, H; magnification 10×, I).

From Joshi NV, Vesey AT, Williams MC, Shah AS, Calvert PA, Craighead FH, Yeoh SE, Wallace W, Salter D, Fletcher AM, van Beek EJ, Flapan AD, Uren NG, Behan MW, Cruden NL, Mills NL, Fox KA, Rudd JH, Dweck MR, Newby DE. ¹⁸F-fluoride positron emission tomography for identification of ruptured and high-risk coronary atherosclerotic plaques: a prospective clinical trial. Lancet. 2014; 383(9918):705–13 . Reprinted with permission from Elsevier (The Lancet, 2014, Vol No.383, 705-13).

Beyond the assessment of luminal stenosis, CT can provide a detailed assessment of plaque morphology. Owing to its high spatial resolution, CT can visualize high-risk features. These include positive remodeling, spotty calcification, high attenuation fibrous plaques, and low-attenuation lipid-rich necrotic plaques [49—51]. Although powerful, CT alone cannot detect biological activity within vascular beds and cannot directly detect developing microcalcification that is beyond its spatial resolution.

Combining CT with PET allows clinicians to take advantage of both imaging modalities. CT can detect areas of arterial calcification and stenosis. PET informs the clinician of biological activity within target vessels by using radiotracers that target specific disease processes such as microcalcification and glycolysis. Combined PET/CT offers a more complete insight into the pathological processes involved in vascular calcification. It enables clinicians to differentiate areas of active biological disease from regions of quiescent established pathology. Areas of luminal stenosis by macrocalcification may not always coincide with zones of high biological activity. The latter is associated with greater plaque vulnerability and subsequent rupture of the thin fibrous cap. Combined PET/CT is thus a powerful tool to differentiate specific disease processes and provide the information required to deliver personalized care.

Positron emission tomography—magnetic resonance imaging

MRI obtains detailed and accurate anatomical images without exposure to ionizing radiation. This makes MRI particularly useful in situations where radiation exposure must be minimized, such as when scanning a pediatric population, pregnant females, or study participants requiring repeated longitudinal assessments.

MRI provides excellent soft-tissue differentiation and improves discrimination of atherosclerotic features. Furthermore, MRI has an excellent temporal resolution [52]. This permits the acquisition of images over predefined time periods to visualize tissues in a mechanically dynamic state. One example is to study myocardial contractility for areas of altered muscle function [53]. Blood flow through the aortic valve and into the ascending aorta can also be visualized [54]. These data can be processed to determine arterial stiffness and areas of vessel wall shear stress.

The administration of gadolinium-based contrast media during MRI can further differentiate soft tissue [55]. In blood vessels, gadolinium-based contrast media can differentiate the blood pool from the adjacent vessel wall, atheroma, and fibrosis. Multicontrast MRI (T1-weighted, T2-weighted, and proton density) offers excellent soft-tissue characterization and allows the constituents of atherosclerotic plaque to be studied without the need for ionizing radiation [56]. MRI is particularly well suited to visualizing larger vessels with limited movements, such as the aorta, carotid, and lower limb arteries.

Combining MRI with PET has proven challenging. However, a significant breakthrough in scanner technology resulted in the development of detectors that can function within MR scanners [57]. PET images typically require image acquisition times from 20 min to 1 h. Combined PET/MRI scanners can obtain PET data simultaneously with MRI images—a strategy that enables accurate coregistration of the MRI and PET datasets. The result is a precise combination of anatomy, tissue characterization, and functional information.

Quantification of PET uptake

The assessment of radiotracer uptake by PET imaging necessitates a uniform approach to allow independent comparisons of scans. The standardized uptake value (SUV) is a widely accepted approach [58]. Standardized uptake values are complex ratios that account for the calibration of PET scanners, the dose of radioactivity injected, and the body weight of the subject.

SUV is often measured within a region of interest and may require bespoke software. SUVs are often reported as max (SUV_{max}) or mean (SUV_{mean}) values. The quantification of radiotracer binding can be adjusted further by calculating the ratio of peak SUV in the area of interest to mean SUV in the blood pool. This ratio is expressed as the tissue-to-background ratio (TBR). The vena cava and the right atrium are reliable areas to measure SUV_{mean} within the blood pool. The TBR is arguably a more reliable unit of radiotracer activity [59].

Imaging high-risk atheromatous features

It is now possible to detect disease processes associated with vessel calcification directly. This approach provides unique insights into plaque features found in association with both micro- and macrocalcification, as well as an understanding of the processes leading to calcium formation.

Structures such as the lipid-rich core, thin fibrous cap, angiogenesis, positive remodeling, and subclinical plaque rupture can be seen in detail using invasive and noninvasive techniques (Table 9.1). PET informs us of the metabolic activity associated with calcifying processes such as glycolysis, hypoxia, and angiogenesis. A combination of smart contrast agents and novel biological radiotracers can detect the cellular composition of tissue and the distribution of macrophages. We discuss later the methods for imaging pathological features related to vessel calcification and how these approaches have improved our understanding of vascular mineralization.

Thin fibrous cap and subclinical plaque rupture

Fibrous cap thinning occurs at an extremely small scale and requires invasive imaging to be detected directly. Optical coherence tomography (OCT) offers exceptionally high-resolution images of the arterial wall and atheromatous plaque. It is a performed using a catheter mounted probe that is passed over a guidewire and placed within the intravascular space adjacent to the atheroma. Using fast frequency-domain analysis, OCT captures detailed images of the adjacent plaque. It can reliably detect thinning of the fibrous cap and identify suspected lesions [60]. The resolution of OCT is so high that some authors report the detection of macrophages within the fibrous cap [61]. These macrophages cause focal deposition of microcalcification within the fibrous cap itself—a marker of disease activity and thought to destabilize atheromatous plaque directly by structurally weakening it.

Detecting plaque rupture can identify culprit lesions in thromboembolic disease. In addition, invasive techniques can visualize thrombus adherent to plaque rupture directly [62−64]. Catheter-based techniques are invasive and require the passage of a guidewire, sometimes across fresh thrombus. The risk of distal embolism with this approach is high. The ability of MRI to differentiate between soft tissues enables assessment of atherosclerotic plaque in a noninvasive and radiation-free

Table 9.1 High-risk plaque features and imaging techniques to detect them.

	Catheter angiography	OCT	NIRS - IVUS	CT	MRI	PET	Radiotracer
Luminal stenosis	✓	✓	✓	✓	✓		
Calcified plaque		✓	✓	✓		✓	¹⁸F-sodium fluoride
Microcalcification							
Thin fibrous cap		✓	✓		✓		
Subclinical plaque rupture		✓	✓				
Lipid-rich necrotic core		✓	✓		MRS		
Macrophage infiltration		✓		✓	USPIO	✓	⁶⁸Ga-DOTATATE, Choline, TSPO
Glycolysis						✓	¹⁸F-FDG
Hypoxia and angiogenesis						✓ ✓	¹⁸F-RGD,¹⁸F-MISO

CT, computed tomography; ¹⁸F-FDG, ¹⁸F-fluorodeoxyglucose; ¹⁸F-MISO, ¹⁸F-fluoromisonidazole; ¹⁸F-RGD, ¹⁸F-arginine–glycine–aspartic tripeptide sequence; ⁶⁸Ga-DOTATATE, [⁶⁸Ga-DOTA⁰-Tyr³]octreotate; IVUS, intravascular ultrasound; MRI, magnetic resonance imaging; MRS, magnetic resonance spectroscopy; NIRS, near-infrared spectroscopy; OCT, optical coherence tomography; PET, positron emission tomography; TSPO, translocator protein ligands; USPIO, ultrasmall paramagnetic particles of iron oxide.

way. T1-weighted high-intensity plaque imaging can detect intraplaque hemorrhage within medium- and large-sized arteries [65,66].

Novel fibrin and platelet-targeted PET radiotracers under development can directly detect fresh thrombus, even in small vessels. This is an exciting prospect. If successful, such tracers will detect fresh thrombus adherent to culprit plaques through PET/CT imaging without the risk of conventional invasive strategies [67,68]. Furthermore, fibrin and platelet PET tracers have the potential to detect plaque rupture in atheroma that may not be causing significant luminal stenosis. This has been a major limitation of conventional morphology-based imaging techniques.

Lipid-rich necrotic core

The lipid-rich necrotic core directly triggers proinflammatory pathways that increase calcium mineralization [69]. Multiplanar reconstruction on CTA of arteries can obtain a cross-section of the vessel perpendicular to its axis. In this reconstructed plane, plaque can be interrogated systematically. Fibrous plaque appears as high attenuation areas, whereas focal necrotic areas are typically of low attenuation [49,50]. The ability of MRI to differentiate between soft-tissue types is particularly useful in seeing the plaque characterization and detecting the lipid-rich core. T1-weighted MRI with gadolinium enhancement can increase the sensitivity of detecting the lipid-rich core [70]. Associated calcification appears hypodense on MRI, whereas fibrous tissue has a low attenuation on T1 and high attenuation on T2-weighted images [71].

Detecting the constituents of the necrotic core allows plaque characterization. Invasive imaging can capture high-resolution images of the lipid-rich pool. Intravascular ultrasound (IVUS) and near-infrared spectroscopy (NIRS) probes capture images from extreme closeness to the intima. NIRS uses wave scatter to produce a gradient map corresponding to the probability of adjacent lipid [72,73]. A lipid core burden index (LCBI) is calculated by an image-processing console contained within modern NIRS machines. The LCBI is the ratio of high lipid content in areas adjacent to the probe against the lipid content in the total study area [74]. Contemporary machines will typically display areas of high lipid content as a yellow gradient over a red ring. Combining IVUS and NIRS provides morphological context to the distribution of lipid within plaque.

Magnetic resonance spectroscopy (MRS) offers a noninvasive technique to detect lipid-rich necrotic cores. In essence, MRS can detect the chemical composition and metabolic state of vascular tissue by detecting a range of atoms. These include 1-Hydrogen (1H), 31-Phosphorus (31P), and 13-Carbon (13C) [75]. MRS can detect cholesterol esters, which are the dominant lipid class in the necrotic core. Regions of interest are identified around the plaque being studied. Chemical shift imaging sequences are used to acquire a spectrum over this region of interest. The amplitudes for specific metabolites, such as lipids, can be measured from acquired spectrums and interpreted as a ratio to the amplitude of intrinsic water.

The final analysis allows detection and quantification of the lipid content within atherosclerotic plaque [76,77].

A large necrotic pool in plaque is a high-risk morphological feature. Imaging now enables us to study the composition of these lipid pools. Detecting the lipid-rich can identify areas of vulnerable plaque in vivo and those at risk of generating calcifying disease.

[18]F-fluoromisonidazole (^{18}F-MISO) is a PET radiotracer that accumulates in viable hypoxic cells due to accumulation of its metabolites in an oxygen-deprived environment [78]. PET scanning can detect ^{18}F-MISO accumulation and produce a map of the distribution of hypoxic tissue [79]. ^{18}F-MISO use is well established in detecting tumor hypoxia [80]. This application may be applicable to directly image myocardial or plaque ischemia. More recently, a novel PET tracer, ^{18}F-Galacto-RGD has been developed. This tracer targets $\alpha_v\beta_3$ receptors directly and can be used with current PET/CT and PET/MRI scanners to detect angiogenesis associated with plaque hypoxia [81].

Macrophage infiltration

Identifying areas of vascular macrophage infiltration is possible by targeting macrophages directly. Activated macrophages express type-2 somatostatin (SST2) receptors [82]. Radiolabeled SST2 analogues can target macrophages directly and overcome the specificity limitations of ^{18}F-FDG. Here, a combination of a somatostatin ligand (-TATE or -NOC) with a DOTA- or NOTA-based "cage" has resulted in the development of various tracers. These cages house a positron-emitting isotope—usually Gallium-68 (^{68}Ga) or Copper-64 (^{64}Cu) [83–85]. Examples of these include ^{68}Ga-DOTATATE ([[68]Gallium-DOTA[0]-Tyr[3]]octreotate) and [68]Gallium-DOTANOC ([^{68}Ga-DOTA[0]-1NaI[3]]octreotide). These agents bind to somatostatin 2 receptors with varying affinities and are being evaluated for optimum uses.

The translocator protein (TSPO) ligands are another promising target for vascular macrophages. TSPO are 18 kDa proteins expressed in the mitochondria of most cells, but markedly upregulated in activated macrophages [86]. This finding is mirrored by histological analysis of culprit plaques abundant in TSPO containing activated macrophages [87]. Novel TSPO ligands, such as 11C-PK11195, can be used as radiotracers in conjunction with PET to isolate macrophage-rich plaque in noninvasive ways [87,88]. Human studies confirm high macrophage-mediated inflammatory activity within culprit plaques that have high TSPO ligand binding [89,90].

"Smart" MRI contrast agents offer an alternative approach to directly target macrophages with the benefit of eliminating exposure to ionizing radiation. One such contrast agent is ultrasmall superparamagnetic particles of iron oxide (USPIOs). These are 30-nm iron oxide containing nanoparticles that are stabilized with low molecular weight dextran [91]. Activated macrophages phagocytose USPIO and remain in the circulation for extended periods. Images are acquired by administering an intravenous dose of USPIO following a baseline T2 and T2*-weighted MRI [92].

Participants are reimaged once USPIO uptake has occurred. This is typically 24−36 h later. Areas rich in USPIO-positive macrophages appear as low-signal intensity on repeat T2 and T2*-weighted MRI [92]. The difference between the two regions of interest can be calculated, and a gradient map applied to highlight areas of increased macrophage infiltration.

Markers of global metabolic activity

^{18}F-Fluorodeoxyglucose (^{18}F-FDG) is a radiolabeled glucose analogue that can be used detected by PET/CT and PET/MRI scanners. Following intravenous administration, ^{18}F-FDG is taken up by metabolically active cells. The immediate metabolite retains the radiolabel and is trapped within cells following phosphorylation. Due to the generalizability of glycolysis in multiple disease states, ^{18}F-FDG has found its place as the most widely used radiotracer in clinical practice [93−95]. In cardiovascular medicine, ^{18}F-FDG is a pioneering radiotracer to detect vessel and myocardial disease.

However, due to the nonspecific uptake of ^{18}F-FDG by tissue, interpretation of results is inconsistent. Furthermore, due to spillover from high glycolytic activity in the myocardium, interpretation of ^{18}F-FDG uptake in adjacent structures is not always possible [40]. One possible workaround is to suppress myocardium uptake using a low-carbohydrate, high-fat diet [96]. Despite these efforts, however, myocardial suppression is ineffective in around a quarter of patients [97]. This limits interpretation of coronary ^{18}F-FDG uptake.

Clinical translation

The overarching objective in clinical practice is to detect disease accurately and stratify the risk of disease progression. Catheter angiography, computed tomographic angiography, and magnetic resonance angiography offer exceptionally high detailed images of the arterial lumen. Invasive imaging techniques provide morphological details about individual plaques. For accessible vessels, ultrasound is an accessible tool to evaluate flow in a target vessel. These modalities form the cornerstone of central and peripheral vascular imaging.

Molecular imaging can now detect specific disease processes. PET can be combined with established anatomical imaging modalities to add further perspective to vascular imaging by detecting areas of metabolically active vascular disease independent of morphological features. The extra dimension of cardiovascular imaging offered by PET provides a platform to direct therapy to metabolically active segments. PET in the vascular system has been shown to detect metabolically active plaques in the coronary, carotid, and peripheral arteries [98−100]. The translation of these techniques into mainstream clinical practice represents the next paradigm shift in cardiovascular imaging.

Mechanisms of vascular occlusion

From a clinical perspective, vascular occlusive disease occurs through specific disease patterns. It is essential to understand these pathological mechanisms to determine the most appropriate imaging modality.

Chronic vascular occlusion from progressive calcification and narrowing of the artery causes a state of long-term organ hypoperfusion. This causes chronic stable ischemia and is responsible for exertional symptoms in stable patients. Ischemia is typically worsened when the metabolic demands of tissue increases—usually under a period of strain. A chronic inability of end organs to meet their metabolic demands results in structural and functional remodeling of the tissue. This initial adaptive response can later become pathological as the organ "overcompensates" over periods of poor metabolic compliance. In the coronary arteries, vessel stenosis causes myocardial ischemia and symptoms of angina on exertion. In the peripheral vasculature, it can lead to intermittent claudication and the development of pain in the calf on walking.

Acute arterial occlusion results from two distinct yet interrelated mechanisms. The first is following plaque rupture and exposure of the lipid-rich necrotic core to luminal blood. The atherosclerotic plaque may cause preexisting vessel stenosis. This plaque is typically highly biologically active with acute-on-chronic inflammatory processes [101]. Rapid thrombus formation at the site of plaque rupture causes complete vessel occlusion at this level. This is the most common mechanism of initial injury that leads to myocardial infarction or lower limb tissue death in acute-on-chronic lower limb ischemia.

In contrast, thrombus may form elsewhere in the body and be carried downstream by blood flow. In the case of the carotid arteries, for instance, acute plaque rupture at the bifurcation of the common carotid artery results in thrombus formation within the internal carotid artery. Blood flow can carry a relatively small thrombus from this site into the anterior cerebral circulation and cause a thromboembolic stroke. Similarly, a thrombus formed in the heart or thoracic aorta can be carried in the bloodstream to previously healthy visceral or limb arteries. Embolic disease causes complete and abrupt occlusion of the target vessel. This may result in catastrophic end-organ tissue loss and necessitates immediate treatment.

Coronary artery disease

The investigation of choice in coronary artery disease is dictated by the underlying pathological mechanism. In chronic coronary disease, imaging can visualize areas of stenosis and quantify the burden of calcification. In the acute setting, imaging can also detect culprit plaques causing vessel occlusion. The information from these imaging modalities can direct treatment and monitor disease progression.

Imaging the coronary vessels has become an increasingly important part of predicting risk in patients with stable chest pain. Traditionally, coronary artery disease risk was estimated using clinical risk prediction tools. In the event of a suspected diagnosis of angina, exercise testing and myocardial perfusion studies were recommended [102]. Diagnostic angiography was an alternative approach to imaging the coronary vessels. However, this approach has fallen out of favor in low- and medium-risk groups. Catheter angiography is invasive, resource heavy, and presents a low but significant risk to the patient. Although angiography is highly sensitive in visualizing luminal stenosis, it is unable to detect areas of subtle calcification within the coronary vessels without extra specialist equipment.

In 2016, NICE updated its guidelines in favor of CT coronary angiography to investigate individuals with stable chest pain. CT coronary angiography is noninvasive and widely available. CT coronary angiography is remarkably sensitive to detect coronary stenosis greater than 50% [103]. Hence, CTCA offers a reliable and low-risk approach to exclude coronary artery disease in low- and medium-risk groups. The SCOT-HEART trial highlighted the benefit of CTCA in patients with stable chest pain [32]. It showed that CT coronary angiography could reclassify the diagnosis of coronary artery disease in up to a quarter of patients and improve the accuracy of diagnosing angina compared to legacy risk-stratification methods [32]. This approach enables the early detection of coronary artery disease and allows better therapeutic targeting. Five-year follow-up results from SCOT-HEART showed that more patients with chest pain were prescribed antianginal medication and prophylactic therapy because of CT coronary angiography. This, perhaps, explains the reduction in coronary heart disease death or nonfatal myocardial infarction seen in the CTCA group [104].

The major risk posed by coronary atheroma is plaque rupture. It follows that cardiovascular events appear to be more closely related to plaque and lipid burden. The more plaque an individual has, the more likely that one will rupture and cause myocardial ischemia. Computed tomography has the advantage of directly visualizing coronary calcification. CT makes it possible to quantify the degree of coronary calcification without giving an additional contrast agent [34]. The resultant coronary artery calcium (CAC) score gives a measure of overall plaque burden and improves our ability to predict future cardiovascular events in individuals at medium and high risk of ischemic heart disease (Framingham risk score $>10\%$). The calcium score is particularly helpful in capturing individuals at high risk of future cardiovascular events. These individuals would otherwise be incorrectly classified using traditional scoring methods [105].

Metabolic imaging in chronic arterial disease

We have robust techniques to visualize the vessel lumen and plaque constituents. Still, these modalities do little to inform the clinician of the metabolic activity within coronary plaque. An area of significant stenosis may be biologically dormant. Conversely, minimally stenotic atherosclerosis may be actively inflamed and at-risk of plaque rupture. Molecular imaging has the potential to bridge this gap in our assessment of coronary artery disease. Positron emission tomography has traditionally struggled to assess the coronary vessels because of the nonspecific uptake of ^{18}F-FDG by the myocardium. Considerable spillover of uptake signals from the myocardium makes the interpretation of coronary artery ^{18}F-FDG uptake extremely challenging.

Novel radiotracers such as ^{68}Ga-DOTATATE [106] may overcome the limitations of FDG. Detecting activated macrophages in atherosclerotic plaque is a far more sensitive approach than detecting global glycolytic activity. ^{68}Ga-DOTA-TATE is a somatostatin type-2 receptor antagonist with a ^{68}Ga radiolabel that directly detects activated macrophages. In clinical studies, histological comparisons confirm preferential binding of ^{68}Ga-DOTA-TATE to plaque rich in CD68-positive macrophages [106]. ^{68}Ga-DOTA-TATE PET is able to identify culprit lesions with good reproducibility and higher sensitivity than ^{18}F-FDG [107]. Macrophage-derived inflammatory activity is strongly implicated in atherosclerotic plaque progression and vessel wall calcification. Hence, directly imaging macrophages in the arterial wall can help to localize areas of focal inflammation and calcifying activity.

Other radiotracers such as ^{18}F-sodium fluoride ([18]F−NaF) are highly specific to areas of microcalcification and are not absorbed by the adjacent myocardium. This makes [18]F−NaF an ideal agent to study biological activity within the coronary arteries. [18]F−NaF PET/CT combines the powerful features of CT coronary angiography and coronary artery calcium (CAC) scoring with the metabolic activity within plaque. In a study of 119 participants, [18]F−NaF PET/CT findings correlated strongly with Framingham risk scores ($P = .011$), prior cardiovascular events ($P = .016$), and underlying angina ($P = .023$) [108]. The ability of [18]F−NaF PET/CT to predict future cardiovascular events remains to be seen. This is the focus of the PRE[18]FFIR study (NCT02278211), which is an ongoing prospective multicenter trial in patients with multivessel disease and recent acute coronary syndrome.

Molecular imaging in acute coronary disease

In acute coronary syndrome, catheter angiography offers a route to simultaneously perform invasive imaging and therapeutic intervention. Indeed, a combination of coronary angiography, optical coherence tomography, IVUS, and NIRS can visualize high-risk features at extreme resolutions (Figs. 9.2 and 9.3). These techniques provide the necessary information to guide therapy for offending plaques. However, this approach is invasive and only useful for individuals requiring percutaneous coronary intervention.

Not all patients with acute coronary syndrome require invasive catheter angiography. In this group, it is difficult to detect culprit plaques. Certainly, assessment of vessel stenosis alone is unable to inform us on the metabolic activity within tissues and not all stenotic lesions are equally at risk of rupture. Combined [18]F−NaF PET with CT angiography offers a noninvasive alternative to obtain complementary information beyond arterial morphology alone. CT coronary angiography can detect luminal irregularities, whereas [18]F−NaF PET can reliably identify IVUS-confirmed culprit plaques [109−111]. The predictive value of high [18]F−NaF uptake by culprit plaques in acute coronary syndrome is currently being studied. A noninvasive approach to detect calcification activity can help to identify culprit lesions and direct intervention.

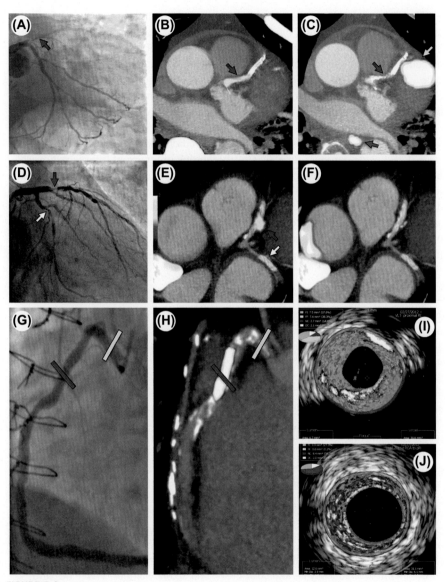

FIGURE 9.2

Focal ^{18}F-fluoride and ^{18}F-fluorodeoxyglucose uptake in patients with myocardial infarction and stable angina. Patient with acute ST-segment elevation myocardial infarction with (A) proximal occlusion (red arrow) of the left anterior descending artery on invasive coronary angiography and (B) intense focal ^{18}F-fluoride (^{18}F-NaF, tissue-to-background ratios, culprit 2·27 versus reference segment 1·09 [108% increase]) uptake (yellow-red) at the site of the culprit plaque (red arrow) on the combined positron emission and computed tomogram (PET-CT). Corresponding ^{18}F-fluorodeoxyglucose PET-CT

Complementary imaging techniques in longitudinal studies can safely detect changes in calcified plaque over time. This information can be used to monitor the response to therapy. These can be used to monitor biological responses to pharmaceutical agents, such as statins and the new proprotein convertase subtilisin/kexin type 9 (PCSK9) inhibitors [112,113]. The ongoing DIAMOND study (NCT02110303) is a prospective trial that aims to determine the effects of dual antiplatelet therapy in patients with high coronary [18]F—NaF PET uptake.

Imaging calcification has become an essential tool in the evaluation of patients with coronary artery disease. It can be used to predict risk in patients with stable disease, detect acute coronary disease, stratify the risk of recurrent future cardiovascular events, and monitor response to therapy. The introduction of molecular imaging to established pathways adds a further dimension to coronary assessment. Multimodality imaging of coronary calcification offers different perspectives to investigate and monitor coronary artery disease. When combined, these powerful imaging techniques empower the clinician with the necessary knowledge to develop bespoke therapeutic strategies and improve the overall cardiovascular health of their patients.

◀───

image (C) showing no uptake at the site of the culprit plaque (^{18}F-FDG, tissue-to-background ratios, 1·63 versus reference segment 1·91 [15% decrease]). Note the significant myocardial uptake overlapping with the coronary artery (yellow arrow) and uptake within the oesophagus (blue arrow). Patient with anterior non-ST-segment elevation myocardial infarction with (D) culprit (red arrow; left anterior descending artery) and bystander non-culprit (white arrow; circumflex artery) lesions on invasive coronary angiography that were both stented during the index admission. Only the culprit lesion had increased ^{18}F-NaF uptake (^{18}F-NaF, tissue-to-background ratios, culprit 2·03 versus reference segment 1·08 [88% increase]) on PET-CT (E) after percutaneous coronary intervention. (F) Corresponding ^{18}F-fluorodeoxyglucose PET-CT showing no uptake either at the culprit (^{18}F-FDG, tissue-to-background ratios, culprit 1·62 versus reference segment 1·49 [9% increase]) or the bystander stented lesion. Note intense uptake within the ascending aorta. In a patient with stable angina with previous coronary artery bypass grafting, invasive coronary angiography (G) showed non-obstructive disease in the right coronary artery. Corresponding PET-CT scan (H) showed a region of increased ^{18}F-NaF activity (positive lesion, red line) in the mid-right coronary artery (tissue-to-background ratio, 3·13) and a region without increased uptake in the proximal vessel (negative lesion, yellow line). Radiofrequency intravascular ultrasound shows that the ^{18}F-NaF negative plaque (I) is principally composed of fibrous and fibrofatty tissue (green) with confluent calcium (white with acoustic shadow) but little evidence of necrosis. On the contrary, the ^{18}F-NaF positive plaque (J) shows high-risk features such as a large necrotic core (red) and microcalcification (white).

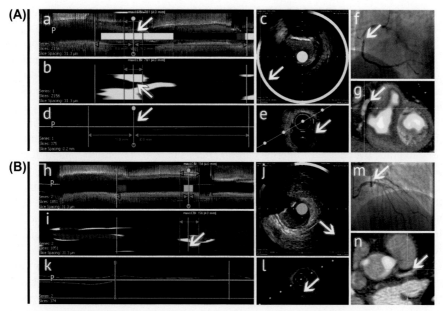

FIGURE 9.3 High-risk features and differences in ^{18}F-sodium fluoride uptake in culprit (A) and non-culprit (B) coronary lesions.

The culprit lesion has a high lipid-core burden index as demonstrated by focal yellow discoloration of the central ribbon (a), topogram (b), and ring (c) on near-infrared spectroscopy—intravascular ultrasound. Optical coherence tomography shows vessel stenosis (d) and marked acoustic artifact from calcified plaque (e). Invasive coronary angiography detects this focal stenosis within the right coronary artery (f). CT angiography confirms a flow-limiting lesion surrounded by dense calcification (g). There is marked focal uptake of ^{18}F-sodium fluoride at the site of plaque rupture (arrow, G). In comparison, nonculprit plaque has a much lower lipid-core burden index denoted by predominantly red near-infrared spectroscopy signals (h–j). The lumen is pinched on optical coherence tomography (k and l). This narrowing is confirmed on catheter (m) and CT angiography (n). There is no ^{18}F-sodium fluoride uptake at the plaque site (arrow, n).

Carotid arteries

The NASCET and ECST trials showed that the degree of internal carotid artery stenosis is an important independent predictor of future cerebrovascular events [114,115]. This is particularly true in men. Managing patients following a cerebrovascular event relies on accurate quantification of the degree of internal carotid artery stenosis. The ultimate goal is to prevent future thromboembolic events. Patients with internal carotid artery disease are stratified as high, medium, and low risk based on the degree of vessel stenosis [38]. Those individuals with symptomatic internal carotid artery stenosis greater than 50% are considered high risk of having a recurrent embolic event. Hence, they are considered for carotid endarterectomy: an open surgical excision of the internal carotid artery atheromatous plaque.

The carotid arteries are large and accessible enough to be studied by ultrasound. Duplex ultrasonography can measure blood flow velocities directly in the common and internal carotid arteries. The proportional rise in blood flow proximal and distal to the plaque is measured at various points of the cardiac cycle. The St Mary's ratio has gained widespread adoption in grading the degree of internal carotid artery stenosis [116]. This is the ratio of peak systolic velocity in the internal carotid artery to the end-diastolic velocity in the distal common carotid artery. The St Mary's ratio is used to grade the degree of internal carotid artery stenosis in medium- and high-risk individuals into deciles.

CT angiography and MR angiography are well-established modalities to image the carotid arteries. They have the benefit of visualizing the lumen along the entire vessel and are unaffected by vessel tortuosity. CT angiography has been instrumental in validating the use of ultrasound as a low-cost, radiation-free alternative. Magnetic resonance imaging of the carotid arteries can characterize atherosclerotic plaque. Late gadolinium-enhanced imaging sequences can visualize the large necrotic core and T1-weighted high-intensity plaque imaging can detect subclinical plaque rupture and intraplaque hemorrhage [117,118]. These are high-risk features suggestive of culprit carotid plaques.

Detecting high-risk metabolic processes in the carotid arteries

Conventional ultrasound, CT angiography, and MRI are powerful imaging modalities and well established in the risk stratification of patients with carotid artery stenosis. However, they do not inform the clinician of metabolic activity within carotid plaque. The detection of upregulated inflammatory and reparative processes within at-risk plaques has the potential to better direct clinical decision-making. A significant proportion of individuals with symptomatic carotid artery disease below the current threshold for surgical intervention are missed by conventional risk-stratification strategies. Similarly, the optimum treatment of asymptomatic carotid artery disease remains controversial because it is not possible to predict whether plaques will cause adverse complications.

Contrast-enhanced ultrasound is a simple and effective way to visualize culprit carotid plaques [119]. A homogenous suspension of inert gas microbubbles (e.g., SF_6) can be administered within the venous system. These microbubbles become trapped within the microvascular structure of carotid plaques. Internal carotid artery lesions with a high degree of abnormal plaque angiogenesis exhibit strong contrast signals on ultrasonography [119]. Contrast-enhanced ultrasound is a radiation-free and low-cost way to identify culprit plaques. It is well suited for individuals that may benefit from repetitive imaging. In the carotid arteries, contrast-enhanced ultrasound can detect neovascularization in culprit lesions with a sensitivity and specificity in excess of 80% [120].

MRI is particularly well suited to imaging the large and stable carotid arterial bed. The high temporal resolution of MRI can detect flow in the carotid vessels, whereas the enhanced soft-tissue differentiation enables the accurate detection of plaque rupture and intraplaque hemorrhage. MRI alone can give important anatomical and functional insights into carotid artery disease. The use of smart contrast agents, such as USPIO, can improve localization of macrophages within plaque [121,122]. Indeed, USPIO have been shown to localize in culprit carotid plaques [123]. The detection of USPIO uptake in carotid plaque requires sequential MRI scanning 24 h apart and bespoke expertize to interpret these images. Patients with symptomatic carotid artery disease frequently undergo urgent endarterectomy and may not be best suited for repeated scans.

PET imaging has the potential to overcome the limitations of USPIO imaging by directly visualizing high-risk metabolic processes simultaneously with MRI. ^{18}F-FDG is the most widely used radiotracer and has been shown to identify culprit carotid lesions. [18]F—NaF binds to microcalcification and can detect upregulated metabolic activity of culprit plaques. [18]F—NaF PET, in particular, can characterize lesions with a lipid-rich necrotic core more accurately than ^{18}F-FDG [41].

A combination of [18]F—NaF PET/MRI holds great promise to deliver a more complete picture of carotid artery disease. The anatomical and functional insights gained from MRI can be combined with information on the metabolic activity within suspect plaque. [18]F—NaF PET/MRI can detect active disease and may also be able to differentiate symptomatic plaques in subthreshold carotid artery stenosis [40,41]. In addition, [18]F—NaF PET/MRI can be used to study the response of carotid plaque to therapy, such as statins and antiplatelet agents. The role of [18]F—NaF in acute carotid artery disease is the focus of ongoing PET/MRI studies in patients requiring carotid endarterectomy (NCT03215550) and those managed with medication alone (NCT03215550).

The easily accessible carotid artery plaque is used frequently for proof-of-concept studies in emerging radiotracers. In carotid artery disease, TSPO PET/CT studies can detect macrophages directly and identify culprit carotid artery lesions with a greater sensitivity than ^{18}F-FDG. Similarly, ^{18}F-MISO and ^{18}F-Galacto-RGD bind to $\alpha_v\beta_3$ integrin surface glycoproteins in response to vascular hypoxia [78,80,124]. ^{18}F-Galacto-RGD, in particular, binds to plaque with increased vasa vasorum density. Both of these high-risk features are directly associated with vascular calcification and carotid plaque instability. The clinical translation of these techniques remains an area of great interest.

Aortic diseases

The aorta is the largest blood vessel in the body and subject to the greatest intravascular pressure changes throughout the cardiac cycle. Consequently, the aorta has a mature tunica media that is rich in smooth muscle cells, collagen, elastin, and components of the extracellular matrix. Complex pathophysiological mechanism driven by exposure to risk factors, such as smoking and hypertension, trigger endothelial-driven remodeling of the aortic media [125]. In connective tissue diseases, the aorta may have a congenital weakening of the medial layer.

Remodeling of the aortic wall is a reparative response and often associated with chronic inflammatory activity. Macrophage infiltration within the aortic media occurs through a leaky endothelium in a patchy distribution. The mechanisms leading to this infiltration are not fully understood but may be related to focal areas of increased endothelial shear stress [126]. A common endpoint of increased aortic inflammation is vessel wall calcification [127]. As is common to calcification throughout the vascular system, microcalcification within the aortic wall ultimately coalesces to established calcified plaque in a process that mirrors atherosclerosis [4].

Abdominal aorta

Aortic remodeling alters the compliance of the aortic wall. Hypertension-driven aortic stiffness produces a positive feedback loop that self-propagates further hypertension. The resultant increase in aortic wall shear stress and stiffness ultimately drives endothelial dysfunction, smooth muscle apoptosis, degeneration of the extracellular matrix, and generalized weakening of the aortic wall. The weakened aorta is prone to dilation under the immense pressures throughout the cardiac cycle. As the aorta dilates, hemodynamic flow within the lumen is altered and intraluminal thrombus deposition occurs. This thrombus, itself, may be biologically active and capable of producing vasoactive substances [128,129]. As the artery progressively dilates and becomes aneurysmal, the risk of aortic rupture increases exponentially [130]. Aortic rupture is a catastrophic endpoint with a mortality in excess of 90% [131].

Aortic aneurysms are most commonly seen in the infrarenal aorta of men and are driven by calcifying vascular disease. Aneurysms of the thoracic and suprarenal aorta are less common. These are often associated with overt or subclinical connective tissue disease. Aortic aneurysmal disease is clinically silent until the aorta ruptures, or rupture is imminent.

In the United Kingdom, the introduction of a one-off ultrasound screening visit for all men in their 56th year of life has transformed the surveillance and treatment of AAA disease [132]. The use of ultrasound to monitor the size of AAA is well established and reproducible. It is noninvasive, inexpensive, and radiation-free. Thus, ultrasound is the ideal modality to repeatedly image individuals under surveillance with AAA. The ultrasound detected anteroposterior diameter is the established way to stratify the risk of aortic rupture [133]. The accepted threshold for intervention is 5.5 cm or a rapid growth in aortic diameter [38]. A CT angiogram adds to ultrasound by visualizing the entire aorta and iliac arteries with exceptional detail.

Despite early detection and detailed morphological analysis, there is no established way to predict the risk of aortic rupture in an individual patient other than the aortic diameter. This method is unreliable [133]. An aneurysm may rupture at subthreshold diameters. Conversely, aneurysms in some individuals can become very large and remain asymptomatic. Aortic surgery carries significant risk to life [134]. There is a need to improve our ability to predict aortic rupture.

One approach is to detect macrophage infiltration within the aortic wall directly. USPIOs are trapped within macrophages and detected using T2*-weighted MRI of the aorta [135,136]. In abdominal aortic aneurysms, high aortic mural USPIO uptake localizes to areas of macrophage infiltration and reflects vessel wall inflammation [137]. In turn, macrophage within the aortic wall are associated with increased rates of aortic expansion and aneurysm events (rupture or repair) [138]. MRI has the additional benefit of differentiating components of intraluminal thrombus from the lumen and aortic wall in high detail (Fig. 9.4B and C).

Microcalcification within the aortic wall is a marker of increased metabolic activity. ^{18}F−NaF PET/CT in AAA disease shows that fluoride binds to aneurysmal aortic wall (Fig. 9.4) with a greater affinity compared to nonaneurysmal aorta [139]. This binding is heterogenous. In a study of 72 participants with AAA, the SOFIA[3] study showed that aneurysms with a high uptake of ^{18}F−NaF demonstrated an increased rate of growth compared to aneurysms with low ^{18}F−NaF uptake [140]. This relationship was independent of aneurysm size and other important clinical risk factors for disease progression (age, gender, body mass index, current smoking, systolic blood pressure). The SOFIA[3] study showed, for the first time, that imaging metabolic pathways in conjunction with traditional anatomical imaging can improve our ability to predict the risk of aortic expansion.

The identification of adverse prognostic indicators in AAA is key to improving our ability to predict the risk of aortic rupture. Morphological features of AAA have been studied extensively. None of these has yielded reliable markers of disease progression. Imaging disease processes represent the next major frontier in stratifying the risk of symptomatic AAA disease [133].

Thoracic aorta

In a similar hypertension-driven mechanism, aortic remodeling causes loss of extracellular architecture and thickening of the aortic media with compensatory vasa vasorum proliferation. Hypertension causes the pressure gradient throughout the cardiac cycle to rise within an increasingly stiff aortic wall [141]. The dense network of vasa vasorum in these conditions is prone to rupture and cause an intramural hematoma [142]. If the arterial endothelium is breached, high-pressure aortic blood flow can strip the endothelium and inner medial layer in the direction of blood flow. Perfusion to critical organs such as the heart, brain, limbs, and viscera may be compromised. This spectrum of diseases is collectively known as acute aortic syndrome [143]. In addition, the aortic wall is weakened and prone to aneurysmal dilatation with the similar consequences of aortic rupture and death.

Similar to aneurysms, there is no way to predict aortic expansion and rupture following acute aortic syndrome other than aortic diameter itself [144]. Aortic expansion in this setting is often more rapid than aneurysms. As the thoracic aorta is most commonly affected, surveillance is performed using CT or MRI [145]. This is a resource-heavy strategy compared to ultrasound surveillance in AAA. Rupture of the thoracic aorta is associated with near certainty of death. Data from international registries show that, despite optimum surveillance and medical treatment, acute aortic syndrome is associated with a mortality rate of nearly a quarter of patients at 3 years [146].

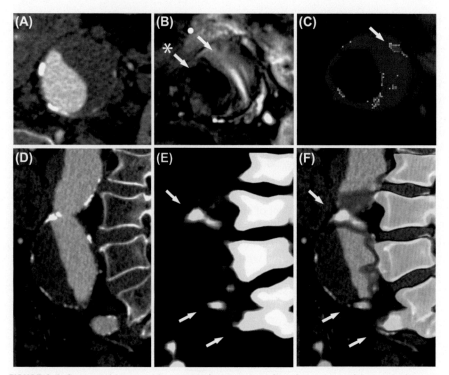

FIGURE 9.4 Computed tomography, magnetic resonance imaging, and positron emission tomography in a patient with an abdominal aortic aneurysm.

Axial view of the aneurysm as seen on computed tomography (A) shows a sac with thrombus and calcified plaque in the aortic wall. T2-weighted magnetic resonance imaging (B) can differentiate between the lumen (*), thrombus (•), and adjacent structures. T2*-magnetic resonance imaging (C) shows high ultrasmall particles of iron oxide uptake (arrow) in the wall of the aneurysm. The sagittal computed tomography view (D) delineates the morphology of the aneurysm. ^{18}F-sodium fluoride uptake (E) seen anterior to the vertebral body (arrows). Superimposing positron emission tomography signals over the computed tomography images (F) confirms high ^{18}F-sodium fluoride uptake at the aneurysm neck, bifurcation, and left common iliac artery (arrows).

PET imaging can help guide our understanding of the metabolic activity within thoracic aortic tissue. ^{18}F-FDG is the only radiotracer used to study disease progression following acute aortic syndrome [147,148]. These case series report increased ^{18}F-FDG uptake at the lacerated edge of the interrupted media and a correlation toward aortic dilatation. This finding suggests ongoing inflammatory activity related to disease progression.

[18]F—NaF PET/CT has been used to predict increased expansion in aneurysms. Aortic expansion following acute aortic syndrome likely occurs through a similar

mechanism. Unpublished in-house data suggest an increased affinity of ^{18}F—NaF binding to thoracic aortic tissue following aortic dissections compared to age-matched healthy individuals. The clinical relevance of ^{18}F—NaF PET/CT to predict aortic expansion and disease progression is currently under investigation.

Peripheral vascular disease

Occlusive vascular disease in the peripheral arteries is slow and progressive. Vascular calcification in the lower limbs adopts a concentric transmural morphology involving the medial layer [4]. Calcification can be circumferential in advanced stages. This pattern is most pronounced in individuals with diabetes mellitus and chronic kidney disease. In addition, the intima can be involved in areas of patchy atherosclerosis. Consequently, chronic hypoperfusion of the lower limb results in starvation of tissue from oxygen-rich blood. Hypoperfusion of the muscle groups in the face of increased metabolic demand causes intermittent claudication. At its most severe form, chronic hypoperfusion can cause ulcers and gangrene. The combination of poor peripheral perfusion and broken skin creates a haven for bacterial infections. Ultimately, there is a significant risk of limb loss [149].

Ultrasound, CT angiography and magnetic resonance angiography are well-established techniques to visualize the peripheral vessels. Duplex ultrasonography has the benefit of detecting flow in the lower limb arteries. The ankle brachial pressure index (ABPI) is a ratio of the pressure required to occlude flow in the pedal vessels compared to the pressure required to do the same in the brachial artery [150,151]. The ABPI is used to grade the severity of lower limb ischemia and used to predict the risk of progressive occlusive disease. An ABPI of less than 0.5 is considered severe limb ischemia [151]. However, this relationship has many caveats. For instance, arteries with extensive circumferential calcification in the calf may remain uncompressible despite high external compression. This gives a falsely elevated ABPI and is particularly true in patients with diabetes mellitus. Microvascular disease and peripheral neuropathy in this group further increases the risk of aggressive tissue loss and infection.

CT angiography and magnetic resonance angiography can produce high-quality images of the arterial lumen and identify areas of stenosis. The information from these modalities is used to plan open or endovascular intervention. However, due to dense calcification in narrow vessels below the knee, both CT and MRI overestimate the degree of vessel stenosis [152]. It is frequently not possible to determine whether narrow calf vessels are patent when dense calcification is present. Catheter diagnostic angiography remains the gold standard to visualize the arterial lumen in the lower limb vessels. This method is invasive and requires exposure to radiation as well as potentially harmful contrast agents.

Metabolic imaging of the peripheral vascular system

Prevention of limb loss is the ultimate goal in the management of peripheral vascular disease. Atherosclerotic plaque rupture in the lower limbs can cause sudden vessel occlusion in acute-on-chronic ischemia. As is true in other parts of the body, plaque rupture is associated with aggressive thrombosis. Treatment in this group is complex [38]. Medical management focuses on antiplatelet therapy and heparin to reduce the risk of propagation of the thrombus. Endovascular interventions are challenging due to the risk of distal embolization of fresh thrombus into smaller vessels. Here, it may cause severe hypoperfusion of a segment of the foot and cause untreatable tissue loss.

Predicting limb loss remains a challenge [153]. Despite best medical management, a significant proportion of patients may develop critical limb ischemia with no reconstructible options. Visualization of the artery lumen is useful in planning surgical or endovascular intervention. However, these alone cannot predict the risk of disease progression at an early stage.

Metabolic imaging using positron emission tomography can aid the process of risk stratification and peripheral vascular disease. Compared to ^{18}F-FDG, ^{18}F—NaF PET/CT imaging corresponded more accurately to areas of active inflammation. ^{18}F—NaF PET/CT identifies areas of microcalcification in the lower limbs [154]. Plaque obtained from the lower limb vessels show elevated ^{18}F—NaF binding to lesions with dense macrophage infiltration [155]. In a clinical study of 409 individuals, femoral ^{18}F—NaF uptake correlated strongly with hypertension, hypercholesterolemia, smoking habit, diabetes mellitus, and prior cardiovascular events [43]. ^{18}F—NaF in femoral artery plaque correlates strongly with established cardiovascular risk factors and histological features of high-risk plaques. The clinical correlation of these findings to predict progressive peripheral vascular disease remains to be validated.

Perfusion of skeletal muscle groups can be performed using ^{15}O-water PET/CT. A PET/CT scan is performed following exercise or the administration of pharmacological stressors. The degree of PET uptake strongly predicted flow reserves in patients with peripheral vascular disease [156]. In addition, ^{15}O-water PET/CT was able to identify muscle groups with poor blood flow and correlate these findings with future amputation levels in patients with severe disease [157]. Animal models have been developed to detect angiogenesis in the peripheral vessels using vascular endothelial growth factor [158] and $\alpha_v\beta_3$ [159] radioligands.

Valvular heart disease

Aortic stenosis (AS) represents a significant health problem in the Western World—it is currently the most prevalent valvular heart disease and, because of the aging population, set to become a growing health burden in the coming decades [160,161].

Calcification is the dominant pathophysiological process leading to aortic stenosis disease progression and represents a complex and intricately regulated process controlled by many governing mediators and pathways. Following lipid deposition and inflammation in the early stages of disease, inflammatory pathways are superseded by unremitting calcification driven by osteogenic differentiation of valvular cells under the influence of many osteogenic pathways and mediators. These include RANK/RANKL/OPG, Notch, Cbfa1/Runx2 transcription factor, and BMP signaling [162—164].

Emerging methods to directly image the process of valvular calcification have provided the ability to improve our understanding of the underlying pathophysiology, grade disease severity, and enable powerful prediction of disease progression. Perhaps consequent to its central importance in AS pathophysiology, quantification of valvular calcification provides the most powerful prediction of disease progression in aortic stenosis, irrespective of the imaging modality utilized. Using a semi-quantitative numerical score on echocardiography, the extent of valvular calcium (from a score of 1 representing no calcium to 4 for severe calcification) has been shown to provide more powerful prediction of disease progression than hemodynamic assessments by echocardiography [165].

However, although echocardiography remains the gold standard for diagnosing AS and grading severity, the emergence of new imaging methods that provide more accurate and sensitive assessments of valvular calcification may enable more powerful prediction of disease progression and become integrated in future clinical care.

One important modality becoming increasingly utilized in valvular heart disease is CT calcium scoring of the aortic valve. CT calcium scoring is an established technique for measuring coronary calcification, and quantification of aortic valve calcium load (CT-AVC) using the same Agatston method has now been implemented in European Guidelines [166]. CT-AVC has been shown to correlate well with calcium volume on explanted valves [167] and accurately grade AS severity. In line with the understanding that women require less calcium to develop severe disease, gender-specific thresholds have been validated and shown to both be accurate in the discrimination of disease severity, and be predictive of clinical outcomes [168,169]. Furthermore, CT-AVC provides important prognostic information in patients with discordant echo results, highlighting an important role for this modality in clinical practice [169,170].

In addition to CT calcium scoring, PET-CT using ^{18}F−NaF has also shown value in detecting disease progression (Fig. 9.5). ^{18}F−NaF binds to hydroxyapatite and therefore detects regions of calcification, with specificity for microcalcific deposits [171]. Indeed, in a study including patients with varying degrees of aortic stenosis severity, regions of increased ^{18}F−NaF uptake appeared to correspond to subsequent microcalcific deposits on CT at 1 year follow-up, correlating with change in calcium score at both 1 and 2 years [172,173]. Whether this imaging modality will add incremental value to CT calcium scoring in risk prediction of disease progression to justify the additional cost and radiation dose in the clinical setting has yet to be determined, but this tool will likely continue to assume an increasing role in the research setting.

In summary, calcification is central to AS pathophysiology and, consequently, directly imaging calcium and understanding the activity of this process holds great value in predicting disease progression and outcomes. This is especially important in a condition such as AS, in which the rate of progression is highly variable and inconsistent and therefore unable to be predicted by a single echocardiographic assessment.

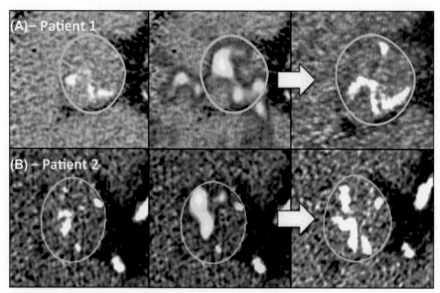

FIGURE 9.5 Change in aortic valve computed tomography (CT) calcium and ^{18}F-sodium fluoride (NaF) positron emission tomography (PET) activity after 1 year.

Coaxial short axis views of the aortic valve from 2 patients with mild aortic stenosis (top and bottom). On baseline CT scans, (left) established regions of macrocalcification appear white. Baseline fused ^{18}F—NaF PET and CT scans (middle) show intense ^{18}F—NaF uptake (red, yellow regions) both overlying and adjacent to existing calcium deposits on the CT. One-year follow-up CT scans (right) demonstrate increased calcium accumulation in much the same distribution as the baseline PET activity.

Adapted from Dweck MR, Jenkins WSA, Vesey AT, Pringle MA, Chin CW, Malley TS et al. ^{18}F—NaF uptake is a marker of active calcification and disease progression in patients with aortic stenosis. Circ Cardiovasc Imaging. 2014; CIRCIMAGING—113 under the Creative Commons License.

Emerging complementary technologies

Related industries outside medical sciences are evolving at an accelerated pace. Expectations from emerging technologies can vary significantly as our understanding of these techniques advances [174]. Emerging complementary technologies in computing power can drive innovation in cardiovascular imaging [175]. Multimodality cardiovascular imaging is essential to detect disease and predict risk. Established imaging techniques such as ultrasound, computed tomography, and MR imaging have reached a plateau of productivity. Cardiovascular PET has seen significant improvements in recent years and holds the promise of complementing established imaging techniques.

With the growth of an ever-connected world, complementary technologies in computing power continue to grow at a disproportionately fast rate [176]. The advent of cloud computing has made power-intensive tasks more accessible. Artificial intelligence, machine learning, and computational fluid dynamics hold the key to analyze large datasets and improve risk prediction. These techniques are now being exploited by data analysts to complement clinical research [177].

Arterial modeling and computational fluid dynamics

Advanced computing allows us to create high definition three-dimensional geometry. These can be manipulated in virtual environments. Contemporary imaging software is able to produce a volume render from stacked CT or MRI images. This is done by assigning a complex gradient map with a separate alpha channel for transparency to specific Hounsfield units in CT data [178]. However, these formats are rarely exportable and have limited use outside of 3D visualization.

Virtual segmentation of blood vessels, viscera, and associated calcification can produce accurate 3D models. Whereas volume rendering produces a gradient, segmentation of organs creates geometric objects with well-defined borders. Modern engineering packages can access these 3D models and simulate blood flow through them [179]. Computational fluid dynamics (CFD) can highlight areas of lamellar or turbulent flow within arteries, along with the distribution of stress on arterial walls caused by calcification [180]. Significant advances continue to be made in modeling the non-Newtonian nature of blood, pulsatile flow, and vessel compliance.

Finite element analysis has already been applied to patterns of microcalcification in the fibrous cap of atherosclerotic plaque by modeling the stress effects caused by large and concentrated calcium-containing vesicles. These preliminary models suggest that microcalcification weakens the fibrous cap directly [2,20]. Finite element analysis models are complex mathematical models with generalized assumptions. The challenge of simulated models is validation against real-world measurements.

Finite element analysis and computational fluid dynamics are resource intensive to develop and execute. Advances in cloud computing allow researchers to offload computing-heavy tasks to powerful remote data-processing centers. Machine learning can automate segmentation and the CFD analysis. Creating a seamless end-to-end pipeline from cross-sectional imaging to CFD analysis is now within reach.

Artificial intelligence and risk prediction

Artificial intelligence encompasses a series of computational techniques that have been designed to mimic the human brains ability to learn from positive and negative reinforcement [177,181]. Supervised machine learning involves two distinct phases: the first involves feeding large datasets into an algorithm as well as the expected result. The computer then "learns" a pattern to connect input variables with the desired outcome through multiple layers of neural networks. The second phase involves testing the learned algorithm. Here, a similar yet unseen dataset is introduced, and the algorithm makes a prediction of outcome. The true correct outcomes are then compared and the accuracy of the system is analyzed [182]. A feedback loop of unsupervised learning can be created as more data is sequentially collected. The fewer degrees of freedom within the expected outcome, the more likely that a machine-learning algorithm will make the correct prediction [182].

The ultimate role of this technique is to make predictions from highly heterogenous datasets. Machine learning is highly scalable and has already proven to process still images and real-time video. Improvements in computational image analysis carry significant prospects to advance cardiovascular imaging. Risk prediction using imaging is already an essential tool for managing individuals with cardiovascular disease. Machine-learning algorithms can automate the detection of high-risk features in medical scans [177].

Artificial intelligence can predict previously unidentified cardiovascular disease risk. A recent Nature publication describes how a deep learning model was trained on retinal data from nearly 300,000 individuals. The inclusion of the neural network model significantly improved the prediction of major adverse cardiac events compared to traditional risk factors alone [183]. Reverse engineering the neural network revealed that the model used anatomical features in the optic disc and retinal blood vessels to make these predictions. Such analysis is difficult to categorize using conventional analysis techniques.

Machine learning can detect subtle differences in images that would otherwise be missed by clinicians. Not only can machine learning automate complex image interpretation, it can detect minute variations and use this information to improve risk prediction. In cardiovascular imaging, the application of these techniques to patterns of calcification in the vascular bed remains to be exploited.

Three-dimensional printing

Virtual 3D models from CT- or MRI-derived imaging can be used to produce a three-dimensional model in the real world with distinct physical properties. 3D printing has seen a marked drop in expense. This has fueled developments that make 3D printing increasingly accessible. Development of new polymers means that the appearance and physical properties of models can be customized. Access to 3D models of calcified blood vessels can improve clinician's understanding of the pattern of disease and plan intervention. In compliant models, procedures can be practiced outside the body before being performed in vivo [184,185].

The next frontier in 3D printing, however, is to replace polymer-based materials with biologically active ones [186]. 3D bioprinting involves harvesting pluripotent stem cells derived from patients themselves. These can be manipulated to differentiate into specific tissue types. The cells are "printed" simultaneously on a collagen-like extracellular mesh to produce biologically compatible blood vessels or heart valves [187].

The chief advantage of autologous bioprosthetics would be to produce immune-system compatible conduits for bypass surgery. Similarly, compatible endoprosthesis printed with autologous stem cells may interact differently with vascular endothelium and avoid aggressive endothelial proliferation. Such technology is still in its infancy. However, proofs of concept have already been shown [188,189]. 3D-printed bioprosthetic implants have immense potential to enhance cardiovascular intervention.

Conclusion

We can already obtain structural and functional information on the vascular tree. Established imaging techniques continue to advance our knowledge and ability to detect disease. However, arterial calcification is varied and complex. Microcalcification is a marker of necrosis within the vessel wall, and its presence in the vascular bed is an adverse prognostic indicator for future cardiovascular events. Microcalcification cannot be detected using anatomical imaging techniques alone.

The emergence of metabolic imaging and its associated radiotracers offers complementary information on specific disease processes within the vessel wall. These give early insight into metabolic processes that ultimately cause morphological changes. ^{18}F-sodium fluoride is a promising biological radiotracer that binds to microcalcification in the vessel wall and can be detected using positron emission tomography. Novel tracers that identify high-risk features within the vessel wall continue to be developed. The clinical application of these techniques is to provide supplementary information to established anatomical-imaging modalities.

Imaging has become an indispensable tool to assess cardiovascular disease and stratify risk in the clinic. A combination of traditional risk factors, anatomical imaging, and biological activity offers a more complete understanding of arterial disease. In the age of precision medicine, this comprehensive understanding will enhance the quality of care offered to patients, improve our ability to predict disease progression, and evaluate the efficacy of therapeutic interventions.

References

[1] Hirsch D, Azoury R, Sarig S, Kruth HS. Colocalization of cholesterol and hydroxyapatite in human atherosclerotic lesions. Calcif Tissue Int 1993;52(2):94−8.

[2] Kelly-Arnold A, Maldonado N, Laudier D, Aikawa E, Cardoso L, Weinbaum S. Revised microcalcification hypothesis for fibrous cap rupture in human coronary arteries. Proc Natl Acad Sci Unit States Am 2013;110(26):10741−6.

[3] Tabas I, Bornfeldt KE. Macrophage phenotype and function in different stages of atherosclerosis. Circ Res 2016;118(4):653−67.

[4] Fuery MA, Liang L, Kaplan FS, Mohler ER. Vascular ossification: pathology, mechanisms, and clinical implications. Bone; 2017. Available from: http://www.sciencedirect.com/science/article/pii/S8756328217302326.

[5] Virmani R, Burke AP, Kolodgie FD, Farb A. Pathology of the thin-cap fibroatheroma: a type of vulnerable plaque. J Interv Cardiol 2003;16(3):267–72.

[6] Vengrenyuk Y, Cardoso L, Weinbaum S. Micro-CT based analysis of a new paradigm for vulnerable plaque rupture: cellular microcalcifications in fibrous caps. Mol Cell Biomech 2008;5(1):37–47.

[7] Nakahara T, Dweck MR, Narula N, Pisapia D, Narula J, Strauss HW. Coronary artery calcification: from mechanism to molecular imaging. JACC Cardiovasc Imaging 2017; 10(5):582–93.

[8] Davaine J-M, Quillard T, Brion R, Lapérine O, Guyomarch B, Merlini T, et al. Osteo-protegerin, pericytes and bone-like vascular calcification are associated with carotid plaque stability. PLoS One 2014;9(9):e107642.

[9] Proudfoot D, Skepper JN, Hegyi L, Bennett MR, Shanahan CM, Weissberg PL. Apoptosis regulates human vascular calcification in vitro: evidence for initiation of vascular calcification by apoptotic bodies. Circ Res 2000;87(11):1055–62.

[10] Aghagolzadeh P, Bachtler M, Bijarnia R, Jackson C, Smith ER, Odermatt A, et al. Calcification of vascular smooth muscle cells is induced by secondary calciprotein particles and enhanced by tumor necrosis factor-α. Atherosclerosis 2016;251:404–14.

[11] Dhore CR, Cleutjens JP, Lutgens E, Cleutjens KB, Geusens PP, Kitslaar PJ, et al. Differential expression of bone matrix regulatory proteins in human atherosclerotic plaques. Arterioscler Thromb Vasc Biol 2001;21(12):1998–2003.

[12] Durham AL, Speer MY, Scatena M, Giachelli CM, Shanahan CM. Role of smooth muscle cells in vascular calcification: implications in atherosclerosis and arterial stiffness. Cardiovasc Res 2018;114(4):590–600.

[13] Hutcheson JD, Maldonado N, Aikawa E. Small entities with large impact: microcalcifications and atherosclerotic plaque vulnerability. Curr Opin Lipidol 2014;25(5):327.

[14] Bischetti S, Scimeca M, Bonanno E, Federici M, Anemona L, Menghini R, et al. Carotid plaque instability is not related to quantity but to elemental composition of calcification. Nutr Metabol Cardiovasc Dis 2017;27(9):768–74.

[15] Bentzon JF, Otsuka F, Virmani R, Falk E. Mechanisms of plaque formation and rupture. Circ Res 2014;114(12):1852–66.

[16] Bobryshev YV, Killingsworth MC, Lord RSA, Grabs AJ. Matrix vesicles in the fibrous cap of atherosclerotic plaque: possible contribution to plaque rupture. J Cell Mol Med 2008;12(5b):2073–82.

[17] Reynolds JL, Joannides AJ, Skepper JN, McNair R, Schurgers LJ, Proudfoot D, et al. Human vascular smooth muscle cells undergo vesicle-mediated calcification in response to changes in extracellular calcium and phosphate concentrations: a potential mechanism for accelerated vascular calcification in ESRD. J Am Soc Nephrol 2004; 15(11):2857–67.

[18] Vengrenyuk Y, Carlier S, Xanthos S, Cardoso L, Ganatos P, Virmani R, et al. A hypothesis for vulnerable plaque rupture due to stress-induced debonding around cellular microcalcifications in thin fibrous caps. Proc Natl Acad Sci Unit States Am 2006;103(40):14678–83.

[19] Bluestein D, Alemu Y, Avrahami I, Gharib M, Dumont K, Ricotta JJ, et al. Influence of microcalcifications on vulnerable plaque mechanics using FSI modeling. J Biomech 2008;41(5):1111–8.

[20] Maldonado N, Kelly-Arnold A, Vengrenyuk Y, Laudier D, Fallon JT, Virmani R, et al. A mechanistic analysis of the role of microcalcifications in atherosclerotic plaque

stability: potential implications for plaque rupture. Am J Physiol Heart Circ Physiol 2012;303(5):H619−28.

[21] Gimbrone MA, García-Cardeña G. Endothelial cell dysfunction and the pathobiology of atherosclerosis. Circ Res 2016;118(4):620−36.

[22] Falk E. Pathogenesis of atherosclerosis. J Am Coll Cardiol 2006;47(8 Supplement): C7−12.

[23] Pirillo A, Bonacina F, Norata GD, Catapano AL. The interplay of lipids, lipoproteins, and immunity in atherosclerosis. Curr Atheroscler Rep 2018;20(3):12.

[24] Goncharov NV, Avdonin PV, Nadeev AD, Zharkikh IL, Jenkins RO. Reactive oxygen species in pathogenesis of atherosclerosis. Curr Pharmaceut Des 2015;21(9): 1134−46.

[25] Sluimer JC, Gasc J-M, Wanroij van JL, Kisters N, Groeneweg M, Gelpke MDS, et al. Hypoxia, hypoxia-inducible transcription factor, and macrophages in human atherosclerotic plaques are correlated with intraplaque angiogenesis. J Am Coll Cardiol 2008;51(13):1258−65.

[26] Winter PM, Morawski AM, Caruthers SD, Fuhrhop RW, Zhang H, Williams TA, et al. Molecular imaging of angiogenesis in early-stage atherosclerosis with $\alpha v \beta 3$-integrin−targeted nanoparticles. Circulation 2003;108(18):2270−4.

[27] Woollard KJ, Geissmann F. Monocytes in atherosclerosis: subsets and functions. Nat Rev Cardiol 2010;7(2):77.

[28] Stary HC, Chandler AB, Dinsmore RE, Fuster V, Glagov S, Insull W, et al. A definition of advanced types of atherosclerotic lesions and a histological classification of atherosclerosis: a report from the committee on vascular lesions of the council on arteriosclerosis, American heart association. Circulation 1995;92(5):1355−74.

[29] Pries AR, Badimon L, Bugiardini R, Camici PG, Dorobantu M, Duncker DJ, et al. Coronary vascular regulation, remodelling, and collateralization: mechanisms and clinical implications on behalf of the working group on coronary pathophysiology and microcirculation. Eur Heart J 2015;36(45):3134−46.

[30] Yamagishi M, Terashima M, Awano K, Kijima M, Nakatani S, Daikoku S, et al. Morphology of vulnerable coronary plaque: insights from follow-up of patients examined by intravascular ultrasound before an acute coronary syndrome. J Am Coll Cardiol 2000;35(1):106−11.

[31] Choi Y, Brunken RC, Hawkins RA, Huang S-C, Buxton DB, Hoh CK, et al. Factors affecting myocardial 2-[F-18]fluoro-2-deoxy-<Emphasis Type="SmallCaps">d</Emphasis>-glucose uptake in positron emission tomography studies of normal humans. Eur J Nucl Med 1993;20(4):308−18.

[32] Investigators S-H. CT coronary angiography in patients with suspected angina due to coronary heart disease (SCOT-HEART): an open-label,parallel-group, multicentre trial. Lancet 2015. Available from: https://ulir.ul.ie/handle/10344/4423.

[33] Doris M, Newby DE. Coronary CT angiography as a diagnostic and prognostic tool: perspectives from the SCOT-heart trial. Curr Cardiol Rep 2016;18.

[34] Greenland P, LaBree L, Azen SP, Doherty TM, Detrano RC. Coronary artery calcium score combined with Framingham score for risk prediction in asymptomatic individuals. J Am Med Assoc 2004;291(2):210−5.

[35] Kanda T, Oba H, Toyoda K, Kitajima K, Furui S. Brain gadolinium deposition after administration of gadolinium-based contrast agents. Jpn J Radiol 2016;34(1): 3−9.

[36] Jasmin NH, Thin MZ, Taylor V, Lythgoe M, Stuckey D. 112 in vivo investigation of intracellular calcium levels in acute myocardial infarction using cardiac T1 mapping-manganese-enhanced MRI. BMJ Publishing Group Ltd and British Cardiovascular Society; 2018.

[37] Gerhard-Herman MD, Gornik HL, Barrett C, Barshes NR, Corriere MA, Drachman DE, et al. 2016 AHA/ACC guideline on the management of patients with lower extremity peripheral artery disease: executive summary: a report of the American College of Cardiology/American Heart Association Task Force on Clinical Practice Guidelines. J Am Coll Cardiol 2017;69(11):1465–508.

[38] Aboyans V, Ricco J-B, Bartelink M-LE, Björck M, Brodmann M, Cohnert T, et al. 2017 ESC guidelines on the diagnosis and treatment of peripheral arterial diseases, in collaboration with the european society for vascular surgery (ESVS) document covering atherosclerotic disease of extracranial carotid and vertebral, mesenteric, renal, upper and lower extremity arteries endorsed by: the European Stroke Organization (ESO) the task force for the diagnosis and treatment of peripheral arterial diseases of the european society of cardiology (ESC) and of the european society for vascular surgery (ESVS). Eur Heart J 2017;39(9):763–816.

[39] Piccini D, Feng L, Bonanno G, Coppo S, Yerly J, Lim RP, et al. Four-dimensional respiratory motion-resolved whole heart coronary MR angiography. Magn Reson Med 2017;77(4):1473–84.

[40] Joshi NV, Vesey AT, Williams MC, Shah ASV, Calvert PA, Craighead FHM, et al. ^{18}F-fluoride positron emission tomography for identification of ruptured and high-risk coronary atherosclerotic plaques: a prospective clinical trial. Lancet 2014;383(9918): 705–13.

[41] Vesey AT, Irkle A, Lewis DY, Skepper JN, Bird JL, Dweck MR, et al. Identifying active vascular microcalcification by ^{18}F-sodium fluoride positron emission tomography. Atherosclerosis 2015;241(1):e14.

[42] Tawakol A, Osborne MT, Fayad ZA. Molecular imaging of atheroma: deciphering how and when to use ^{18}F-sodium fluoride and ^{18}F-fluorodeoxyglucose. Circ Cardiovasc Imaging 2017;10(3). e006183.

[43] Janssen T, Bannas P, Herrmann J, Veldhoen S, Busch JD, Treszl A, et al. Association of linear ^{18}F-sodium fluoride accumulation in femoral arteries as a measure of diffuse calcification with cardiovascular risk factors: a PET/CT study. J Nucl Cardiol 2013; 20(4):569–77.

[44] Rubeaux M, Joshi NV, Dweck MR, Fletcher A, Motwani M, Thomson LE, et al. Motion correction of ^{18}F-NaF PET for imaging coronary atherosclerotic plaques. J Nucl Med 2016;57(1):54–9.

[45] Pawade TA, Cartlidge TRG, Jenkins WSA, Adamson PD, Robson P, Lucatelli C, et al. Optimization and reproducibility of aortic valve ^{18}F-fluoride positron emission tomography in patients with aortic stenosis. Circ Cardiovasc Imaging 2016;9(10):e005131.

[46] Zhang Y, Li H, Jia Y, Yang P, Zhao F, Wang W, et al. Noninvasive assessment of carotid plaques calcification by ^{18}F-sodium fluoride accumulation: correlation with pathology. J Stroke Cerebrovasc Dis 2018;27(7):1796–801.

[47] Moneta GL, Edwards JM, Chitwood RW, Taylor LM, Lee RW, Cummings CA, et al. Correlation of North American Symptomatic Carotid Endarterectomy Trial (NASCET) angiographic definition of 70% to 99% internal carotid artery stenosis with duplex scanning. J Vasc Surg 1993;17(1):152–9.

[48] Josephson SA, Bryant SO, Mak HK, Johnston SC, Dillon WP, Smith WS. Evaluation of carotid stenosis using CT angiography in the initial evaluation of stroke and TIA. Neurology 2004;63(3):457—60.

[49] Divakaran S, Cheezum MK, Hulten EA, Bittencourt MS, Silverman MG, Nasir K, et al. Use of cardiac CT and calcium scoring for detecting coronary plaque: implications on prognosis and patient management. Br J Radiol 2014;88(1046):20140594.

[50] Motoyama S, Ito H, Sarai M, Kondo T, Kawai H, Nagahara Y, et al. Plaque characterization by coronary computed tomography angiography and the likelihood of acute coronary events in mid-term follow-up. J Am Coll Cardiol 2015;66(4):337—46.

[51] Maurovich-Horvat P, Ferencik M, Voros S, Merkely B, Hoffmann U. Comprehensive plaque assessment by coronary CT angiography. Nat Rev Cardiol 2014;11(7): 390—402.

[52] Tsao J, Kozerke S. MRI temporal acceleration techniques. J Magn Reson Imaging 2012;36(3):543—60.

[53] Epstein FH. MRI of left ventricular function. J Nucl Cardiol 2007;14(5):729—44.

[54] Garcia J, Jarvis KB, Schnell S, Malaisrie S, Clennon C, Collins JD, et al. 4D flow MRI of the aorta demonstrates age- and gender-related differences in aortic size and blood flow velocity in healthy subjects. J Cardiovasc Magn Reson 2015;17(1):P39.

[55] Ambale-Venkatesh B, Lima JAC. Cardiac MRI: a central prognostic tool in myocardial fibrosis. Nat Rev Cardiol 2015;12(1):18—29.

[56] Hatsukami TS, Ross R, Polissar NL, Yuan C. Visualization of fibrous cap thickness and rupture in human atherosclerotic carotid plaque in vivo with high-resolution magnetic resonance imaging. Circulation 2000;102(9):959—64.

[57] Vandenberghe S, Marsden PK. PET-MRI: a review of challenges and solutions in the development of integrated multimodality imaging. Phys Med Biol 2015;60(4):R115.

[58] Westerterp M, Pruim J, Oyen W, Hoekstra O, Paans A, Visser E, et al. Quantification of FDG PET studies using standardised uptake values in multi-centre trials: effects of image reconstruction, resolution and ROI definition parameters. Eur J Nucl Med Mol Imaging 2007;34(3):392—404.

[59] Cao A, Pandya AK, Serhatkulu GK, Weber RE, Dai H, Thakur JS, et al. A robust method for automated background subtraction of tissue fluorescence. J Raman Spectrosc 2007;38(9):1199—205.

[60] Jang I-K, Tearney GJ, MacNeill B, Takano M, Moselewski F, Iftima N, et al. In vivo characterization of coronary atherosclerotic plaque by use of optical coherence tomography. Circulation 2005;111(12):1551—5.

[61] Tearney GJ, Yabushita H, Houser SL, Aretz HT, Jang I-K, Schlendorf KH, et al. Quantification of macrophage content in atherosclerotic plaques by optical coherence tomography. Circulation 2003;107(1):113—9.

[62] Kataoka Y, Puri R, Hammadah M, Duggal B, Uno K, Kapadia SR, et al. Spotty calcification and plaque vulnerability in vivo: frequency-domain optical coherence tomography analysis. Cardiovasc Diagn Ther 2014;4(6):460—9.

[63] Costopoulos C, Brown AJ, Teng Z, Hoole SP, West NEJ, Samady H, et al. Intravascular ultrasound and optical coherence tomography imaging of coronary atherosclerosis. Int J Cardiovasc Imaging 2016;32(1):189—200.

[64] Newby DE, Fox KAA. Invasive assessment of the coronary circulation: intravascular ultrasound and Doppler. Br J Clin Pharmacol 2002;53(6):561—75.

[65] Kerwin WS, Zhao X, Yuan C, Hatsukami TS, Maravilla KR, Underhill HR, et al. Contrast-enhanced MRI of carotid atherosclerosis: dependence on contrast agent. J Magn Reson Imaging 2009;30(1):35−40.

[66] Teng Z, Sadat U, Huang Y, Young VE, Graves MJ, Lu J, et al. In vivo MRI-based 3D mechanical stress−strain profiles of carotid plaques with Juxtaluminal plaque haemorrhage: an exploratory study for the mechanism of subsequent cerebrovascular events. Eur J Vasc Endovasc Surg 2011;42(4):427−33.

[67] Oliveira BL, Blasi F, Rietz TA, Rotile NJ, Day H, Caravan P. Multimodal molecular imaging reveals high target uptake and specificity of [111]In- and ^{68}Ga-labeled fibrin-binding probes for thrombus detection in rats. J Nucl Med 2015;56(10):1587−92.

[68] Ziegler M, Alt K, Paterson BM, Kanellakis P, Bobik A, Donnelly PS, et al. Highly sensitive detection of minimal cardiac ischemia using positron emission tomography imaging of activated platelets. Sci Rep 2016;6:38161.

[69] Ota H, Yu W, Underhill HR, Oikawa M, Dong L, Zhao X, et al. Hemorrhage and large lipid-rich necrotic cores are independently associated with thin or ruptured fibrous caps: an in vivo 3T MRI study. Arterioscler Thromb Vasc Biol 2009;29(10):1696−701.

[70] Xia J, Yin A, Li Z, Liu X, Peng X, Xie N. Quantitative analysis of lipid-rich necrotic core in carotid atherosclerotic plaques by in vivo magnetic resonance imaging and clinical outcomes. Med Sci Monit 2017;23:2745−50.

[71] Chai JT, Biasiolli L, Li L, Alkhalil M, Galassi F, Darby C, et al. Quantification of lipid-rich core in carotid atherosclerosis using magnetic resonance T2 mapping: relation to clinical presentation. JACC Cardiovasc Imaging 2017;10(7):747−56.

[72] Waxman S, Dixon SR, L'Allier P, Moses JW, Petersen JL, Cutlip D, et al. In vivo validation of a catheter-based near-infrared spectroscopy system for detection of lipid core coronary plaques: initial results of the SPECTACL study. JACC Cardiovasc Imaging 2009;2(7):858−68.

[73] Brugaletta S, Garcia-Garcia HM, Serruys PW, Boer S de, Ligthart J, Gomez-Lara J, et al. NIRS and IVUS for characterization of atherosclerosis in patients undergoing coronary angiography. JACC Cardiovasc Imaging 2011;4(6):647−55.

[74] Schuurman A-S, Vroegindewey M, Kardys I, Oemrawsingh RM, Cheng JM, de Boer S, et al. Near-infrared spectroscopy-derived lipid core burden index predicts adverse cardiovascular outcome in patients with coronary artery disease during long-term follow-up. Eur Heart J 2018;39(4):295−302.

[75] Neubauer S. Cardiac magnetic resonance spectroscopy. Curr Cardiol Rep 2003;5(1):75−82.

[76] Xin L, Lanz B, Lei H, Gruetter R. Assessment of metabolic fluxes in the mouse brain in vivo using 1H-[13C] NMR spectroscopy at 14.1 Tesla. J Cereb Blood Flow Metab 2015;35(5):759−65.

[77] Deelchand DK, Van de Moortele P-F, Adriany G, Iltis I, Andersen P, Strupp JP, et al. In vivo 1H NMR spectroscopy of the human brain at 9.4 T: initial results. J Magn Reson 2010;206(1):74−80.

[78] Mateo J, Izquierdo-Garcia D, Badimon JJ, Fayad ZA, Fuster V. Noninvasive assessment of hypoxia in rabbit advanced atherosclerosis using ^{18}F-fluoromisonidazole positron emission tomographic imaging. Circ Cardiovasc Imaging 2014;7(2):312−20.

[79] Takasawa M, Beech JS, Fryer TD, Hong YT, Hughes JL, Igase K, et al. Imaging of brain hypoxia in permanent and temporary middle cerebral artery occlusion in the rat using ^{18}F-fluoromisonidazole and positron emission tomography: a pilot study. J Cereb Blood Flow Metab 2007;27(4):679−89.

[80] Rajendran JG, Krohn KA. F-18 fluoromisonidazole for imaging tumor hypoxia: imaging the microenvironment for personalized cancer therapy. Semin Nucl Med 2015; 45(2):151–62.

[81] Haubner R, Kuhnast B, Mang C, Weber WA, Kessler H, Wester H-J, et al. [18F] Galacto-RGD: synthesis, radiolabeling, metabolic stability, and radiation dose estimates. Bioconjug Chem 2004;15(1):61–9.

[82] Dalm VASH, van Hagen PM, van Koetsveld PM, Achilefu S, Houtsmuller AB, Pols DHJ, et al. Expression of somatostatin, cortistatin, and somatostatin receptors in human monocytes, macrophages, and dendritic cells. Am J Physiol Endocrinol Metab 2003;285(2):E344–53.

[83] Malmberg C, Ripa RS, Johnbeck CB, Knigge U, Langer SW, Mortensen J, et al. 64Cu-DOTATATE for noninvasive assessment of atherosclerosis in large arteries and its correlation with risk factors: Head-to-Head comparison with 68Ga-DOTATOC in 60 patients. J Nucl Med 2015;56(12):1895–900.

[84] Kim H, Lee S-J, Davies-Venn C, Kim JS, Yang BY, Yao Z, et al. 64Cu-DOTA as a surrogate positron analog of Gd-DOTA for cardiac fibrosis detection with PET: pharmacokinetic study in a rat model of chronic MI. Nucl Med Commun 2016;37(2):188–96.

[85] Ruf J, Schiefer J, Furth C, Kosiek O, Kropf S, Heuck F, et al. 68Ga-DOTATOC PET/CT of neuroendocrine tumors: spotlight on the CT phases of a triple-phase protocol. J Nucl Med 2011;52(5):697–704.

[86] Largeau B, Dupont A-C, Guilloteau D, Santiago-Ribeiro M-J, Arlicot N. TSPO PET imaging: from microglial activation to peripheral sterile inflammatory diseases? [Internet] Contrast Media Mol Imaging 2017;25:6592139. Available from: https://www.hindawi.com/journals/cmmi/2017/6592139/abs/.

[87] Gaemperli O, Shalhoub J, Owen DRJ, Lamare F, Johansson S, Fouladi N, et al. Imaging intraplaque inflammation in carotid atherosclerosis with 11C-PK11195 positron emission tomography/computed tomography. Eur Heart J 2012;33(15):1902–10.

[88] Boutin H, Murray K, Pradillo J, Maroy R, Smigova A, Gerhard A, et al. 18F-GE-180: a novel TSPO radiotracer compared to 11C-R-PK11195 in a preclinical model of stroke. Eur J Nucl Med Mol Imaging 2015;42(3):503–11.

[89] Cuhlmann S, Gsell W, Van der Heiden K, Habib J, Tremoleda JL, Khalil M, et al. In vivo mapping of vascular inflammation using the translocator protein tracer 18F-FEDAA1106. Mol Imaging 2014;13 (101120118).

[90] Tiwari AK, Yui J, Fujinaga M, Kumata K, Shimoda Y, Yamasaki T, et al. Characterization of a novel acetamidobenzoxazolone-based PET ligand for translocator protein (18 kDa) imaging of neuroinflammation in the brain. J Neurochem 2014;129(4): 712–20.

[91] Kooi ME, Cappendijk VC, Cleutjens KBJM, Kessels AGH, Kitslaar PJ, Borgers M, et al. Accumulation of ultrasmall superparamagnetic particles of iron oxide in human atherosclerotic plaques can be detected by in vivo magnetic resonance imaging. Circulation 2003;107(19):2453–8.

[92] Richards JMJ, Semple SI, MacGillivray TJ, Gray C, Langrish JP, Williams M, et al. Abdominal aortic aneurysm growth predicted by uptake of ultrasmall superparamagnetic particles of iron oxide: a pilot study. Circ Cardiovasc Imaging 2011;4(3):274–81.

[93] Youssef G, Leung E, Mylonas I, Nery P, Williams K, Wisenberg G, et al. The use of 18F-FDG PET in the diagnosis of cardiac sarcoidosis: a systematic review and metaanalysis including the Ontario experience. J Nucl Med 2012;53(2):241–8.

[94] Wykrzykowska J, Lehman S, Williams G, Parker JA, Palmer MR, Varkey S, et al. Imaging of inflamed and vulnerable plaque in coronary arteries with ^{18}F-FDG PET/CT in patients with suppression of myocardial uptake using a low-carbohydrate, high-fat preparation. J Nucl Med 2009;50(4):563−8.

[95] Hoh CK. Clinical use of FDG PET. Nucl Med Biol 2007;34(7):737−42.

[96] Williams G, Kolodny GM. Suppression of myocardial ^{18}F-FDG uptake by preparing patients with a high-fat, low-carbohydrate diet. Am J Roentgenol 2008;190(2): W151−6.

[97] Harisankar CNB, Mittal BR, Agrawal KL, Abrar ML, Bhattacharya A. Utility of high fat and low carbohydrate diet in suppressing myocardial FDG uptake. J Nucl Cardiol 2011;18(5):926.

[98] Teague HL, Ahlman MA, Alavi A, Wagner DD, Lichtman AH, Nahrendorf M, et al. Unraveling vascular inflammation: from immunology to imaging. J Am Coll Cardiol 2017;70(11):1403−12.

[99] Adamson PD, Moss AJ, Doris MK, Dweck M, Newby D. Molecular imaging of coronary atherosclerosis: quantification and reproducibility of coronary ^{18}F-fluoride positron emission tomography. J Am Coll Cardiol 2018;71(11, Supplement):A1468.

[100] McBride OMB, Joshi NV, Robson JMJ, MacGillivray TJ, Gray CD, Fletcher AM, et al. Positron emission tomography and magnetic resonance imaging of cellular inflammation in patients with abdominal aortic aneurysms. Eur J Vasc Endovasc Surg 2016; 51(4):518−26.

[101] Newby AC. Metalloproteinase production from macrophages − a perfect storm leading to atherosclerotic plaque rupture and myocardial infarction. Exp Physiol 2016; 101(11):1327−37.

[102] ESC Guidelines for the diagnosis and treatment of acute and chronic heart failure 2008‡ - Dickstein - 2008 - European Journal of Heart Failure - Wiley Online Library Available from: https://onlinelibrary.wiley.com/doi/abs/10.1016/j.ejheart.2008.08. 005.

[103] Pugliese F, Mollet NRA, Runza G, Mieghem C van, Meijboom WB, Malagutti P, et al. Diagnostic accuracy of non-invasive 64-slice CT coronary angiography in patients with stable angina pectoris. Eur Radiol 2006;16(3):575−82.

[104] Coronary CT angiography and 5-year risk of myocardial infarction. N Engl J Med 2018;379:924−33.

[105] Polonsky TS, McClelland RL, Jorgensen NW, Bild DE, Burke GL, Guerci AD, et al. Coronary artery calcium score and risk classification for coronary heart disease prediction. J Am Med Assoc 2010;303(16):1610−6.

[106] Tarkin JM, Joshi FR, Evans NR, Chowdhury MM, Figg NL, Shah AV, et al. Detection of atherosclerotic inflammation by ^{68}Ga-DOTATATE PET compared to [^{18}F]FDG PET imaging. J Am Coll Cardiol 2017;69(14):1774−91.

[107] Li X, Samnick S, Lapa C, Israel I, Buck AK, Kreissl MC, et al. ^{68}Ga-DOTATATE PET/ CT for the detection of inflammation of large arteries: correlation with ^{18}F-FDG, calcium burden and risk factors. EJNMMI Res 2012;2(1):52.

[108] Dweck MR, Joshi FR, Newby DE, Rudd JH. Noninvasive imaging in cardiovascular therapy: the promise of coronary arterial ^{18}F-sodium fluoride uptake as a marker of plaque biology. Expert Rev Cardiovasc Ther 2012;10(9):1075−7.

[109] Dweck MR, Joshi FR, Newby DE, Rudd JHF. Noninvasive imaging in cardiovascular therapy: the promise of coronary arterial ^{18}F-sodium fluoride uptake as a marker of plaque biology. Expert Rev Cardiovasc Ther 2012;10(9):1075−7.

[110] Dweck MR, Chow MWL, Joshi NV, Williams MC, Jones C, Fletcher AM, et al. Coronary arterial ^{18}F-sodium fluoride uptake: a novel marker of plaque biology. J Am Coll Cardiol 2012;59(17):1539—48.

[111] Joshi NV, Vesey A, Newby DE, Dweck MR. Will ^{18}F-sodium fluoride PET-CT imaging be the magic bullet for identifying vulnerable coronary atherosclerotic plaques? Curr Cardiol Rep 2014;16(9):521.

[112] Giugliano RP, Pedersen TR, Park J-G, De Ferrari GM, Gaciong ZA, Ceska R, et al. Clinical efficacy and safety of achieving very low LDL-cholesterol concentrations with the PCSK9 inhibitor evolocumab: a prespecified secondary analysis of the FOURIER trial. Lancet 2017;390(10106):1962—71.

[113] Stroes E, Colquhoun D, Sullivan D, Civeira F, Rosenson RS, Watts GF, et al. Anti-PCSK9 antibody effectively lowers cholesterol in patients with statin intolerance: the GAUSS-2 randomized, placebo-controlled phase 3 clinical trial of evolocumab. J Am Coll Cardiol 2014;63(23):2541—8.

[114] Gasecki AP, Eliasziw M, Ferguson GG, Hachinski V, Barnett HJM. Long-term prognosis and effect of endarterectomy in patients with symptomatic severe carotid stenosis and contralateral carotid stenosis or occlusion: results from NASCET. J Neurosurg 1995;83(5):778—82.

[115] Randomised trial of endarterectomy for recently symptomatic carotid stenosis: final results of the MRC European Carotid Surgery Trial (ECST). Lancet 1998; 351(9113):1379—87.

[116] Matz O, Nikoubashman O, Rajkumar P, Keuler A, Wiesmann M, Schulz JB, et al. Grading of proximal internal carotid artery (ICA) stenosis by Doppler/duplex ultrasound (DUS) and computed tomographic angiography (CTA): correlation and interrater reliability in real-life practice. Acta Neurol Belg 2017;117(1):183—8.

[117] Chu B, Kampschulte A, Ferguson MS, Kerwin WS, Yarnykh VL, O'Brien KD, et al. Hemorrhage in the atherosclerotic carotid plaque: a high-resolution MRI study. Stroke 2004;35(5):1079—84.

[118] Viereck J, Ruberg FL, Qiao Y, Perez AS, Detwiller K, Johnstone M, et al. MRI of atherothrombosis associated with plaque rupture. Arterioscler Thromb Vasc Biol 2005; 25(1):240—5.

[119] Huang R, Abdelmoneim SS, Ball CA, Nhola LF, Farrell AM, Feinstein S, et al. Detection of carotid atherosclerotic plaque neovascularization using contrast enhanced ultrasound: a systematic review and meta-analysis of diagnostic accuracy studies. J Am Soc Echocardiogr 2016;29(6):491—502.

[120] Staub D, Patel MB, Tibrewala A, Ludden D, Johnson M, Espinosa P, et al. Vasa vasorum and plaque neovascularization on contrast-enhanced carotid ultrasound imaging correlates with cardiovascular disease and past cardiovascular events. Stroke 2010; 41(1):41—7.

[121] Tang TY, Howarth SP, Miller SR, Graves MJ, U-King-Im J-M, Li Z-Y, et al. Correlation of carotid atheromatous plaque inflammation using USPIO-enhanced MR imaging with degree of luminal stenosis. Stroke 2008;39(7):2144—7.

[122] Tang TY, Howarth SPS, Li ZY, Miller SR, Graves MJ, U-King-Im JM, et al. Correlation of carotid atheromatous plaque inflammation with biomechanical stress: utility of USPIO enhanced MR imaging and finite element analysis. Atherosclerosis 2008; 196(2):879—87.

[123] Trivedi RA, Mallawarachi C, U-King-Im J-M, Graves MJ, Horsley J, Goddard MJ, et al. Identifying inflamed carotid plaques using in vivo USPIO-enhanced MR imaging to label plaque macrophages. Arterioscler Thromb Vasc Biol 2006;26(7):1601−6.

[124] Laitinen I, Saraste A, Weidl E, Poethko T, Weber AW, Nekolla SG, et al. Evaluation of $\alpha v \beta 3$ integrin-targeted positron emission tomography tracer ^{18}F-Galacto-RGD for imaging of vascular inflammation in atherosclerotic Mice. Circ Cardiovasc Imaging 2009;2(4):331−8.

[125] Kaess BM, Rong J, Larson MG, Hamburg NM, Vita JA, Levy D, et al. Aortic stiffness, blood pressure progression, and incident hypertension. J Am Med Assoc 2012;308(9): 875−81.

[126] Prado CM, Rossi MA. Circumferential wall tension due to hypertension plays a pivotal role in aorta remodelling. Int J Exp Pathol 2006;87(6):425−36.

[127] McEniery CM, McDonnell BJ, So A, Aitken S, Bolton CE, Munnery M, et al. Aortic calcification is associated with aortic stiffness and isolated systolic hypertension in healthy individuals. Hypertension 2009;53(3):524−31.

[128] Vorp DA, Lee PC, Wang DH, Makaroun MS, Nemoto EM, Ogawa S, et al. Association of intraluminal thrombus in abdominal aortic aneurysm with local hypoxia and wall weakening. J Vasc Surg 2001;34(2):291−9.

[129] Kazi M, Thyberg J, Religa P, Roy J, Eriksson P, Hedin U, et al. Influence of intraluminal thrombus on structural and cellular composition of abdominal aortic aneurysm wall. J Vasc Surg 2003;38(6):1283−92.

[130] Fillinger MF, Marra SP, Raghavan ML, Kennedy FE. Prediction of rupture risk in abdominal aortic aneurysm during observation: wall stress versus diameter. J Vasc Surg 2003;37(4):724−32.

[131] Johansson G, Swedenborg J. Ruptured abdominal aortic aneurysms: a study of incidence and mortality. BJS 1986;73(2):101−3.

[132] Thompson SG, Ashton HA, Gao L, Scott RAP. Screening men for abdominal aortic aneurysm: 10 year mortality and cost effectiveness results from the randomised Multicentre Aneurysm Screening Study. BMJ 2009;338:b2307.

[133] Forsythe RO, Newby DE, Robson JMJ. Monitoring the biological activity of abdominal aortic aneurysms beyond ultrasound. Heart 2016;102(11):817−24.

[134] Birkmeyer JD, Siewers AE, Finlayson EV, Stukel TA, Lucas FL, Batista I, et al. Hospital volume and surgical mortality in the United States. N Engl J Med 2002;346(15): 1128−37.

[135] Alam SR, Stirrat C, Richards J, Mirsadraee S, Semple SIK, Tse G, et al. Vascular and plaque imaging with ultrasmall superparamagnetic particles of iron oxide. J Cardiovasc Magn Reson 2015;17(1).

[136] McBride OMB, Berry C, Burns P, Chalmers RTA, Doyle B, Forsythe R, et al. MRI using ultrasmall superparamagnetic particles of iron oxide in patients under surveillance for abdominal aortic aneurysms to predict rupture or surgical repair: MRI for abdominal aortic aneurysms to predict rupture or surgery—the MA3RS study. Open Heart 2015;2(1).

[137] Brophy CM, Reilly JM, Smith GW, Tilson MD. The role of inflammation in nonspecific abdominal aortic aneurysm disease. Ann Vasc Surg 1991;5(3):229−33.

[138] Choke E, Cockerill G, Wilson WRW, Sayed S, Dawson J, Loftus I, et al. A review of biological factors implicated in abdominal aortic aneurysm rupture. Eur J Vasc Endovasc Surg 2005;30(3):227−44.

[139] Forsythe RO. Microcalcification predicts abdominal aortic aneurysm expansion and repair: the ^{18}F-sodium fluoride imaging in abdominal aortic aneurysms (SoFIA3) study. J Vasc Surg 2017;65(6):24S–5S.

[140] Forsythe RO, Dweck MR, McBride OMB, Vesey AT, Semple SI, Shah ASV, et al. ^{18}F–Sodium fluoride uptake in abdominal aortic aneurysms: the SoFIA3 study. J Am Coll Cardiol 2018;71(5):513–23.

[141] Gaddum NR, Keehn L, Guilcher A, Gomez A, Brett S, Beerbaum P, et al. Altered dependence of aortic pulse wave velocity on transmural pressure in hypertension revealing structural change in the aortic wall. Hypertension 2015;65(2):362–9.

[142] Sano M, Unno N, Sasaki T, Baba S, Sugisawa R, Tanaka H, et al. Topologic distributions of vasa vasorum and lymphatic vasa vasorum in the aortic adventitia– Implications for the prevalence of aortic diseases. Atherosclerosis 2016;247:127–34.

[143] Gawinecka J, Schönrath F, von Eckardstein A. Acute aortic dissection: pathogenesis, risk factors and diagnosis. Swiss Med Wkly 2017;147:w14489.

[144] Boudoulas KD, Triposkiadis F, Stefanadis C, Boudoulas H. Aortic size and aortic dissection: does one size fit all? Cardiology 2018;139(3):147–50.

[145] members AF, Erbel R, Aboyans V, Boileau C, Bossone E, Bartolomeo RD, et al. 2014 ESC Guidelines on the diagnosis and treatment of aortic diseases: document covering acute and chronic aortic diseases of the thoracic and abdominal aorta of the adult the task force for the diagnosis and treatment of aortic diseases of the European Society of Cardiology (ESC). Eur Heart J 2014;35(41):2873–926.

[146] Evangelista A, Isselbacher EM, Bossone E, Gleason TG, Eusanio MD, Sechtem U, et al. Insights from the international registry of acute aortic dissection: a 20-year experience of collaborative clinical research. Circulation 2018;137(17):1846–60.

[147] Kuehl H, Eggebrecht H, Boes T, Antoch G, Rosenbaum S, Ladd S, et al. Detection of inflammation in patients with acute aortic syndrome: comparison of FDG-PET/CT imaging and serological markers of inflammation. Heart 2008;94(11):1472–7.

[148] Gorla R, Erbel R, Kuehl H, Kahlert P, Tsagakis K, Jakob H, et al. Prognostic value of ^{18}F-fluorodeoxyglucose PET-CT imaging in acute aortic syndromes: comparison with serological biomarkers of inflammation. Int J Cardiovasc Imaging 2015;31(8):1677–85.

[149] Mills Sr JL, Conte MS, Armstrong DG, Pomposelli FB, Schanzer A, Sidawy AN, et al. The society for vascular surgery lower extremity threatened limb classification system: risk stratification based on wound, ischemia, and foot infection (WIfI). J Vasc Surg 2014;59(1):220–34.

[150] Lew E, Nicolosi N, Botek G. Lower extremity amputation risk factors associated with elevated ankle brachial indices and radiographic arterial calcification. J Foot Ankle Surg 2015;54(3):473–7.

[151] Mallipeddi VP, Kullo I, Afzal N, Kalra M, Lewis B, Scott C, et al. Association of ankle brachial index with limb outcomes in peripheral artery disease: a community-based study. J Am Coll Cardiol 2017;69(11 Supplement):2083.

[152] Lim JC, Ranatunga D, Owen A, Spelman T, Galea M, Chuen J, et al. Multidetector (64+) computed tomography angiography of the lower limb in symptomatic peripheral arterial disease: assessment of image quality and accuracy in a tertiary care setting. J Comput Assist Tomogr 2017;41(2):327.

[153] Brownrigg JRW, Hinchliffe RJ, Apelqvist J, Boyko EJ, Fitridge R, Mills JL, et al. Performance of prognostic markers in the prediction of wound healing or amputation

among patients with foot ulcers in diabetes: a systematic review. Diabetes Metab Res Rev 2016;32:128–35.

[154] Derlin T, Tóth Z, Papp L, Wisotzki C, Apostolova I, Habermann CR, et al. Correlation of inflammation assessed by ^{18}F-FDG PET, active mineral deposition assessed by ^{18}F-fluoride PET, and vascular calcification in atherosclerotic plaque: a dual-tracer PET/CT study. J Nucl Med 2011;52(7):1020–7.

[155] Stacy MR, Sinusas AJ. Novel applications of radionuclide imaging in peripheral vascular disease. Cardiol Clin 2016;34(1):167–77.

[156] Schmidt MA, Chakrabarti A, Shamim-Uzzaman Q, Kaciroti N, Koeppe RA, Rajagopalan S. Calf flow reserve with H(2)(15)O PET as a quantifiable index of lower extremity flow. J Nucl Med 2003;44(6):915–9.

[157] Scremin OU, Figoni SF, Norman K, Scremin AME, Kunkel CF, Opava-Rutter D, et al. Preamputation evaluation of lower-limb skeletal muscle perfusion with H(2) (15)O positron emission tomography. Am J Phys Med Rehabil 2010;89(6):473–86.

[158] Lu E, Wagner WR, Schellenberger U, Abraham JA, Klibanov AL, Woulfe SR, et al. Targeted in vivo labeling of receptors for vascular endothelial growth factor: approach to identification of ischemic tissue. Circulation 2003;108(1):97–103.

[159] Jeong JM, Hong MK, Chang YS, Lee Y-S, Kim YJ, Cheon GJ, et al. Preparation of a promising angiogenesis PET imaging agent: ^{68}Ga-labeled c(RGDyK)-isothiocyanato-benzyl-1,4,7-triazacyclononane-1,4,7-triacetic acid and feasibility studies in mice. J Nucl Med 2008;49(5):830–6.

[160] Nkomo VT, Gardin JM, Skelton TN, Gottdiener JS, Scott CG, Enriquez-Sarano M. Burden of valvular heart diseases: a population-based study. Lancet 2006; 368(9540):1005–11.

[161] Iung B, Baron G, Butchart EG, Delahaye F, Gohlke-Bärwolf C, Levang OW, et al. A prospective survey of patients with valvular heart disease in Europe: the Euro heart survey on valvular heart disease. Eur Heart J 2003;24(13):1231–43.

[162] Pawade TA, Newby DE, Dweck MR. Calcification in aortic stenosis: the skeleton key. J Am Coll Cardiol 2015;66(5):561–77.

[163] Dweck MR, Boon NA, Newby DE. Calcific aortic stenosis: a disease of the valve and the myocardium. J Am Coll Cardiol 2012;60(19):1854–63.

[164] Dweck MR, Pawade TA, Newby DE. Aortic stenosis begets aortic stenosis: between a rock and a hard place? Heart 2015;101(12):919–20.

[165] Rosenhek R, Binder T, Porenta G, Lang I, Christ G, Schemper M, et al. Predictors of outcome in severe, asymptomatic aortic stenosis. N Engl J Med 2000;343(9):611–7.

[166] Baumgartner H, Falk V, Bax JJ, De Bonis M, Hamm C, Holm PJ, et al. 2017 ESC/EACTS Guidelines for the management of valvular heart disease. Eur Heart J 2017; 38(36):2739–91.

[167] Messika-Zeitoun D, Aubry M-C, Detaint D, Bielak LF, Peyser PA, Sheedy PF, et al. Evaluation and clinical implications of aortic valve calcification measured by electron-beam computed tomography. Circulation 2004;110(3):356–62.

[168] Clavel M-A, Pibarot P, Messika-Zeitoun D, Capoulade R, Malouf J, Aggarval S, et al. Impact of aortic valve calcification, as measured by MDCT, on survival in patients with aortic stenosis: results of an international registry study. J Am Coll Cardiol 2014;64(12):1202–13.

[169] Clavel M-A, Messika-Zeitoun D, Pibarot P, Aggarwal SR, Malouf J, Araoz PA, et al. The complex nature of discordant severe calcified aortic valve disease grading: new

insights from combined Doppler echocardiographic and computed tomographic study. J Am Coll Cardiol 2013;62(24):2329–38.

[170] Pawade T, Clavel M-A, Tribouilloy C, Dreyfus J, Mathieu T, Tastet L, et al. Computed tomography aortic valve calcium scoring in patients with aortic stenosis. Circ Cardiovasc Imaging 2018;11(3):e007146.

[171] Irkle A, Vesey AT, Lewis DY, Skepper JN, Bird JLE, Dweck MR, et al. Identifying active vascular microcalcification by (18)F-sodium fluoride positron emission tomography. Nat Commun 2015;6:7495.

[172] Dweck MR, Jenkins WSA, Vesey AT, Pringle MAH, Chin CWL, Malley TS, et al. ^{18}F-Sodium fluoride uptake is a marker of active calcification and disease progression in patients with aortic stenosis. Circ Cardiovasc Imaging 2014;7(2).

[173] Jenkins WSA, Vesey AT, Shah ASV, Pawade TA, Chin CWL, White AC, et al. Valvular ^{18}F-fluoride and ^{18}F-fluorodeoxyglucose uptake predict disease progression and clinical outcome in patients with aortic stenosis. J Am Coll Cardiol 2015;66(10):1200–1.

[174] Lajoie EW, Bridges L. Innovation decisions: using the Gartner hype cycle. Libr Leader Manag 2014;28(4).

[175] John Walker S. Big data: a revolution that will transform how we live, work, and think. Taylor & Francis; 2014.

[176] Denning PJ, Lewis TG. Exponential laws of computing growth. Commun ACM 2016; 60(1):54–65.

[177] Krittanawong C, Zhang H, Wang Z, Aydar M, Kitai T. Artificial intelligence in precision cardiovascular medicine. J Am Coll Cardiol 2017;69(21):2657–64.

[178] Fishman EK, Ney DR, Heath DG, Corl FM, Horton KM, Johnson PT. Volume rendering versus maximum intensity projection in CT angiography: what works best, when, and why. Radiographics 2006;26(3):905–22.

[179] Dappa E, Higashigaito K, Fornaro J, Leschka S, Wildermuth S, Alkadhi H. Cinematic rendering—an alternative to volume rendering for 3D computed tomography imaging. Insights Imaging 2016;7(6):849–56.

[180] Morris PD, Narracott A, von Tengg-Kobligk H, Soto DAS, Hsiao S, Lungu A, et al. Computational fluid dynamics modelling in cardiovascular medicine. Heart 2016; 102(1):18–28.

[181] Bench-Capon TJM, Dunne PE. Argumentation in artificial intelligence. Artif Intell 2007;171(10):619–41.

[182] Liang M, Hu X. Recurrent convolutional neural network for object recognition. In: Proceedings of the IEEE Conference on computer vision and pattern recognition; 2015. p. 3367–75.

[183] Poplin R, Varadarajan AV, Blumer K, Liu Y, McConnell MV, Corrado GS, et al. Prediction of cardiovascular risk factors from retinal fundus photographs via deep learning. Nat Biomed Eng 2018;2(3):158–64.

[184] O'Reilly MK, Reese S, Herlihy T, Geoghegan T, Cantwell CP, Feeney RN, et al. Fabrication and assessment of 3D printed anatomical models of the lower limb for anatomical teaching and femoral vessel access training in medicine. Anat Sci Educ 2016;9(1): 71–9.

[185] Lim KHA, Loo ZY, Goldie SJ, Adams JW, McMenamin PG. Use of 3D printed models in medical education: a randomized control trial comparing 3D prints versus cadaveric materials for learning external cardiac anatomy. Anat Sci Educ 2016;9(3):213–21.

[186] Kang H-W, Lee SJ, Ko IK, Kengla C, Yoo JJ, Atala A. A 3D bioprinting system to pro-duce human-scale tissue constructs with structural integrity. Nat Biotechnol 2016; 34(3):312.

[187] Murphy SV, Atala A. 3D bioprinting of tissues and organs. Nat Biotechnol 2014;32(8): 773—85.

[188] Jia W, Gungor-Ozkerim PS, Zhang YS, Yue K, Zhu K, Liu W, et al. Direct 3D bio-printing of perfusable vascular constructs using a blend bioink. Biomaterials 2016; 106:58—68.

[189] 3D Bioprinting of heterogeneous aortic valve conduits with alginate/gelatin hydrogels - Duan - 2013 - Journal of Biomedical Materials Research Part A - Wiley Online Li-brary. Available from: https://onlinelibrary.wiley.com/doi/abs/10.1002/jbm.a.34420.

Index

'*Note:* Page numbers followed by "t" indicate tables and "f" indicate figures.'

Printed in the United States
By Bookmasters